Annals of Mathematics Studies

Number 98

K-THEORY OF FORMS

BY

ANTHONY BAK

PRINCETON UNIVERSITY PRESS

AND

UNIVERSITY OF TOKYO PRESS

PRINCETON, NEW JERSEY

1981

Published in Japan exclusively by
University of Tokyo Press;
In other parts of the world by
Princeton University Press
41 William Street
Princeton, New Jersey

Printed in the United States of America
by Princeton University Press, Princeton, New Jersey

Library of Congress Cataloging in Publication data will
be found on the last printed page of this book

In memory

of my father

TABLE OF CONTENTS

K-Theory of Forms

§1. INTRODUCTION

A. *General remarks*

This book contains a unified treatment of basic materials for the theories of quadratic, even hermitian, and hermitian forms and their 'classical' algebraic K-theories. In the last ten years, there has been a great deal of activity in the areas above, especially in the quadratic case which has been influenced by geometric surgery and the problem of computing the surgery obstruction groups. One precise aim of the text is to provide fundamental materials for application in [3], [4], [8], and [9] to the problem above. Accordingly, special attention is paid to free modules with preferred bases (based modules) which are important to the application above. A broader aim is to provide a reference for access between the areas above, so that an individual working in one area can learn easily what has been done in a neighboring area and to what extent the techniques carry over. In some cases, the foundational materials carry over on a one-to-one basis so that results in one area can be equivalent to those in another. We take care also to compare K-theory and Witt groups arising from different situations; thus in §11, it is shown how these groups are affected in going from quadratic to even hermitian to hermitian forms and in §8, it is shown how these groups are affected in going from projective modules to free modules to based modules. The procedure above of 'shifting gears' is useful not only for comparison purposes, but also for verification purposes, since analogous results in different settings can present different degrees of difficulty to verify.

The ingredient used to unify the theories of quadratic and hermitian forms is the form parameter Λ. However, Λ serves not only a unifying role, but also solves certain deficiencies in the theories of even hermitian and hermitian forms. For example, in the setting of even hermitian or hermitian forms, there is no reduction theorem modulo a nilpotent ideal; but in the refined context of Λ-quadratic and Λ-hermitian modules, there is such a theorem. We comment further on this in §1D.

A reader acquainted with the literature will realize that special cases of many of the results here have appeared already in the literature. Four papers of special merit in this regard are the foundational papers of H. Bass [11], A. Ranicke [22], and C. T. C. Wall [36], and the paper of R. Sharpe [26] on the quadratic Steinberg group. I was especially influenced by the paper of H. Bass and I would like to thank him for access to his early manuscripts.

Next, I would like to comment on the origins of the text and in particular, on the concept of a form parameter. The concept itself was required to classify [1] normal subgroups of unitary groups of even hermitian forms, but its unification role was not immediately evident to me. This occurred only after C. T. C. Wall suggested that I try to establish stability theorems for quadratic forms which were analogous to those I had established for even hermitian forms. In subsequent investigations to develop a K-theory of even hermitian and hermitian forms, it became evident that the form parameter was also necessary if one wanted to obtain results, such as Mayer-Vietoris sequences, which were analogous to known results for projective modules. These requirements plus the convenience mentioned above of shifting gears gave rise to the material presented in this book. The manuscript itself was written during a visit to the Université de Genève in 1972-73 and the book was announced in [4].

It is a pleasure for me to express my gratitude to H. Bass who gave very generously of his time and advice during the preparation of my thesis [1], a portion of which appears in §1B, §2, and §3. His influence is continually evident in this book. I would also like to thank the Université de Genève and my host there M. Kervaire for their hospitality during the writing of the manuscript and W.-C. Hsiang of the Annals Studies for his patient cooperation in the publication of the book.

B. *Quadratic modules*

The novelty in our definitions is the introduction of the form parameter. Let A be a ring with involution $a \mapsto \bar{a}$; thus $\bar{\bar{a}} = a$ and $\overline{ab} = \bar{b}\bar{a}$. Let $\lambda \in \text{center}(A)$ such that $\lambda\bar{\lambda} = 1$. A *form parameter* Λ is an additive subgroup of A such that

1. $\{a - \lambda\bar{a} \mid a \in A\} \subset \Lambda \subset \{a \mid a \in A, a = -\lambda\bar{a}\}$
2. $a\Lambda\bar{a} \subset \Lambda$ for all $a \in A$.

We denote the minimum (resp. maximum) choice of Λ often by min (resp. max). The pair

$$(A, \Lambda)$$

is called a *form ring*. If we wish to emphasize the symmetry λ we shall write ${}^{\lambda}(A, \Lambda)$ in place of (A, Λ). A homomorphism ${}^{\lambda}(A, \Lambda) \to {}^{\lambda'}(A', \Lambda')$ of form rings is a homomorphism $f: A \to A'$ of rings with involution such that $f(\lambda) = \lambda'$ and $f(\Lambda) \subset \Lambda'$. It is easy to check that all the K-theory groups we shall construct define functors from the category of form rings to the category of abelian groups.

Let M be a right A-module. A *sesquilinear form* on M is a biadditive map

$$B : M \times M \to A$$

such that $B(ma, nb) = \bar{a}B(m, n)b$. We denote the additive group of sesquilinear forms on M by $\text{Sesq}(M)$. $\text{Sesq}(M)$ has an involution $B \mapsto \bar{B}$ defined by $\bar{B}(m, n) = \overline{B(n, m)}$. B is called *λ-hermitian* if $B = \lambda\bar{B}$, and *even λ-hermitian* if $B = C + \lambda\bar{C}$ for some $C \in \text{Sesq}(M)$.

A *Λ-quadratic module* is a pair

$$(M, B)$$

where M is a right A-module and $B \in \text{Sesq}(M)$. To define a morphism of Λ-quadratic modules we associate to (M, B) a

$$\Lambda\text{-}\textit{quadratic form}\ q_B : M \to A/\Lambda,$$
$$m \mapsto [B(m, m)]$$

and a

$$\lambda\textit{-hermitian form} \; < , >_B : M \times M \to A .$$
$$< , >_B = B + \lambda\overline{B} .$$

Then a morphism $(M, B) \to (M', B')$ of Λ-quadratic modules is a linear map $M \to M'$ which preserves the associated quadratic and hermitian forms. We say that two Λ-quadratic modules (M, B) and (M', B') are equal (and write $(M, B) = (M', B')$) if (M, B) and (M', B') have the same quadratic and hermitian forms. Note that this does *not* imply $B = B'$.

Classically, a quadratic form is a 0-quadratic module. This requires of course that $\lambda = 1$ (because $1 - \lambda \in \Lambda$) and that the involution be trivial (because $a - \overline{a} \in \Lambda$). In order to remove the triviality restriction on the involution, Tits [30] gives a definition of a quadratic form which is nearly equivalent to a min-quadratic module. Precise equivalence occurs when the underlying modules are finitely generated and projective (see 9.6). The next result shows that even λ-hermitian forms are a special case of Λ-quadratic forms.

THEOREM 1.1. *A linear map* $M \to M'$ *is a morphism* $(M, B) \to (M', B')$ *of max-quadratic modules if and only if it preserves the associated even* λ*-hermitian forms.*

Proof. Let $f : M \to M'$ be a linear map such that $<m, n>_B = <f(m), f(n)>_{B'}$ for all m, $n \in M$. The equation implies that $B(m, m) - B'(f(m), f(m)) = -\lambda\overline{B}(m, m) + \lambda\overline{B}'(f(m), f(m))$. Thus, by definition, $B(m, m) - B'(f(m), f(m)) \in$ max. Hence, $q_B(m) = q_{B'}(f(m))$.

We define the *orthogonal sum* of two Λ-quadratic modules by $(M, B) \perp (M', B') = (M \oplus M', B \oplus B')$. We say that (M, B) is *nonsingular* if M is finitely generated, projective and if the map $M \to M^* = \mathrm{Hom}_A(M, A)$, $m \mapsto <m, \;>_B$, is an isomorphism.

The most important example of a nonsingular quadratic module is the hyperbolic module. If P is a finitely generated, projective, right A-module, we define the *hyperbolic module*

$$H(P) = (P \oplus P^*, B_P)$$

where $B_P((p,f),(q,g)) = f(q)$. $P^* = \mathrm{Hom}_A(P,A)$ is the right A-module on which the action of A is defined by $(f\cdot a)(p) = \overline{a}(f(p))$. One can check easily that there is a canonical isomorphism $H(P\oplus Q) \cong H(P) \perp H(Q)$. $H(A)$ is called the *hyperbolic plane*.

Recall now the definitions of the algebraic K-theory groups $K_0(A)$ and $K_1(A)$ defined in [10, IX §1]. The dual operator on right A-modules, $M \mapsto M^*$ (if M is finitely generated, projective then the canonical map $M \to M^{**}$, $m \mapsto (f \mapsto f(m))$, is an isomorphism), and the conjugate transpose operator on matrices, $(a_{ij}) \mapsto {}^t(\overline{a}_{ij})$, induce respectively involutions, i.e. actions of $Z/2Z$, on $K_0(A)$ and $K_1(A)$. Let X and Y be involution invariant subgroups respectively of $K_0(A)$ and $K_1(A)$. For convenience, we shall assume that X contains the class of the free module A and that Y contains the classes of the matrices -1 and $-\lambda$. We define the categories

$$Q^\lambda(A,\Lambda)_X$$

$$Q^\lambda(A,\Lambda)_{\text{based}-Y}$$

as follows. The objects of $Q^\lambda(A,\Lambda)_X$ are all nonsingular Λ-quadratic modules (M,B) such that the class of M in $K_0(A)/X$ vanishes. Morphisms in $Q^\lambda(A,\Lambda)_X$ are defined analogously to morphisms in $Q^\lambda(A,\Lambda)$. The objects of $Q^\lambda(A,\Lambda)_{\text{based}-Y}$ are all nonsingular Λ-quadratic modules (M,B) such that M is a free module with a preferred (distinguished) basis $e_1,\cdots,e_m$ such that the $m\times m$-matrix $(\langle e_i,e_j\rangle_B)$ vanishes in $K_1(A)/Y$. A morphism $f:(M,B)\to(M',B')$ is an isomorphism of Λ-quadratic modules such that the preferred bases on M and M′ have the same number of elements and such that the matrix determined by f and the preferred bases vanishes in $K_1(A)/Y$. The condition that -1 vanishes in $K_1(A)/Y$ guarantees that the operation of orthogonal sum in $Q^\lambda(A,\Lambda)_{\text{based}-Y}$ is commutative up to isomorphism. The *standard* preferred basis for the underlying module $A\oplus A^*$ of $H(A)$ is $e=(1,0)$, $f=(0,\text{identity})$. If $H(\Lambda)$ has the standard preferred basis then we denote it by

$$H(A)_{\text{based}}.$$

$H(A)_{\text{based}}$ is called the *based hyperbolic plane*. The condition that $-\lambda$ vanishes in $K_1(A)/Y$ guarantees that $H(A)_{\text{based}} \in Q^{\lambda}(A,\Lambda)_{\text{based}-Y}$. When we do not wish to emphasize the symmetry λ, we shall drop the superscript λ. We shall write often

$$Q(A,\Lambda)$$

$$Q(A,\Lambda)_{\text{free}}$$

in place of $Q(A,\Lambda)_{K_0(A)}$ and $Q(A,\Lambda)_{\text{based}-K_1(A)}$.

The following variations of $Q(A,\Lambda)_{\text{based}-Y}$ are sometimes convenient to have. The second variation below is perhaps the one which has the most utility, especially for problems of current interest. Let Y be an involution invariant subgroup of $K_1(A)$ such that λ vanishes in $K_1(A)/Y$. Define

$$Q(A,\Lambda)_{\text{even-based}-Y}$$

analogously to $Q(A,\Lambda)_{\text{based}-Y}$ with the added restriction that a preferred basis has an even number of elements. Notice that the condition that -1 vanishes in $K_1(A)/Y$ is not needed any more to guarantee that the product is commutative. The condition that λ vanishes in $K_1(A)/Y$ guarantees that $H(A)_{\text{based}} \in Q(A,\Lambda)_{\text{even-based}-Y}$. Suppose now that (M,B) is any nonsingular, quadratic module with a preferred basis $e_1,\cdots,e_{2m}$. The $2m\times 2m$ matrix

$$\operatorname{discr}(M,B) = \left(\begin{array}{c|c} & \begin{matrix}\lambda & & \\ & \ddots & \\ & & \lambda\end{matrix} \\ \hline \begin{matrix}1 & & \\ & \ddots & \\ & & 1\end{matrix} & \end{array}\right)^{-1} (\langle e_i, e_j\rangle_B)$$

is called the *discriminant* of (M,B). If Y is an involution invariant subgroup of $K_1(A)$, we define

$$Q(A,\Lambda)_{\text{discr-based-}Y}$$

analogously to $Q(A,\Lambda)_{\text{based-}Y}$, except that we assume the $\operatorname{discr}(M,B)$ vanishes in $K_1(A)/Y$ instead of the matrix $(\langle e_i, e_j\rangle_B)$. It is automatic that the product in $Q(A,\Lambda)_{\text{discr-based-}Y}$ is commutative and that $H(A)_{\text{based}} \in Q(A,\Lambda)_{\text{discr-based-}Y}$.

We recall that if C is a category with product $\perp$ [10, VII §1] such that the isomorphism classes $[M]$ of objects M of C form a set then

$$K_0C$$

is the free abelian group on the isomorphism classes $[M]$ modulo the relations $[M]+[N]=[M\perp N]$.

Define the *Grothendieck groups*

$$\begin{aligned}
KQ_0(A,\Lambda)_X &= K_0Q(A,\Lambda)\\
KQ_0(A,\Lambda)_{\text{based-}Y} &= K_0Q(A,\Lambda)_{\text{based-}Y}\\
KQ_0(A,\Lambda)_{\text{even-based-}Y} &= K_0Q(A,\Lambda)_{\text{even-based-}Y}\\
KQ_0(A,\Lambda)_{\text{discr-based-}Y} &= K_0Q(A,\Lambda)_{\text{discr-based-}Y}.
\end{aligned}$$

Define the *Witt groups*

$$\begin{aligned}
WQ_0(A,\Lambda)_X &= KQ_0(A,\Lambda)_X/\{H(P) \mid [P]\in X\}\\
WQ_0(A,\Lambda)_{\text{based-}Y} &= KQ_0(A,\Lambda)_{\text{based-}Y}/[H(A)_{\text{based}}]\\
WQ_0(A,\Lambda)_{\text{even-based-}Y} &= KQ_0(A,\Lambda)_{\text{even-based-}Y}/[H(A)_{\text{based}}]\\
WQ_0(A,\Lambda)_{\text{discr-based-}Y} &= KQ_0(A,\Lambda)_{\text{discr-based-}Y}/[H(A)_{\text{based}}].
\end{aligned}$$

Let

$$P(A)$$

denote the category with product $\oplus$ of finitely generated, projective, right A-modules where only isomorphisms are allowed as morphisms. The construction above of the hyperbolic module leads to a product preserving functor

$$\begin{aligned} \Lambda - \mathbf{H} : \bar{P}(A) &\to \bar{Q}(A,\Lambda) \\ P &\mapsto \mathbf{H}(P) \\ \sigma : P \to Q &\mapsto \sigma \oplus \sigma^{*-1} \end{aligned}$$

called the Λ-*hyperbolic functor*. Since $\Lambda - \mathbf{H}$ is product preserving, it induces a homomorphism

$$\mathbf{H} : K_0(A) \to KQ_0(A,\Lambda)$$

called the *hyperbolic map*. From the definitions above, it is clear that $WQ_0(A,\Lambda) = \operatorname{coker} \mathbf{H} : K_0(A) \to KQ_0(A,\Lambda)$.

Suppose that Γ is another form parameter defined with respect to λ such that

$$\Lambda \subseteqq \Gamma .$$

We want to determine how the groups above are affected when Λ is replaced by Γ. If $x \in \Gamma$ and $a \in A$ then the rules $x \mapsto ax\bar{a}$ and $x \mapsto \bar{a}xa$ define respectively left and right actions of A on the quotient Γ/Λ. Let

$$S(\Gamma/\Lambda) = (\Gamma/\Lambda \otimes_A \Gamma/\Lambda)/\{a \otimes b - b \otimes a,\ a \otimes b - a \otimes ba\bar{b}\} .$$

If $K_0(A,\Lambda)$ denotes any one of the Grothendieck or Witt groups defined above then there is a canonical map $S(\Gamma/\Lambda) \to K_0(A,\Lambda)$, $[a \otimes b] \mapsto [A \oplus A, \begin{pmatrix} a & 0 \\ 1 & b \end{pmatrix}] - [A \oplus A, \begin{pmatrix} 0 & 0 \\ 1 & 0 \end{pmatrix}]$, where in the based cases it is assumed that $A \oplus A$ has the preferred basis $\{(1,0),(0,1)\}$. Call A *trace noetherian* if it is noetherian as a module over the subring generated by 1 and all $a + \bar{a}$ such that $a \in \operatorname{center}(A)$. Any order [29] of characteristic $\neq 2$ or any A such that $A/2A$ is finite satisfies the condition above.

THEOREM 1.2. *If A is trace noetherian or if A is semilocal and complete modulo its Jacobson radical then the sequence below is split exact*

$$0 \to S(\Gamma/\Lambda) \to K_0(A,\Lambda) \to K_0(A,\Gamma) \to 0 .$$

The theorem is proved in §11.

C. *Hermitian modules*

Let A, λ, and Λ be as in B.

Let M be a right A-module. A *Λ-hermitian form* B on M is a $-\lambda$-hermitian form such that $B(m,m) \in \Lambda$ for all $m \in M$. A *Λ-hermitian module* is a pair

$$(M, B) .$$

A morphism $(M,B) \to (M',B')$ of Λ-hermitian modules is a linear map $M \to M'$ which preserves the Λ-hermitian forms. The *orthogonal sum* of two Λ-hermitian modules is defined by $(M,B) \perp (M',B') = (M \oplus M', B \oplus B')$. A Λ-hermitian module (M,B) is called *nonsingular* if M is finitely generated, projective and if the map $M \mapsto M^* = \mathrm{Hom}_A(M,A)$, $m \mapsto B(m,\ )$, is an isomorphism.

Clearly the expression max-hermitian form has the same meaning as the expression $-\lambda$-hermitian form. The corresponding result for min-hermitian forms is

THEOREM 1.3. *If* M *is finitely generated, projective then the expression min-hermitian form on* M *has the same meaning as the expression even* $-\lambda$*-hermitian form on* M.

Proof. Suppose B is a min-hermitian form on M. Choose N such that $M \oplus N \cong A^n$. Let 0 denote the trivial form on N and set $B' = B \oplus 0$. Pick a basis $e_1, \cdots, e_n$ for A^n. The $n \times n$-matrix $(B'(e_i, e_j))$ is $-\lambda$-hermitian, i.e. $(B'(e_i,e_j)) = -\lambda^t(\overline{B'(e_i,e_j)})$, and since B' is min-hermitian, the diagonal coefficients $B'(e_i,e_i) \in \min$. Hence, $B' = C' + (-\lambda)\overline{C}'$ for some sesquilinear form C' on M, and if $C = C'$ restricted to M then $B = C + (-\lambda)\overline{C}$. Thus, B is even. Conversely, if B is an even $-\lambda$-hermitian form then by definition $B = C - \lambda\overline{C}$ for some sesquilinear form C. Thus, $B(m,m) = C(m,m) - \lambda\overline{C(m,m)} \in \min$. Thus, B is a min-hermitian form.

The most important example of a nonsingular, Λ-hermitian module is the Λ-metabolic module. If P is a finitely generated, projective, right

A-module then we call

$$\Lambda - M(P) = (P \oplus P^*, B)$$

Λ-metabolic if for all $p \in P$ and $f, f_1 \in P^*$, we have $B(f, p) = f(p)$, $B(f, f_1) = 0$, and $B(p, p) \in \Lambda$. $P^* = \mathrm{Hom}_A(P, A)$ is the right A-module on which the action of A is defined by $(fa)(p) = \overline{a}(f(p))$. If $B(p, p_1) = 0$ for all $p, p_1 \in P$ then we call $\Lambda - M(P)$ *hyperbolic* and denote it by

$$H(P) .$$

One can check easily that there is a canonical isomorphism $\Lambda - M(P \oplus Q) \cong \Lambda - M(P) \perp \Lambda - M(Q)$. It should be noted that the notation $\Lambda - M(P)$ does not uniquely determine the Λ-hermitian form B on $P \oplus P^*$. This deficit will be removed later in the section when we introduce the Λ-metabolic functor on the category $S(A, \Lambda)$.

Recall now the algebraic K-theory groups $K_0(A)$ and $K_1(A)$, and the involutions defined on these groups in §1B. Let X and Y be involution invariant subgroups respectively of $K_0(A)$ and $K_1(A)$. For convenience, we shall assume that X contains the class of the free module A and that Y contains the classes of the matrices -1 and $-\lambda$. We define the categories

$$H^{-\lambda}(A, \Lambda)_X$$

$$H^{-\lambda}(A, \Lambda)_{\text{based}-Y}$$

as follows. If in the definition in §1B of $Q^\lambda(A, \Lambda)_X$, one replaces the word quadratic by the word hermitian then one has the definition of $H^{-\lambda}(A, \Lambda)_X$. If in the definition of $Q^\lambda(A, \Lambda)_{\text{based}-Y}$, one replaces the word quadratic by the word hermitian and the matrix $(\langle e_i, e_j \rangle_B)$ by the matrix $(B(e_i, e_j))$ then one has the definition of $H^{-\lambda}(A, \Lambda)_{\text{based}-Y}$. The condition that -1 vanishes in $K_1(A)/Y$ guarantees that the operation of orthogonal sum in $H^{-\lambda}(A, \Lambda)_{\text{based}-Y}$ is commutative up to isomorphism. The *standard* preferred basis for the underlying module $A \oplus A^*$ of a metabolic module $\Lambda - M(A)$ is $e = (1, 0)$, $f = (0, \text{identity})$. If $\Lambda - M(A)$ has the standard preferred basis then we denote it by

$$\Lambda - M(A)_{\text{based}} .$$

$\Lambda - M(\Lambda)_{\text{based}}$ is called a *based* Λ*-metabolic plane*. The condition that λ vanishes in $K_1(A)/Y$ guarantees that each $\Lambda - M(A)_{\text{based}} \in H^{-\lambda}(A, \Lambda)_{\text{based}-Y}$. When we do not wish to emphasize the symmetry $-\lambda$, we drop the superscript $-\lambda$. We shall write often

$$H(A, \Lambda)$$

$$H(A, \Lambda)_{\text{free}}$$

in place of $H(A, \Lambda)_{K_0(A)}$ and $H(A, \Lambda)_{\text{based}-K_1(A)}$.

The following variations of $H(A, \Lambda)_{\text{based}-Y}$ are sometimes convenient to have. The second variation below is perhaps the one which has the most utility, especially for problems of current interest. Let Y be an involution invariant subgroup of $K_1(A)$ such that λ vanishes in $K_1(A)/Y$. Define

$$H(A, \Lambda)_{\text{even-based}-Y}$$

analogously to $H(A, \Lambda)_{\text{based}-Y}$ with the added restriction that a preferred basis has an even number of elements. Notice that the condition that -1 vanishes in $K_1(A)/Y$ is not needed any more to guarantee that the product is commutative. The condition that λ vanishes in $K_1(A)/Y$ guarantees that each $\Lambda - M(A)_{\text{based}} \in H(A, \Lambda)_{\text{even-based}-Y}$. Suppose now that (M, B) is any nonsingular hermitian module with a preferred basis $e_1, \cdots, e_{2m}$. The $2m \times 2m$ matrix

$$\operatorname{discr}(M, B) = \left(\begin{array}{ccc|ccc} & & & -\lambda & & \\ & & & & \ddots & \\ & & & & & -\lambda \\ \hline 1 & & & & & \\ & \ddots & & & & \\ & & 1 & & & \end{array}\right)^{-1} (B(e_i, e_j))$$

is called the *discriminant* of (M, B). If Y is an involution invariant subgroup of $K_1(A)$, we define

$$H(A, \Lambda)_{\text{discr-based}-Y}$$

analogously to $H(A, \Lambda)_{\text{based}-Y}$, except that we assume the $\operatorname{discr}(M, B)$

vanishes in $K_1(A)/Y$ instead of the matrix $(B(e_i, e_j))$. It is automatic that the product in $H(A,\Lambda)_{\text{discr-based-Y}}$ is commutative and that each $\Lambda - M(A)_{\text{based}} \in H(A,\Lambda)_{\text{discr-based-Y}}$.

The definition of K_0 of a category with product is recalled in §1B. Define the *Grothendieck groups*

$$\begin{aligned}
KH_0(A,\Lambda)_X &= K_0H(A,\Lambda)\\
KH_0(A,\Lambda)_{\text{based-Y}} &= K_0H(A,\Lambda)_{\text{based-Y}}\\
KH_0(A,\Lambda)_{\text{even-based-Y}} &= K_0H(A,\Lambda)_{\text{even-based-Y}}\\
KH_0(A,\Lambda)_{\text{discr-based-Y}} &= K_0H(A,\Lambda)_{\text{discr-based-Y}}.
\end{aligned}$$

Define the *Witt groups*

$$\begin{aligned}
WH_0(A,\Lambda)_X &= KH_0(A,\Lambda)_X/\{H(P) \mid [P] \in X\}\\
WH_0(A,\Lambda)_{\text{based-Y}} &= KH_0(A,\Lambda)_{\text{based-Y}}/[H(A)_{\text{based}}]\\
WH_0(A,\Lambda)_{\text{even-based-Y}} &= KH_0(A,\Lambda)_{\text{even-based-Y}}/[H(A)_{\text{based}}]\\
WH_0(A,\Lambda)_{\text{discr-based-Y}} &= KH_0(A,\Lambda)_{\text{discr-based-Y}}/[H(A)_{\text{based}}].
\end{aligned}$$

It is worth noting that it will follow from 2.11 that each Λ-metabolic module $\Lambda - M(P)$ such that $[P] \in X$ vanishes in $WH_0(A,\Lambda)_X$ and from 2.12 that each based Λ-metabolic plane $\Lambda - M(A)_{\text{based}}$ vanishes in any of the based Witt groups.

Define the category with product

$$S(A,\Lambda)_X$$

as follows. Its objects are pairs (P,α) where P is a finitely generated, projective, right A-module which vanishes in $K_0(A)/X$ and $\alpha : P \to P^*$ is a homomorphism such that $\alpha = -\lambda\alpha^*$ and $\alpha(p)(p) \in \Lambda$ for all $p \in P$ (we identify P with its double dual P^{**} via the canonical map $P \to P^{**}$, $p \mapsto (f \mapsto \overline{f(p)})$). A morphism $(P,\alpha) \to (Q,\beta)$ is an A-linear isomorphism $\sigma: P \to Q$ such that $\sigma^*\beta\sigma = \alpha$. The product is defined by $(P,\alpha) \perp (Q,\beta) = (P \oplus Q, \alpha \oplus \beta)$. The construction above of the Λ-metabolic module leads to a product preserving functor

$$\begin{aligned}\Lambda - \mathbf{M} : S(A,\Lambda) &\to \mathbf{H}(A,\Lambda)\\ (P,\alpha) &\mapsto \left(P\oplus P^*, \begin{pmatrix}\alpha & -\lambda I_{P^*}\\ I_P & 0\end{pmatrix}\right)\\ (\sigma:(P,\alpha)\to(Q,\beta)) &\mapsto \sigma\oplus\sigma^{*-1}\end{aligned}$$

called the Λ-*metabolic functor*. Since $\Lambda - \mathbf{M}$ is product preserving, it induces a homomorphism

$$\mathbf{M} : K_0S(A,\Lambda) \to KH_0(A,\Lambda)$$

called the *metabolic map*. From the remark following the definition of the Witt groups, it is clear that $WH_0(A,\Lambda) = \operatorname{coker} \mathbf{M} : K_0S(A,\Lambda) \to KH_0(A,\Lambda)$.

If $K_0(A,\Lambda)$ denotes any one of the Grothendieck or Witt groups defined above and if Γ is another form parameter such that $\Lambda \subseteq \Gamma$ then the kernel and cokernel of the canonical map $K_0(A,\Lambda) \to K_0(A,\Gamma)$ are more difficult to handle than the analogous kernel and cokernel in §1B. We shall leave our results in this direction till §11.

D. *Necessity for refined definitions*

Let A and Λ be as in §1B. Let $\mathfrak{q}$ be an involution invariant ideal of A such that A is complete in the $\mathfrak{q}$-adic topology. By a result [10, III 2.12b)] of H. Bass, the canonical map $K_0(A) \to K_0(A/\mathfrak{q})$ is an isomorphism, and by a result [33] of C. T. C. Wall, the canonical map $KQ_0(A, \min) \to KQ_0(A/\mathfrak{q}, \min)$ is an isomorphism. However, the same conclusion fails for the canonical maps $KQ_0(A, \max) \to KQ_0(A/\mathfrak{q}, \max)$ and $KH_0(A, \max) \to KH_0(A/\mathfrak{q}, \max)$. The remedy is to allow intermediate kinds of forms to occur; namely, to replace the maximum form parameter on $A/\mathfrak{q}$ by the image Γ of the maximum form parameter on A. Γ is not necessarily the maximum form parameter on $A/\mathfrak{q}$. In general, if Λ is a form parameter on A and if Γ is its image in $A/\mathfrak{q}$ (resp. $A/\mathfrak{q}^n$) then the canonical map $KQ_0(A,\Lambda) \to KQ_0(A/\mathfrak{q},\Gamma)$ (resp. $KH_0(A,\Lambda) \to KH_0(A/\mathfrak{q}^n,\Gamma)$ is an isomorphism (resp. isomorphism for n suitably large). The results are proved in §10.

If A is a ring with involution and if $\lambda \in \text{center}(A)$ such that $\lambda\bar{\lambda} = 1$, let $m(A)$ denote either the minimum or maximum form parameter with respect to λ on A. Let

$$(1) \qquad \begin{array}{ccc} A & \longrightarrow & A_2 \\ \downarrow & & \downarrow \\ A_1 & \longrightarrow & \hat{A} \end{array}$$

be a fibre product square of rings with involution and let

$$(2) \qquad \begin{array}{ccc} m(A) & \longrightarrow & m(A_2) \\ \downarrow & & \downarrow \\ m(A_1) & \longrightarrow & m(\hat{A}) \end{array}$$

be the corresponding square of form parameters. With certain conditions on (1), such as the surjectivity or density of the map $A_2 \to \hat{A}$, we would like to associate to (1) an exact Mayer-Vietoris sequence (6.32) of either KQ-groups or KH-groups. If the conditions on (1) imply the corresponding conditions on (2) then one can do this. However, when the corresponding conditions on (2) do not hold, one can often remedy the situation by replacing some of the m's by suitable other form parameters such that the resulting square of form parameters satisfies the corresponding conditions and obtain thereby an exact Mayer-Vietoris sequence involving either KQ-groups or KH-groups with 'nontrivial' form parameters. The results are found in §6 and §7.

§2. HYPERBOLIC AND METABOLIC MODULES

The definitions of hyperbolic and metabolic modules appear in §1B and C.

Let $\mathbb{C}$ be a category with product $\perp$. A subcategory $\mathbb{C}' \subset \mathbb{C}$ will be called *cofinal* if given an object $M \in \mathbb{C}$, there exists an object $N \in \mathbb{C}$ and objects $M_1', \cdots, M_n' \in \mathbb{C}'$ such that $M \perp N \simeq M_1' \perp \cdots \perp M_n'$. An object of $\mathbb{C}$ is called *cofinal* if the subcategory it defines is cofinal. A product preserving functor $F : \mathbb{C}'' \to \mathbb{C}$ of categories with product is called *cofinal* if image(F) is a cofinal subcategory of $\mathbb{C}$. The purpose of this section is to prove that the hyperbolic and metabolic planes $\Lambda - \mathbf{H}(A)$ and $\Lambda - \mathbf{M}(A)$ are cofinal respectively in $\mathbf{Q}(A, \Lambda)$ and $\mathbf{H}(A, \Lambda)$.

Let (M, B) be a quadratic (resp. hermitian module). Two elements $m, n \in M$ are called *orthogonal* if $< m, n >_B = 0$ (resp. $B(m, n) = 0$). Let U be a submodule of M. Let $U^{\perp} = \{m \in M | m$ is orthogonal to every element of $U\}$. U is called *totally isotropic* if q_B and $< , >_B$ are trivial on U (resp. B is trivial on U).

LEMMA 2.1. *Suppose* (M, B) *is a* Λ*-quadratic (resp.* Λ*-hermitian) module. If* $M = U + V$ *and if each element of* U *is orthogonal to each element of* V *then the canonical map* $(U, B|_U) \perp (V, B|_V) \to (M, B)$, $(u, v) \mapsto u + v$, *is an isomorphism.*

Whenever the hypotheses of 2.1 are satisfied we shall write $(M, B) = U \perp V$.

Proof. The principle behind the isomorphism in 2.1 for quadratic modules is the following. If B and $B' \in \mathrm{Sesq}(M)$ such that $B - B' = C - \lambda \overline{C}$ for some C, then B and B' determine the same Λ-quadratic module; namely, the identity map $M \to M$ defines an isomorphism $(M, B) \to (M', B')$, because $q_B = q_{B'}$ and $< , >_B = < , >_{B'}$. To prove the lemma for quadratic modules, one defines $C \in \mathrm{Sesq}(M)$ such that if both $m, n \in U$ or both $m, n \in V$, then $C(m,n)=0$, and such that if $u \in U$ and $v \in V$ then $C(u,v)=B(u,v)$ and $C(v,u)=0$. Then $B|_U \oplus B|_V + C - \lambda\bar{C} = B$.

The proof of 2.1 in the hermitian case is clear.

LEMMA 2.2. *Suppose that* (M,B) *is a hermitian module and that* (U,C) *is a subspace. If the map* $M \to M^*$, $m \mapsto B(m,\ )$, *is an isomorphism then the map* $U \to U^*$, $u \mapsto C(u,\ )$, *is an isomorphism if and only if* $M = U \perp U^{\perp}$.

Proof. Clearly $U \cap U^{\perp} = 0$. Furthermore, if $m \in M$ then we can choose $u \in U$ such that the linear functionals $B(m,\ )$ and $B(u,\ )$ agree on U. Hence, $m-u \in U^{\perp}$. Thus, $M = U \perp U^{\perp}$.

COROLLARY 2.3. *If* (M,B) *is a nonsingular* Λ*-quadratic (resp.* Λ*-hermitian) module then a subspace* (U,C) *of* (M,B) *is nonsingular if and only if* $M = U \perp U^{\perp}$.

COROLLARY 2.4. *Suppose* (M,B) *is a* Λ*-quadratic (resp.* Λ*-hermitian) module such that the map* $M \to M^*$, $m \mapsto \langle m,\ \rangle_B$ *(resp.* $m \mapsto B(m,\ )$*) is an isomorphism. Let* (M',B') *be another* Λ*-quadratic (resp.* Λ*-hermitian) module. Then any morphism* $f: (M,B) \to (M',B')$ *is injective and its image is an orthogonal summand of* (M',B').

Proof. We prove only the quadratic case. The hermitian case is handled similarly. One begins by noting that $fm = 0 \Longrightarrow \langle fm,\ \rangle_{B'} = 0 \Longrightarrow \langle m,\ \rangle_B = 0 \Longrightarrow m = 0$. Hence, f is injective. Furthermore, applying 2.2 to the subspace $(f(M), \langle\ ,\ \rangle_{B'}|_{f(M)}) \subseteq (M',B')$, one can conclude that $f(M)$ is an orthogonal summand of (M',B').

LEMMA 2.5. *Suppose* (M,B) *is a* Λ*-quadratic (resp.* Λ*-hermitian) module such that the map* $M \to M^*$, $m \mapsto \langle m,\ \rangle_B$ *(resp.* $m \mapsto B(m,\ )$*), is an isomorphism. If* A *is finitely generated as a module over its center and if* M *is finitely generated over* A *then any endomorphism of* (M,B) *is an automorphism.*

Proof. By 2.4, any endomorphism f is injective and its image is a direct summand of M. Thus, $M \cong M \oplus N$ for some N. To complete the proof, it suffices to show that $N = 0$. If $\mathfrak{p}$ is a maximal ideal in the center(A), let $A_{\mathfrak{p}}$ denote A localized at $\mathfrak{p}$ and let $A'_{\mathfrak{p}} = A_{\mathfrak{p}}/\text{Jacobson radical}(A_{\mathfrak{p}})$. By a well-known principle [10, III 4.3], $N = 0 \iff N \otimes_A A_{\mathfrak{p}} = 0$ for all $\mathfrak{p}$, and by Nakayama's lemma [10, III 2.2], $N \otimes_A A_{\mathfrak{p}} = 0 \iff N \otimes_A A'_{\mathfrak{p}} = 0$. But, since $A'_{\mathfrak{p}}$ is semisimple, it follows from the isomorphism $M \otimes_A A'_{\mathfrak{p}} \cong M \otimes_A A'_{\mathfrak{p}} \oplus N \otimes_A A'_{\mathfrak{p}}$ that $N \otimes_A A'_{\mathfrak{p}} = 0$.

LEMMA 2.6. *Suppose* (M, B) *is a nonsingular* Λ*-hermitian module. If* U *is a totally isotropic direct summand of* M *then*

$$(M, B) \cong U^{\perp}/U \perp \Lambda - \mathbf{M}(U).$$

Proof. The hermitian form B on $U^{\perp}$ induces in a canonical way a hermitian form on $U^{\perp}/U$. Since U is a direct summand of M, it follows that U is projective. Hence, U^* is projective. Thus, the exact sequence $0 \to U^{\perp} \to M \to U^* \to 0$ splits, and $U^{\perp}$ is a direct summand of M. Write $M = U^{\perp} \oplus V$. The map $\phi : V \to U^*$, $v \mapsto B(v, \ )$ is an isomorphism because B is nonsingular. Let $i : U \to U^{**}$, $u \mapsto (f \mapsto \overline{f(u)})$, be the canonical identification of U with U^{**} (remember we make U^* and U^{**} into right A-modules via the involution on A). Then $\phi^* i : U \to V^*$ is also an isomorphism. But $\phi^* i(u) = \overline{B}(\ , u) = -\overline{\lambda} B(u, \ )$. Hence, $U \to V^*$, $u \mapsto B(u, \ )$ is an isomorphism. Thus, $U \oplus V \to (U \oplus V)^*$, $u + v \mapsto B(u+v, \ )$ is an isomorphism (because $U \to U^*$, $u \mapsto B(u, \ )$, is trivial) and $(U \oplus V, B|_{U \oplus V})$ is nonsingular. By 2.1, we can pick an orthogonal complement U' to $U \oplus V$. Clearly $U' \subset U^{\perp}$, and $U' \oplus U = U^{\perp}$. The last equation implies $U' \cong U^{\perp}/U$, and hence, $(M, B) \cong U^{\perp}/U \perp (U \oplus V, B|_{U \oplus V})$. The assertion of the lemma follows now from 2.8 below.

LEMMA 2.7. *Suppose* (M, B) *is a nonsingular* Λ*-quadratic module. Let* U *be a totally isotropic direct summand of* M. *Suppose that* $U^{\perp} = U' \oplus U$ *and let* $B' = B|_{U'}$. *Then*

$$(M, B) \cong (U', B') \perp \Lambda - \mathbf{H}(U).$$

Proof. As in the proof of 2.6, we can write $M = U^{\perp} \oplus V$ and show that $(U \oplus V, B|_{U \oplus V})$ is nonsingular. By 2.2, we can pick an orthogonal complement (U'', B'') to $(U \oplus V, B_{U\oplus V})$. Clearly, $U'' \subset U^{\perp}$ and $U'' \oplus U = U^{\perp}$. The composite $U' \to U'' \oplus U \xrightarrow{\text{projection on } U''} U''$ induces an isomorphism $(U', B') \cong (U'', B'')$ of Λ-quadratic modules. Hence, $(M, B) \cong (U', B') \perp (U \oplus V, B|_{U\oplus V})$. The lemma follows now from 2.8 below.

LEMMA 2.8. *Suppose* (M, B) *is a nonsingular* Λ*-quadratic (resp.* Λ*-hermitian) module. Then* (M, B) *is* Λ*-hyperbolic (resp.* Λ*-metabolic) if and only if* M *contains a totally isotropic direct summand* U *such that* $U = U^{\perp}$, *in which case* $(M, B) \cong \Lambda - \mathbf{H}(U)$ *(resp.* $\Lambda - \mathbf{M}(U)$*).*

Proof. We consider first the case of quadratic modules. The first task is to find a totally isotropic direct complement to U. Write $M = U \oplus V$. Every direct complement to U in M is of the form $\{v + h(v) | v \in V, h: V \to U\}$. Choose h such that $-B(m, n) = \langle hm, n\rangle_B$ for all $m, n \in M$ (we can do this since $\langle\ ,\ \rangle_B$ is nonsingular, see [19]). Then $q_B(v + h(v)) \equiv B(v, v) + B(hv, v) + B(v, hv) \equiv B(v, v) + (B(hv, v) + \lambda\overline{B(v, hv)}) + (-\lambda\overline{B(v, hv)} + B(v, hv)) \equiv B(v, v) + \langle hv, v\rangle_B = 0$, and $\langle v + hv, w + hw\rangle_B = \langle v, w\rangle_B + \langle hv, w\rangle_B + \langle v, hw\rangle_B = (B(v, w) + \langle hv, w\rangle_B) + (\lambda\overline{B(w, v)} + \lambda\overline{\langle hw, v\rangle_B}) = 0 + 0 = 0$.

Now we suppose that $M = U \oplus V$ and that V is totally isotropic. Let $\Lambda - \mathbf{H}(U) = (U \oplus U^*, C)$. Define $U \oplus V \to U \oplus V^*$, $u + v \mapsto u + \phi(v)$ where $\phi: V \to U^*$, $v \mapsto \langle v,\ \rangle_B$. Then $q_C(u + \phi(v)) \equiv \phi(v)u \equiv \langle v, u\rangle_B \equiv (B(u, v) + \lambda\overline{B(v, u)}) + (-\lambda\overline{B(v, u)} + B(v, u)) \equiv B(u, v) + B(v, u) \equiv q_B(u + v)$ and $\langle u + \phi(v), u_1 + \phi(v_1)\rangle_C = \langle v, u_1\rangle_B + \lambda\overline{\langle v_1, u\rangle_B} = \langle v, u_1\rangle_B + \langle u, v_1\rangle_B = \langle u + v, u_1 + v_1\rangle_B$.

We consider next the case of hermitian modules. Write $M = U \oplus V$. Define $U \oplus V \to U \oplus U^*$, $u + v \mapsto u + \phi(v)$, where $\phi: V \to U^*$, $v \mapsto B(v,\ )$. Let $\Lambda - \mathbf{M}(U) = (U \oplus U^*, C)$ where $C(f, g) = B(\phi^{-1}f, \phi^{-1}g)$ for all $f, g \in U^*$. Then $\langle u + \phi(v), u_1 + \phi(v_1)\rangle_C = B(v, u_1) + \lambda\overline{B(v_1, u)} + B(v, v_1) = B(v, u_1) + B(u, v_1) + B(v, v_1) = B(u + v, u_1 + v_1)$.

LEMMA 2.9. *Suppose* (M, B) *is a nonsingular* Λ*-quadratic (resp.* Λ*-hermitian) module. Then*

$$(M,B) \perp (M,-B) \cong \Lambda - \mathbf{H}(M) \text{ (resp. } \Lambda - \mathbf{M}(M)).$$

Proof. The diagonal subspace $U = \{(m, m \mid m \in M\}$ of $(M,B) \perp (M,-B)$ is a totally isotropic direct summand such that $U = U^{\perp}$. Hence, the lemma follows from 2.8.

COROLLARY 2.10.

a) $\Lambda - \mathbf{H}(A)$ *is cofinal in* $Q(A,\Lambda)_X$.

b) $\Lambda - \mathbf{H}(A)$ *with a fixed preferred basis is cofinal in* $Q(A,\Lambda)_{\text{based}-Y}$, $Q(A,\Lambda)_{\text{even-based}-Y}$, *and* $Q(A,\Lambda)_{\text{discr-based}-Y}$.

c) *The metabolic modules* $\Lambda - \mathbf{M}(A)$ *are cofinal in* $Q(A,\Lambda)_X$.

d) *Fix a preferred basis for* $A \oplus A^*$ *and give this preferred basis to each* Λ*-metabolic module* $\Lambda - \mathbf{M}(A)$. *The resulting based metabolic modules are cofinal in* $H(A,\Lambda)_{\text{based}-Y}$, $H(A,\Lambda)_{\text{even-based}-Y}$, *and* $H(A,\Lambda)_{\text{discr-based}-Y}$.

Proof. a) and c) follow from 2.9.

b) We consider the based-Y case. The other cases are handled similarly. Suppose $(M,B) \in Q(A,\Lambda)_{\text{based}-Y}$. Forgetting for a moment the preferred bases for M and $\Lambda - \mathbf{H}(A)$, we can find a free Λ-quadratic module (M_1, B_1) such that $(M,B) \perp (M_1,B_1) \cong \overset{n}{\perp} \Lambda - \mathbf{H}(A)$. Let σ denote this isomorphism. Pick a preferred basis for M_1. With respect to the resulting preferred basis for $M \oplus M_1$ and the preferred basis for $\overset{n}{\perp} \Lambda - \mathbf{H}(A)$, σ determines an element, say x, of $K_1(A)$. Let $N = M \oplus M_1$ but choose for N a preferred basis such that the isomorphism $\sigma\colon N \to \overset{n}{\perp} \Lambda - \mathbf{H}(A)$ determines the element x^{-1} of $K_1(A)$. Then $\sigma \perp \sigma\colon ((M,B) \perp (M_1,B_1)) \perp (N, B \oplus B_1) \to \overset{2n}{\perp} \Lambda - \mathbf{H}(A)$ is an isomorphism of based-Y quadratic modules.

d) is proved similarly to b).

LEMMA 2.11. *For notational purposes, recall that* $\mathbf{M}$ *is a functor* $S(A,\Lambda) \to \mathbf{H}(A,\Lambda)$, $(P,\alpha) \mapsto \left(P \oplus P^*, \begin{pmatrix} \alpha & -\lambda I_{P^*} \\ I_P & 0 \end{pmatrix}\right)$. *The assertion is that there is an isomorphism*

$$M(P,\alpha) \perp M(P,-\alpha) \;\cong\; M(P,\alpha) \perp H(P)\,.$$

Proof. By definition $H(P) = M(P,0)$. Identify canonically $M(P,\alpha) \perp M(P,-\alpha) = M(P \oplus P, \alpha \oplus -\alpha)$ and $M(P,\alpha) \perp H(P) = M(P \oplus P, \alpha \oplus 0)$. Then the map

$$\left(\begin{array}{cc|cc} I & & & \\ & I & & \\ \hline & & I & \\ -\bar{\lambda}\alpha & & & I \end{array}\right)\left(\begin{array}{cc|cc} I & -I & & \\ & I & & \\ \hline & & I & \\ & & I & I \end{array}\right)$$

$: P \oplus P \oplus (P \oplus P)^* \to P \oplus P \oplus (P \oplus P)^*$ defines an isomorphism $M(P \oplus P, \alpha \oplus -\alpha) \to M(P \oplus P, \alpha \oplus 0)$.

COROLLARY 2.12. *Pick two preferred bases for* A^n *and give* $(A^n)^*$ *the corresponding dual bases. Give one of the resulting preferred bases on* $A^n \oplus (A^n)^*$ *to* $M(A^n, \alpha)$ *and give the other preferred basis on* $A^n \oplus (A^n)^*$ *to both* $M(A^n, -\alpha)$ *and* $H(A^n)$. *Then, in any of the based categories in Corollary 2.10d), there is an isomorphism*

$$M(A^n,\alpha) \perp M(A^n,-\alpha) \;\cong\; M(A^n,\alpha) \perp H(A^n)\,.$$

Proof. If $\rho : A^n \oplus (A^n)^* \oplus A^n \oplus (A^n)^* \to A^n \oplus A^n \oplus (A^n \oplus A^n)^*$,

$$\rho = \begin{pmatrix} I & & & \\ & 0 & I & \\ & I & 0 & \\ & & & I \end{pmatrix}$$

and if $\sigma : A^n \oplus A^n \oplus (A^n \oplus A^n)^* \to A^n \oplus A^n \oplus (A^n \oplus A^n)^*$,

$$\sigma = \left(\begin{array}{cc|cc} I & & & \\ & I & & \\ \hline & & I & \\ -\bar{\lambda}\alpha & & & I \end{array}\right)\left(\begin{array}{cc|cc} I & -I & & \\ & I & & \\ \hline & & I & \\ & & I & I \end{array}\right)$$

then $\rho^{-1}\sigma\rho$ defines an isomorphism $M(A^n,\alpha) \perp M(A^n,-\alpha) \to M(A^n,\alpha) \perp H(A^n)$.

LEMMA 2.13. *Give* A^n *and* A^{2n} *preferred bases. With respect to these bases, let* β *and* $\begin{pmatrix} \alpha & -\lambda\bar{\beta} \\ \beta & 0 \end{pmatrix}$ *denote matrices corresponding to nonsingular* Λ*-hermitian forms on respectively* A^n *and* A^{2n}. *Let* $M(A^n, \alpha)$ *have a preferred basis as in Corollary 2.12. Then in any of the based categories in Corollary 2.10d), there is an isomorphism*

$$M(A^n,\alpha) \perp (A^n,\beta) \cong \left(A^n \oplus A^n, \begin{pmatrix} \alpha & -\lambda\bar{\beta} \\ \beta & 0 \end{pmatrix}\right) \perp (A^n,\beta).$$

Proof. The matrix $\begin{pmatrix} I & & \\ & \beta & \\ & & \beta^{-1} \end{pmatrix}$ defines an isomorphism

$$M(A^n,\alpha) \perp (A^n,\beta) \xrightarrow{\cong} \left(A^n \oplus A^n, \begin{pmatrix} \alpha & -\lambda\bar{\beta} \\ \beta & 0 \end{pmatrix}\right) \perp (A^n,\beta).$$

§3. AUTOMORPHISM GROUPS OF NONSINGULAR MODULES

Let ${}^{\lambda}(A,\Lambda)$ be a form ring. If σ is a matrix (σ_{ij}) with coefficients $\sigma_{ij} \in A$, let $\overline{\sigma}$ denote the conjugate transpose of σ, i.e. $\overline{\sigma}$ = transpose $(\overline{\sigma}_{ij})$.

If we pick a basis for the free right A-module A^n and the dual basis for $(A^n)^*$ then we can identify the group $\mathrm{Aut}(\Lambda\text{-}H(A^n))$ with a subgroup

$$GQ_{2n}(A,\Lambda)$$

of $GL_{2n}(A)$ called the *general* Λ*-quadratic group.*

LEMMA 3.1. *A* $2n \times 2n$ *matrix* $\begin{pmatrix} \alpha & \beta \\ \gamma & \delta \end{pmatrix} \in GL_{2n}(A)$ *belongs to* $GQ_{2n}(A,\Lambda)$ *if and only if*

i) $\begin{pmatrix} \alpha & \beta \\ \gamma & \delta \end{pmatrix}^{-1} = \begin{pmatrix} \overline{\delta} & \lambda\overline{\beta} \\ \overline{\lambda}\overline{\gamma} & \overline{\alpha} \end{pmatrix}$

ii) *The diagonal coefficients of* $\overline{\gamma}\alpha$ *and* $\overline{\delta}\beta$ *lie in* Λ.

3.1 is an immediate consequence of 3.4 below.

A matrix σ with the properties that $\sigma = -\lambda\overline{\sigma}$ and the diagonal coefficients of σ lie in Λ is called Λ*-hermitian.* It follows from 3.1 i) and ii) that the matrices $\overline{\gamma}\alpha$ and $\overline{\delta}\beta$ in 3.1 ii) are Λ-hermitian.

COROLLARY 3.2. a) *A* $2n \times 2n$ *matrix* $\begin{pmatrix} \alpha & \beta \\ \gamma & \delta \end{pmatrix} \in GL_{2n}(A)$ *belongs to* $GQ_{2n}(A,\Lambda)$ *if and only if*

i) $\overline{\delta}\alpha + \lambda\overline{\beta}\gamma = 1$

$\overline{\delta}\beta + \lambda\overline{\beta}\delta = 0$

$\overline{\lambda}\overline{\gamma}\alpha + \overline{\alpha}\gamma = 0$

ii) *the diagonal coefficients of* $\overline{\gamma}\alpha$ *and* $\overline{\delta}\beta$ *lie in* Λ.

b) *A* $2n \times 2n$ *matrix* $\begin{pmatrix} \alpha & \beta \\ \gamma & \delta \end{pmatrix} \in GL_{2n}(A)$ *belongs to* $GQ_{2n}(A, \Lambda)$ *if and only if*

i) $\alpha\bar{\delta} + \bar{\lambda}\beta\bar{\gamma} = 1$

$\gamma\bar{\delta} + \bar{\lambda}\delta\bar{\gamma} = 0$

$\lambda\alpha\bar{\beta} + \beta\bar{\alpha} = 0$

ii) *the diagonal coefficients of* $\bar{\gamma}\alpha$ *and* $\bar{\delta}\beta$ *lie in* Λ.

c) *A* $2n \times 2n$ *matrix* $\begin{pmatrix} \alpha & \beta \\ \gamma & \delta \end{pmatrix}$ *belongs to* $GQ_{2n}(A, \Lambda)$ *if and only if*

i) *the equations of* a)(i) *and* b)(i) *hold*

ii) *the diagonal coefficients of* $\bar{\gamma}\alpha$ *and* $\bar{\delta}\beta$ *lie in* Λ.

Proof. a) Clearly, (ii) is equivalent to 3.1 (ii). Furthermore, the equations in (i) are equivalent to the assertion that $\begin{pmatrix} \bar{\delta} & \lambda\bar{\beta} \\ \bar{\lambda}\bar{\gamma} & \bar{\alpha} \end{pmatrix}$ is a left inverse to $\begin{pmatrix} \alpha & \beta \\ \gamma & \delta \end{pmatrix}$. Thus, a) follows from 3.1.

b) Clearly, (ii) is equivalent to 3.1 (ii). Furthermore, the equations in (i) are equivalent to the assertion that $\begin{pmatrix} \bar{\delta} & \lambda\bar{\beta} \\ \bar{\lambda}\bar{\gamma} & \bar{\alpha} \end{pmatrix}$ is a right inverse to $\begin{pmatrix} \alpha & \beta \\ \gamma & \delta \end{pmatrix}$. Thus, b) follows from 3.1.

c) is deduced easily from the proofs of a) and b).

COROLLARY 3.3. *If* A *is finitely generated as a module over its center then a* $2n \times 2n$ *matrix* $\begin{pmatrix} \alpha & \beta \\ \gamma & \delta \end{pmatrix}$ *belongs to* $GQ_{2n}(A, \Lambda)$ *if and only if*

i) *either the equations of* 3.2 a)(i) *or* 3.2 b)(i) *hold*

ii) *the diagonal coefficients of* $\bar{\gamma}\alpha$ *and* $\bar{\delta}\beta$ *lie in* Λ.

Proof. If $\begin{pmatrix} \alpha & \beta \\ \gamma & \delta \end{pmatrix}$ lies in $GQ_{2n}(A, \Lambda)$ then it follows from 3.2 that (i) and (ii) are valid. Conversely, suppose that (i) and (ii) are valid. Suppose that (i) means that 3.2 a)(i) holds. 3.2 a)(i) is equivalent to the condition that $\overline{\begin{pmatrix} \alpha & \beta \\ \gamma & \delta \end{pmatrix}}\begin{pmatrix} 0 & \lambda I \\ I & 0 \end{pmatrix}\begin{pmatrix} \alpha & \beta \\ \gamma & \delta \end{pmatrix} = \begin{pmatrix} 0 & \lambda I \\ I & 0 \end{pmatrix}$. Thus, $\begin{pmatrix} \alpha & \beta \\ \gamma & \delta \end{pmatrix}$ is an endomorphism of $H(A^n)$. Thus, by 2.5, $\begin{pmatrix} \alpha & \beta \\ \gamma & \delta \end{pmatrix} \in GL_{2n}(A)$. Thus, the hypotheses of 3.2 a) are satisfied. Thus, $\begin{pmatrix} \alpha & \beta \\ \gamma & \delta \end{pmatrix} \in GQ_{2n}(A,\Lambda)$. Similarly, if (i) means that 3.2 b)(i) holds then one can show that the hypotheses of 3.2 b) are satisfied and conclude that $\begin{pmatrix} \alpha & \beta \\ \gamma & \delta \end{pmatrix} \in GQ_{2n}(A,\Lambda)$.

LEMMA 3.4. *Pick a basis for* A^n *and pick the dual basis for* $(A^n)^*$. *Let* $\phi \in \mathrm{Aut}_A(A^n \oplus (A^n)^*)$ *and let* $\begin{pmatrix} \alpha & \beta \\ \gamma & \delta \end{pmatrix} \in GL_{2n}(A)$ *be the matrix associated with* ϕ. *Then*

i) $$\begin{pmatrix} \alpha & \beta \\ \gamma & \delta \end{pmatrix}^{-1} = \begin{pmatrix} \overline{\delta} & \lambda\overline{\beta} \\ \overline{\lambda\gamma} & \overline{\alpha} \end{pmatrix} \Longleftrightarrow$$

$$\overline{\begin{pmatrix} \alpha & \beta \\ \gamma & \delta \end{pmatrix}}\begin{pmatrix} 0 & \lambda I \\ I & 0 \end{pmatrix}\begin{pmatrix} \alpha & \beta \\ \gamma & \delta \end{pmatrix} = \begin{pmatrix} 0 & \lambda I \\ I & 0 \end{pmatrix} \Longleftrightarrow$$

ϕ *preserves the hermitian form associated to* $\Lambda - \mathbf{H}(A^n)$.

ii) *If* ϕ *satisfies the conditions in* (i) *then the diagonal coefficients of* $\overline{\gamma}\alpha$ *and* $\overline{\delta}\beta$ *lie in* $\Lambda \Longleftrightarrow \phi$ *preserves the quadratic form associated to* $\Lambda - \mathbf{H}(A^n)$.

Proof. i) The matrix of the hermitian form associated to $\Lambda - \mathbf{H}(A^n)$ is $\begin{pmatrix} 0 & \lambda I \\ I & 0 \end{pmatrix}$. i) follows.

ii) By definition, ϕ preserves the Λ-quadratic form on $\Lambda - \mathbf{H}(A^n) \Longleftrightarrow$ the matrices $\overline{\begin{pmatrix} \alpha & \beta \\ \gamma & \delta \end{pmatrix}}\begin{pmatrix} 0 & 0 \\ I & 0 \end{pmatrix}\begin{pmatrix} \alpha & \beta \\ \gamma & \delta \end{pmatrix}$ and $\begin{pmatrix} 0 & 0 \\ I & 0 \end{pmatrix}$ have the same associated Λ-quadratic form. Proceeding straightforwardly, one can verify that the two matrices above have the same associated Λ-quadratic form $\Longleftrightarrow$ their difference $\overline{\begin{pmatrix} \alpha & \beta \\ \gamma & \delta \end{pmatrix}}\begin{pmatrix} 0 & 0 \\ I & 0 \end{pmatrix}\begin{pmatrix} \alpha & \beta \\ \gamma & \delta \end{pmatrix} - \begin{pmatrix} 0 & 0 \\ I & 0 \end{pmatrix}$ is Λ-hermitian. Multiplying out, one obtains the difference above is the matrix $\begin{pmatrix} \overline{\gamma}\alpha & \overline{\gamma}\beta \\ \overline{\delta}\alpha - I & \overline{\delta}\beta \end{pmatrix}$. Since ϕ satisfies (i), it follows that the equations of 3.2 a)(i) hold. Thus, $\begin{pmatrix} \overline{\gamma}\alpha & \overline{\gamma}\beta \\ \overline{\delta}\alpha - I & \overline{\delta}\beta \end{pmatrix} = \begin{pmatrix} \overline{\gamma}\alpha & \overline{\gamma}\beta \\ -\lambda\overline{\beta}\gamma & \overline{\delta}\beta \end{pmatrix}$ and the matrices $\overline{\gamma}\alpha$ and $\overline{\delta}\beta$ are max-hermitian, i.e. $\overline{\gamma}\alpha = -\lambda\overline{\overline{\gamma}\alpha}$ and $\overline{\delta}\beta = -\lambda\overline{\overline{\delta}\beta}$. Thus, $\begin{pmatrix} \overline{\gamma}\alpha & \overline{\gamma}\beta \\ -\lambda\overline{\beta}\gamma & \overline{\delta}\beta \end{pmatrix}$ is Λ-hermitian $\Longleftrightarrow$ the diagonal coefficients of $\overline{\gamma}\alpha$ and $\overline{\delta}\beta$ lie in Λ.

The equations 3.2 show that if α and δ are invertible $n\times n$ matrices then $\begin{pmatrix} \alpha & 0 \\ 0 & \delta \end{pmatrix}$ belongs to $GQ_{2n}(A,\Lambda) \Longleftrightarrow \delta = \overline{\alpha}^{-1}$. The matrix $\begin{pmatrix} \alpha & 0 \\ 0 & \overline{\alpha}^{-1} \end{pmatrix}$ is called a *hyperbolic matrix* and is denoted often by

$$H(\alpha) = \begin{pmatrix} \alpha & 0 \\ 0 & \overline{\alpha}^{-1} \end{pmatrix}.$$

The equations 3.2 show that if β (resp. γ) is an $n\times n$ matrix then $\begin{pmatrix} I & \beta \\ 0 & I \end{pmatrix}$ (resp. $\begin{pmatrix} I & 0 \\ \gamma & I \end{pmatrix}$) belongs to $GQ_{2n}(A,\Lambda) \Longleftrightarrow \beta$ is Λ-hermitian (resp. γ is $\overline{\Lambda}$-hermitian). We let

$$EQ_{2n}(A,\Lambda)$$

denote the subgroup of $GQ_{2n}(A,\Lambda)$ generated by all $\begin{pmatrix} I & \beta \\ 0 & I \end{pmatrix}$, $\begin{pmatrix} I & 0 \\ \gamma & I \end{pmatrix}$, and $\begin{pmatrix} \varepsilon & 0 \\ 0 & \overline{\varepsilon}^{-1} \end{pmatrix}$ such that β and γ are as above and ε is an elementary matrix (defined below). $EQ_{2n}(A,\Lambda)$ is called the *elementary Λ-quadratic group*.

$EQ_{2n}(A,\Lambda)$ has a set of generators called *elementary Λ-quadratic matrices* which we describe next. Let i and j be two integers between 1 and n and let $a \in A$. An $n\times n$ matrix is called *elementary* if it is of the kind

$$\varepsilon_{ij}(a) = \begin{pmatrix} 1 & & & a_{ij} \\ & \cdot & & \\ & & \cdot & \\ & & & 1 \end{pmatrix} \quad (i \neq j).$$

The notation $\begin{pmatrix} 1 & & & a_{ij} \\ & \cdot & & \\ & & \cdot & \\ & & & 1 \end{pmatrix}$ denotes the matrix with a as (i, j)'th coefficient, the identity as diagonal coefficients, and 0 as all other coefficients. We call a $2n \times 2n$ matrix *elementary* Λ*-quadratic* if it is one of

$$H(\varepsilon_{ij}(a)) = \left(\begin{array}{ccc|ccc} 1 & \cdot & a_{ij} & & & \\ & \cdot & & & & \\ & & 1 & & & \\ \hline & & & 1 & & \\ & & & & \cdot & \\ & & & & & \cdot \\ & & & & & \cdot \\ & & & -\bar{a}_{ji} & & 1 \end{array}\right) \quad (i \neq j)$$

$$\varepsilon_{i,n+j}(a) = \left(\begin{array}{ccc|ccc} 1 & & & 0 & \cdot & a_{ij} \\ & \cdot & & & \cdot & \\ & & \cdot \; 1 & -\lambda\bar{a}_{ji} & \cdot & 0 \\ \hline & & & 1 & \cdot & \\ & & & & \cdot & \\ & & & & \cdot & 1 \end{array}\right) \quad (i \neq j)$$

$$\varepsilon_{n+i,j}(a) = \left(\begin{array}{ccc|ccc} 1 & \cdot & & & & \\ & \cdot & & & & \\ & & \cdot \; 1 & & & \\ \hline 0 & \cdot & a_{ij} & 1 & \cdot & \\ & \cdot & & & \cdot & \\ -\lambda\bar{a}_{ji} & \cdot & 0 & & \cdot & 1 \end{array}\right) \quad (i \neq j)$$

$$\varepsilon_{i,n+i}(a) = \left(\begin{array}{ccc|ccc} 1 & & & 0 & & \\ & \cdot & & & \cdot & \\ & & & & 0 & \\ & \cdot & & & a_{ii} & \\ & & & & \cdot \; 0 & \\ & & \cdot & & & \cdot \\ & & 1 & & & 0 \\ \hline & & & 1 & & \\ & & & & \cdot & \\ & & & & & \cdot \\ & & & & & \cdot \; 1 \end{array}\right) \quad (a \in \Lambda)$$

$$\varepsilon_{n+i,i}(a) = \left(\begin{array}{ccc|ccc} 1 & & & & & \\ & \ddots & & & & \\ & & 1 & & & \\ \hline 0 & & & 1 & & \\ & a_{ii} & & & \ddots & \\ & & 0 & & & 1 \end{array}\right) (a \in \bar{\Lambda}) .$$

The first matrix above is called a *hyperbolic elementary* Λ*-quadratic matrix*. The elementary Λ-quadratic matrices satisfy the following identities which are similar to the standard identities [10, V 1.2 (a), (b)] for elementary matrices. Later, we shall give a longer list of identities sufficient to define the functor KQ_2. If σ and ρ are two elements of a group, we let $[\sigma, \rho] = \sigma^{-1}\rho^{-1}\sigma\rho$ denote their *commutator*.

3.5 (a) *If* $\tau(a) = \varepsilon_{rs}(a)$ *is an elementary* Λ*-quadratic matrix then* $\tau(a)\tau(b) = \tau(a+b)$.

If i, j, *and* k *are distinct then*

(b) $[H(\varepsilon_{ij}(a)), \varepsilon_{j,n+k}(b)] = \varepsilon_{i\,n+k}(ab)$

$[H(\varepsilon_{ij}(a)), \varepsilon_{j,n+j}(b)] = \varepsilon_{i\,n+j}(ab)\,\varepsilon_{j,n+j}(ab\bar{a})$

(c) $[H(\varepsilon_{ij}(a)), \varepsilon_{n+i,k}(b)] = \varepsilon_{n+j,k}(-\bar{a}b)$

$[H(\varepsilon_{ij}(a)), \varepsilon_{n+i,i}(b)] = \varepsilon_{n+j,i}(-\bar{a}b)\,\varepsilon_{n+i,i}(-\bar{a}ba)$

(d) $[\varepsilon_{i,n+j}(a), \varepsilon_{n+i,k}(b)] = H(\varepsilon_{jk}(-\lambda\bar{a}b))$

$[\varepsilon_{i,n+j}(a), \varepsilon_{n+j,j}(b)] = H(\varepsilon_{ij}(ab))\,\varepsilon_{i,n+i}(-\lambda ab\bar{a})$.

3.5 follows with a little computation from the following matrix identities.

3.6 (a) $\left[\begin{pmatrix} \varepsilon & 0 \\ 0 & \bar{\varepsilon}^{-1} \end{pmatrix}, \begin{pmatrix} I & \beta \\ 0 & I \end{pmatrix}\right] = \begin{pmatrix} I & \beta - \varepsilon^{-1}\beta\bar{\varepsilon}^{-1} \\ 0 & I \end{pmatrix}$

(b) $\left[\begin{pmatrix} \varepsilon & 0 \\ 0 & \bar{\varepsilon}^{-1} \end{pmatrix}, \begin{pmatrix} I & 0 \\ \gamma & I \end{pmatrix}\right] = \begin{pmatrix} I & 0 \\ \gamma - \bar{\varepsilon}\gamma\varepsilon & I \end{pmatrix}$

(c) $\left[\begin{pmatrix} I & \beta \\ 0 & I \end{pmatrix}, \begin{pmatrix} I & 0 \\ \gamma & I \end{pmatrix}\right] = \begin{pmatrix} I + \beta\gamma + \beta\gamma\beta\gamma & \beta\gamma\beta \\ -\gamma\beta\gamma & I - \gamma\beta \end{pmatrix}$.

Let $\mathfrak{q}$ be an involution invariant ideal of A. We define

$$GQ_{2n}(A,\Lambda,\mathfrak{q}) = \ker(GQ_{2n}(A,\Lambda) \to GQ_{2n}(A/\mathfrak{q},\Lambda/\mathfrak{q}))$$

where $\Lambda/\mathfrak{q}$ denotes the image of Λ in $A/\mathfrak{q}$. We define

$$EQ_{2n}(A,\Lambda,\mathfrak{q})$$

to be the normal subgroup of $EQ_{2n}(A,\Lambda)$ generated by all elementary Λ-quadratic matrices where the off diagonal coefficients lie in $\mathfrak{q}$. $GQ_{2n}(A,\Lambda,\mathfrak{q})$ (resp. $EQ_{2n}(A,\Lambda,\mathfrak{q})$) is called the *relative* or *congruence subgroup* (resp. *relative elementary subgroup*) *of level* $\mathfrak{q}$. Clearly, $GQ_{2n}(A,\Lambda,A) = GQ_{2n}(A,\Lambda)$ and $EQ_{2n}(A,\Lambda,\Lambda) = EQ_{2n}(A,\Lambda)$.

If G is a group and $H \subset G$ is a normal subgroup then the *mixed commutator group* $[G,H]$ of G and H is the subgroup of G generated by all commutators $[g,h]$ such that $g \in G$ and $h \in H$. The group $[G,G]$ is the *commutator subgroup* of G. G is called *connected* or *perfect* if $G = [G,G]$.

The next result is deduced routinely from 3.5 and 3.6. Details will be left to the reader.

3.7. *If* $n \geq 3$ *then*

(a) $EQ_{2n}(A,\Lambda,\mathfrak{q})$ *is generated as a normal subgroup of* $EQ_{2n}(A,\Lambda)$ *by all*

$$\begin{pmatrix} I & \beta \\ 0 & I \end{pmatrix} \quad \text{and} \quad \begin{pmatrix} I & 0 \\ \gamma & I \end{pmatrix}$$

where α *and* $\beta \equiv 0 \bmod \mathfrak{q}$.

(b) $EQ_{2n}(A,\Lambda,\mathfrak{q}) = [EQ_{2n}(A,\Lambda,\mathfrak{q}), EQ_{2n}(A,\Lambda)]$. *In particular,* $EQ_{2n}(A,\Lambda)$ *is perfect.*

There is a natural embedding

$$GQ_{2n}(A,\Lambda) \to GQ_{2(n+1)}(A,\Lambda), \begin{pmatrix} \alpha & \beta \\ \gamma & \delta \end{pmatrix} \mapsto \left(\begin{array}{cc|cc} \alpha & 0 & \beta & 0 \\ 0 & 1 & 0 & 0 \\ \hline \gamma & 0 & \delta & 0 \\ 0 & 0 & 0 & 1 \end{array}\right).$$

We define

$$GQ(A,\Lambda,q) = \varinjlim_{n} GQ_{2n}(A,\Lambda,q)$$

$$EQ(A,\Lambda,q) = \varinjlim_{n} EQ_{2n}(A,\Lambda,q).$$

QUADRATIC WHITEHEAD LEMMA 3.8. *Suppose*

$$\begin{pmatrix} \alpha & \beta \\ \gamma & \delta \end{pmatrix} \text{ and } \begin{pmatrix} A & B \\ C & D \end{pmatrix} \in GQ_{2n}(A,\Lambda).$$

Then in $GQ_{4n}(A,\Lambda)$ *we have*

$$\left(\begin{array}{cc|cc} \alpha & & \beta & \\ & I & & \\ \hline \gamma & & \delta & \\ & & & I \end{array}\right)\left(\begin{array}{cc|cc} A & & B & \\ & I & & \\ \hline C & & D & \\ & & & I \end{array}\right) =$$

$$\left(\begin{array}{cc|cc} I & & & -\beta \\ \bar{\delta} & I & \lambda\bar{\beta} & \\ \hline & & I & -\delta \\ & & & I \end{array}\right)\left(\begin{array}{cc|cc} & I & & \\ -I & & & \\ \hline & & & I \\ & & -I & \end{array}\right)\left(\begin{array}{cc|cc} A & & B & \\ & \alpha & & \beta \\ \hline C & & D & \\ & \gamma & & \delta \end{array}\right)\left(\begin{array}{cc|cc} I & & & -\lambda\bar{B} \\ A & I & B & \\ \hline & & I & -\bar{A} \\ & & & I \end{array}\right)\left(\begin{array}{cc|cc} I & -\bar{D} & & \\ & I & & \\ \hline & -\overline{\lambda C} & I & \\ C & & D & I \end{array}\right).$$

Proof. Straightforward computation.

COROLLARY 3.9. $[GQ(A,\Lambda,q), GQ(A,\Lambda)] = EQ(A,\Lambda,q)$.

Proof. Consider the absolute case (i.e. $q = A$) first. 3.5 shows that the left-hand side contains the right-hand side. Conversely, suppose

$$P = \begin{pmatrix} A & B \\ C & D \end{pmatrix} \in GQ_{2n}(A,\Lambda)$$

$$P \perp I = \text{image of } P \text{ in } GQ_{4n}(A,\Lambda)$$

$$I \perp P = \left(\begin{array}{cc|cc} I & & & \\ & A & & B \\ \hline & & I & \\ & C & & D \end{array}\right) .$$

3.8 implies $(P \perp I)^{-1}(I \perp P) \in EQ_{4n}(A,\Lambda)$. Thus, if $\pi \in GQ_{2n}(A,\Lambda)$ we obtain (for suitable E, E_1, $E_2 \in EQ_{4n}(A,\Lambda)$) $(\pi \perp I)(P \perp I) = (\pi \perp I)(I \perp P)$ $E = (I \perp P)(I \perp \pi)E_1E = (I \perp (P\pi))E_1E = (P\pi \perp I)E_2E_1E = (P \perp I)(\pi \perp I)E_2E_1E$.

We consider the general case next. We use the relativization procedure described in §4C and identify $G = GQ(A \ltimes q, \Lambda \ltimes \Lambda \cap q)$ with $GQ(A,\Lambda) \ltimes GQ(A,\Lambda,q)$ and $E = EQ(A \ltimes q, \Lambda \ltimes \Lambda \cap q)$ with $EQ(A,\Lambda) \ltimes EQ(A,\Lambda,q)$. From the absolute case, we obtain that $[G,G] = E$. But the standard commutator formulas show that $[G,G] = [GQ(A,\Lambda), GQ(A,\Lambda)] \ltimes [GQ(A,\Lambda), GQ(A,\Lambda,q)]$. Hence $EQ(A,\Lambda,q) = [GQ(A,\Lambda), GQ(A,\Lambda,q)]$.

The next theorem extends a result of Sharpe [26, §5] by eliminating the assumptions that λ be a root of unity and that the form parameter $\Lambda = \min$. In fact, if one examines Sharpe's proof, one sees that he does not use the latter assumption. Our proof simplifies the matrix computations in Sharpe's paper.

Let

$$\omega_4 = \left(\begin{array}{cc|cc} & & & -1 \\ & & \lambda & \\ \hline & -\bar{\lambda} & & \\ 1 & & & \end{array}\right) = \left(\begin{array}{cc|cc} 1 & & & -1 \\ & 1 & \lambda & \\ \hline & & 1 & \\ & & & 1 \end{array}\right)\left(\begin{array}{cc|cc} 1 & & & \\ & 1 & & \\ \hline & -\bar{\lambda} & 1 & \\ 1 & & & 1 \end{array}\right)\left(\begin{array}{cc|cc} 1 & & & -1 \\ & 1 & \lambda & \\ \hline & & 1 & \\ & & & 1 \end{array}\right)$$

$$\omega_{4n} = \omega_4 \perp \cdots \perp \omega_4 \quad (n \text{ times}).$$

$\omega_{4n} \in EQ_{4n}(A,\Lambda)$.

THEOREM 3.10. *Every element of* $EQ(A,\Lambda)$ *can be expressed as the product below of 5 elements from* $EQ_{4n}(A,\Lambda)$ *where* n *is suitably large and* E *is a product of* $n \times n$ *elementary matrices.*

$$\begin{pmatrix} I & U \\ 0 & I \end{pmatrix} \omega_{4n} \begin{pmatrix} I & B \\ 0 & I \end{pmatrix}\begin{pmatrix} I & 0 \\ L & I \end{pmatrix}\begin{pmatrix} E & 0 \\ 0 & \bar{E}^{-1} \end{pmatrix} .$$

Proof. Let $\varepsilon_4^+ = \left(\begin{array}{cc|cc} 1 & & & 1 \\ & 1 & -\lambda & \\ \hline & & 1 & \\ & & & 1 \end{array}\right)$ and $\varepsilon_4^- = \left(\begin{array}{cc|cc} 1 & & & \\ & 1 & & \\ \hline & \bar{\lambda} & 1 & \\ -1 & & & 1 \end{array}\right)$.

If $\varepsilon_{4m}^+ = \varepsilon_4^+ \perp \cdots \perp \varepsilon_4^+$ (m times) and $\varepsilon_{4m}^- = \varepsilon_4^- \perp \cdots \perp \varepsilon_4^-$ (m times) then the identities preceding 3.10 show that $\varepsilon_{4m}^+ \omega_{4m} \varepsilon_{4m}^+ \varepsilon_{4m}^- = 1$. By stabilizing a 5-term decomposition $u\,\omega_{4n}\,b\,\ell\,a \in EQ_{4n}(A,\Lambda)$ we mean replacing it by $(u \perp \varepsilon_{4m}^+)\omega_{4(n+m)}(b \perp \varepsilon_{4m}^+)(\ell \perp \varepsilon_{4m}^-)(a \perp I_{4m}) \in EQ_{4(n+m)}(A,\Lambda)$. However, by *stabilizing* an element $\begin{pmatrix} \alpha & \beta \\ \gamma & \delta \end{pmatrix} \in E_{2n}(A,\Lambda)$ we mean replacing it by its canonical image in $EQ_{2(n+m)}(A,\Lambda)$.

Suppose we have an element $x \in EQ(A,\Lambda)$ which has a 5-term decomposition $x = u\omega_{4n}b\ell a$ as in the theorem. Since $EQ(A,\Lambda)$ is generated by matrices of the kind

$$\varepsilon_{4m}^+(U) = \begin{pmatrix} I_{2m} & U \\ 0 & I_{2m} \end{pmatrix} \quad \text{and} \quad \varepsilon_{4m}^-(L) = \begin{pmatrix} I_{2m} & 0 \\ L & I_{2m} \end{pmatrix}$$

it suffices to show that $\varepsilon_{4m}^+(U)x$ and $\varepsilon_{4m}(L)x$ have 5-term decompositions. After stabilizing the 5-term decomposition of x we can assume that $m = n$. Then $\varepsilon_{4m}^+(U)x = (\varepsilon_{4m}^+(U)u)\omega_{4n}b\ell a$ which is a 5-term decomposition for $\varepsilon_{4m}^+(U)x$. For a suitable U, we can write $\varepsilon_{4m}^-(L) = \omega_{4m}^{-1}\varepsilon_{4m}^+(U)\omega_{4m}$. Note that $\omega_{4m}^{-1} = -\omega_{4m}$. Hence, to complete the proof of the theorem, it suffices to show that if $m \le n$ then $\omega_{4m}x$ has a 5-term decomposition. But this is an immediate consequence of the next lemma.

REDUCTION LEMMA 3.11. *Let* $0 < m \le n$. *Let*

$$I = I_{2n}, \quad 1 = I_{2m} \ \textit{or} \ I_{2(n-m)}$$

$$\tau = \begin{pmatrix} 0 & -1 \\ \lambda & 0 \end{pmatrix} \oplus \cdots \oplus \begin{pmatrix} 0 & -1 \\ \lambda & \sigma \end{pmatrix} \quad (n \ \textit{times})$$

$$\pi = \begin{pmatrix} 0 & -1 \\ \lambda & 0 \end{pmatrix} \oplus \cdots \oplus \begin{pmatrix} 0 & -1 \\ \lambda & 0 \end{pmatrix} \quad (\text{m } \textit{times})$$

$$\pi' = \begin{pmatrix} 0 & \overline{\lambda} \\ -1 & 0 \end{pmatrix} \oplus \cdots \oplus \begin{pmatrix} 0 & \overline{\lambda} \\ -1 & 0 \end{pmatrix} \quad ((\text{n}-\text{m}) \ \textit{times}) .$$

Let

$$\begin{pmatrix} I & U \\ 0 & I \end{pmatrix}, \begin{pmatrix} I & B \\ 0 & I \end{pmatrix} \in EQ_{4n}(A, \Lambda)$$

Let

$$U = \begin{pmatrix} U_1 & U_2 \\ -\lambda \overline{U}_2 & U_4 \end{pmatrix}$$

where U_1 *and* U_4 *are respectively* $2m \times 2m$ *and* $2(n-m) \times 2(n-m)$ *matrices. Then*

$$\left(\begin{array}{ccc|ccc} 0 & & & \pi & 0 & \\ & 1 & & 0 & 0 & \\ & & I & & & \\ \hline -\overline{\pi} & 0 & & 0 & & \\ 0 & 0 & & & 1 & \\ & & & & & I \end{array}\right) \left(\begin{array}{cc|cc} I & & U & \\ & I & & \\ \hline & & I & \\ & & & I \end{array}\right) \left(\begin{array}{cc|cc} & & \tau & \\ & I & & \\ \hline -\overline{\tau} & & 0 & \\ & & & I \end{array}\right) \left(\begin{array}{cc|cc} I & & B & \\ & I & & \\ \hline & & I & \\ & & & I \end{array}\right) =$$

$$\left\{ \left(\begin{array}{ccc|ccc} 1 & & & 0 & 0 & \\ \overline{U}_2 \pi & 1 & & 0 & U_4 & \\ & & I & & & \\ \hline & & & 1 & -\overline{\pi} U_2 & \\ & & & & 1 & \\ & & & & & I \end{array}\right) \left(\begin{array}{ccc|cccc} I & & & & & -\pi & 0 \\ & & & & & 0 & 0 \\ 0 & 0 & I & -\pi & 0 & & \\ 0 & 1 & & 0 & 0 & & \\ \hline & & & I & & 0 & 0 \\ & & & & & 0 & -1 \\ & & & & & I & \end{array}\right) \left(\begin{array}{cc|cc} I & & & \\ -I & & & \\ \hline & & & I \\ & & -I & \end{array}\right) \left(\begin{array}{cc|cc} I & & U_1 0 & \\ & & 0\ 0 & \\ & I & & \\ \hline & & I & \\ & & & I \end{array}\right) \right.$$

$$
\left.\begin{matrix} \\ I_{4n}\perp 1_{4m}\perp \varepsilon^{+}_{4(n-m)} \\ \\ \end{matrix}\right\} \omega_{8n} \left\{ I_{4n}\perp 1_{4m}\perp \varepsilon^{+}_{4(n-m)} \begin{pmatrix} I & \begin{smallmatrix} 0 & 0 \\ 0 & -\tau\pi' \end{smallmatrix} & & \tau \\ \begin{smallmatrix} -U_1\bar{\tau} & 0 \\ 0 & 0 \end{smallmatrix} & I & \tau & \\ \hline & & I & \begin{smallmatrix} \tau\bar{U}_1 & 0 \\ 0 & 0 \end{smallmatrix} \\ & & \begin{smallmatrix} 0 & 0 \\ 0 & \pi'\tau \end{smallmatrix} & I \end{pmatrix} \right.
$$

$$
\left.\left(\begin{array}{cc|cc} I & & B & \\ & I & & \\ \hline & & I & \\ & & & I \end{array}\right)\right\} \left\{ I_{4n}\perp 1_{4m}\perp \varepsilon^{-}_{4(n-m)} \left(\begin{array}{cc|cc} I & B\bar{\tau} & & \\ & I & & \\ \hline & -\bar{\tau} & I & \\ -\bar{\tau} & & -\bar{\tau}B & I \end{array}\right)\right\}
$$

Proof. Since

$$
\left(\begin{array}{ccc|ccc} 0 & & & \pi & 0 & \\ & 1 & & 0 & 0 & \\ & & I & & & \\ \hline -\bar{\pi} & 0 & & 0 & & \\ 0 & 0 & & & 1 & \\ & & & & & I \end{array}\right) \left(\begin{array}{ccc|c} I & 0 & U_2 & \\ & -\lambda\bar{U}_2 & U_4 & \\ & & I & \\ \hline & & & I \\ & & & \\ & & & I \end{array}\right) \left(\begin{array}{ccc|ccc} 0 & & & \pi & 0 & \\ & 1 & & 0 & 0 & \\ & & I & & & \\ \hline -\bar{\pi} & 0 & & 0 & & \\ 0 & 0 & & & 1 & \\ & & & & & I \end{array}\right)^{-1} =
$$

$$
\left(\begin{array}{ccc|ccc} 1 & & & 0 & 0 & \\ \bar{U}_2\pi & 1 & & 0 & U_4 & \\ & & I & & & \\ \hline & & & 1 & -\bar{\pi}U_2 & \\ & & & & 1 & \\ & & & & & I \end{array}\right), \text{ we may assume } U_2 = U_4 = 0 .
$$

$$
\left(\begin{array}{ccc|ccc} 0 & & & \pi & 0 & \\ & 1 & & 0 & 0 & \\ & & I & & & \\ \hline -\bar{\pi} & 0 & & 0 & & \\ 0 & 0 & & & 1 & \\ & & & & & I \end{array}\right) \left(\begin{array}{cc|cc} I & & \begin{smallmatrix} U_1 & 0 \\ 0 & 0 \end{smallmatrix} & \\ & I & & \\ \hline & & I & \\ & & & I \end{array}\right) \left(\begin{array}{cc|cc} 0 & & \tau & \\ & I & & \\ \hline -\bar{\tau} & & 0 & \\ & & & I \end{array}\right) \left(\begin{array}{cc|cc} I & & B & \\ & I & & \\ \hline & & I & \\ & & & I \end{array}\right) =
$$

(by 3.8, and letting ϕ = product of last 2 matrices)

$$\left(\begin{array}{ccc|ccc} I & & & & \begin{smallmatrix}-\pi & 0\\ 0 & 0\end{smallmatrix} & \\ & \begin{smallmatrix}0 & 0\\ 0 & 1\end{smallmatrix} & & \begin{smallmatrix}-\pi & 0\\ 0 & 0\end{smallmatrix} & & \\ & & I & & & \\ \hline & & & I & & \\ & & & & \begin{smallmatrix}0 & 0\\ 0 & -1\end{smallmatrix} & \\ & & & & & I \end{array}\right)\left(\begin{array}{cc|cc} & I & & \\ -I & & & \\ \hline & & & I \\ & & -I & \end{array}\right)\left(\begin{array}{cc|cc} I & & \begin{smallmatrix}U_1 & 0\\ 0 & 0\end{smallmatrix} & \\ & I & & \\ \hline & & I & \\ & & & I \end{array}\right)$$

$$\left(\begin{array}{ccc|ccc} I & & & & & \\ & 0 & & & \pi & 0 \\ & & 1 & & 0 & 0 \\ \hline & & & I & & \\ -\bar\pi & 0 & & & 0 & \\ 0 & 0 & & & & 1 \end{array}\right)\left(\begin{array}{cc|cc} I & & \begin{smallmatrix}U_1 & 0\\ 0 & 0\end{smallmatrix} & \begin{smallmatrix}-\lambda\bar U_1 & 0\\ 0 & 0\end{smallmatrix} \\ I & I & & \\ \hline & & I & -I \\ & & & I \end{array}\right)\left(\begin{array}{cc|cc} I & -I & & \\ & I & & \\ \hline & & I & \\ & & I & I \end{array}\right)\phi =$$

(letting ϕ_1 = product of first 3 matrices, and passing $\left(\begin{array}{c|c} & \tau \\ \hline -\bar\tau & \end{array}\right)$ by the sixth and fifth matrices)

$$\phi_1\left(\begin{array}{cc|cc} \begin{smallmatrix}0 & 0\\ 0 & 1\end{smallmatrix} & & \tau & \begin{smallmatrix}\pi & 0\\ 0 & 0\end{smallmatrix} \\ \hline -\bar\tau & & & \\ \begin{smallmatrix}-\bar\pi & 0\\ 0 & 0\end{smallmatrix} & & & \begin{smallmatrix}0 & 0\\ 0 & 1\end{smallmatrix} \end{array}\right)\left(\begin{array}{cc|cc} I & & & \tau \\ \begin{smallmatrix}-U_1\bar\tau & 0\\ 0 & 0\end{smallmatrix} & I & \tau & \\ \hline & & & \begin{smallmatrix}\tau\bar U_1 & 0\\ 0 & 0\end{smallmatrix} \\ & & & I \end{array}\right)$$

$$\left(\begin{array}{cc|cc} I & & & \\ & I & & \\ \hline & -\bar\tau & I & \\ -\bar\tau & & & I \end{array}\right)\left(\begin{array}{cc|cc} I & & B & \\ & I & & \\ \hline & & I & \\ & & & I \end{array}\right) =$$

(passing the last matrix by the one preceding it, and then using the identity $1_{4(n-m)} = \varepsilon^+_{4(n-m)}\,\omega_{4(n-m)}\,\varepsilon^+_{4(n-m)}\,\varepsilon^-_{4(n-m)}$)

$$\phi_1(I_{4n}\perp 1_{4m}\perp \varepsilon^+_{4(n-m)})\,\omega_{8n}(I_{4n}\perp 1_{4m}\perp \varepsilon_{4(n-m)})(I_{4n}\perp 1_{4m}\perp \varepsilon^-_{4(n-m)})$$

$$\left(\begin{array}{ccc|ccc} & I & & & & \tau \\ -\bar{U}_1\tau & 0 & I & \tau & & \\ 0 & 0 & & & & \\ \hline & & & I & \tau\bar{U}_1 & 0 \\ & & & & 0 & 0 \\ & & & & & I \end{array}\right)\left(\begin{array}{cc|cc} I & & B & \\ & I & & \\ \hline & & I & \\ & & & I \end{array}\right)\left(\begin{array}{cc|cc} I & B\bar{\tau} & & \\ & I & & \\ \hline & -\bar{\tau} & I & \\ -\bar{\tau} & & -\bar{\tau}B & I \end{array}\right) =$$

(passing $I_{4n} \perp 1_{4m} \perp \varepsilon_{4(n-m)}^{-}$ by the 2 matrices immediately to the right of it) the right-hand side of the identity in the lemma.

The following special case of the reduction lemma is useful for applications.

COROLLARY 3.12. *Let*

$$I = I_{2n}$$

$$\pi = \begin{pmatrix} 0 & -1 \\ \lambda & 0 \end{pmatrix} \oplus \cdots \oplus \begin{pmatrix} 0 & -1 \\ \lambda & 0 \end{pmatrix} \quad (n \textit{ times})$$

$$\omega_{4n} = \begin{pmatrix} 0 & \pi \\ -\pi^{-1} & 0 \end{pmatrix}$$

$$\omega_{8n} = \omega_{4n} \oplus \omega_{4n}$$

$$\begin{pmatrix} I & \beta \\ & I \end{pmatrix}, \begin{pmatrix} I & U \\ & I \end{pmatrix}, \begin{pmatrix} I & B \\ & I \end{pmatrix} \in EQ_{4n}(A, \Lambda) .$$

Then

$$\begin{pmatrix} I & \beta \\ & I \end{pmatrix} \omega_{4n} \begin{pmatrix} I & U \\ & I \end{pmatrix} \omega_{4n} \begin{pmatrix} I & B \\ & I \end{pmatrix} =$$

$$\left(\begin{array}{cc|cc} & I & & \\ -I & & & \\ \hline & & & I \\ & & -I & \end{array}\right)\left(\begin{array}{cc|cc} I & U\bar{\pi} & & \\ & I & & \\ \hline & & I & \\ & & -\bar{U} & I \end{array}\right)\left(\begin{array}{cc|cc} I & & -U+U\bar{\pi}\beta\pi\bar{U} & \pi-U\bar{\pi}\beta \\ & I & \pi-\beta\pi\bar{U} & \beta \\ \hline & & I & \\ & & & I \end{array}\right)$$

$$\omega_{8n}\left(\begin{array}{cc|cc} I & & B & \pi \\ & I & \pi & U \\ \hline & & I & \\ & & & I \end{array}\right)\left(\begin{array}{cc|cc} I & & & \\ & I & & \\ \hline & -\bar{\pi} & I & \\ -\bar{\pi} & \bar{\pi}B\bar{\pi} & & I \end{array}\right)\left(\begin{array}{cc|cc} I & B\bar{\pi} & & \\ & I & & \\ \hline & & I & \\ & & -\bar{\pi}B & I \end{array}\right).$$

Proof. $\begin{pmatrix} I & \beta \\ & I \end{pmatrix}\omega_{4n}\begin{pmatrix} I & U \\ & I \end{pmatrix}\omega_{4n}\begin{pmatrix} I & B \\ & I \end{pmatrix}$ = (by 3.11)

$$\left(\begin{array}{cc|cc} I & & \beta & \\ & I & & \\ \hline & & I & \\ & & & I \end{array}\right)\left(\begin{array}{cc|cc} I & & & -\pi \\ & I & -\pi & \\ \hline & & I & \\ & & & I \end{array}\right)\left(\begin{array}{cc|cc} & I & & \\ -I & & & \\ \hline & & & I \\ & & -I & \end{array}\right)\left(\begin{array}{cc|cc} I & & U & \\ & I & & \\ \hline & & I & \\ & & & I \end{array}\right)$$

$$\omega_{8n}\left(\begin{array}{cc|cc} I & & & \\ -U\bar{\pi} & I & & \\ \hline & & I & \pi\bar{U} \\ & & & I \end{array}\right)\left(\begin{array}{cc|cc} I & & & \pi \\ & I & \pi & U \\ \hline & & I & \\ & & & I \end{array}\right)\left(\begin{array}{cc|cc} I & & B & \\ & I & & \\ \hline & & I & \\ & & & I \end{array}\right)$$

$$\left(\begin{array}{cc|cc} I & & & \\ & I & & \\ \hline & -\bar{\pi} & I & \\ -\bar{\pi} & \bar{\pi}B\bar{\pi} & & I \end{array}\right)\left(\begin{array}{cc|cc} I & B\bar{\pi} & & \\ & I & & \\ \hline & & I & \\ & & -\bar{\pi}B & I \end{array}\right) = \text{(moving} \left(\begin{array}{cc|cc} & I & & \\ -I & & & \\ \hline & & & I \\ & & -I & \end{array}\right)$$

two places to the left and consolidating terms)

$$\left(\begin{array}{cc|cc} & I & & \\ -I & & & \\ \hline & & & I \\ & & -I & \end{array}\right)\left(\begin{array}{cc|cc} I & & U & \pi \\ & I & \pi & \beta \\ \hline & & I & \\ & & & I \end{array}\right)\omega_{8n}\left(\begin{array}{cc|cc} I & & & \\ -U\bar{\pi} & I & & \\ \hline & & I & \pi\bar{U} \\ & & & I \end{array}\right)$$

$$\left(\begin{array}{cc|cc} I & & B & \pi \\ & I & \pi & U \\ \hline & & I & \\ & & & I \end{array}\right)\left(\begin{array}{cc|cc} I & & & \\ & I & & \\ \hline & -\bar{\pi} & I & \\ -\bar{\pi} & \bar{\pi}B\bar{\pi} & & I \end{array}\right)\left(\begin{array}{cc|cc} I & B\bar{\pi} & & \\ & I & & \\ \hline & & I & \\ & & -\bar{\pi}B & I \end{array}\right).$$

If one moves ω_{8n} one place to the right then one changes

$$\left(\begin{array}{cc|cc} I & & & \\ -U\bar{\pi} & I & & \\ \hline & & I & \pi\bar{U} \\ & & & I \end{array}\right) \text{ to } \left(\begin{array}{cc|cc} I & U\bar{\pi} & & \\ & I & & \\ \hline & & I & \\ & & -\pi\bar{U} & I \end{array}\right).$$ If one now moves

$$\left(\begin{array}{cc|cc} I & U\bar{\pi} & & \\ & I & & \\ \hline & & I & \\ & & -\pi\bar{U} & I \end{array}\right)$$ one place to the left then one obtains the right-hand side of the equation in the corollary.

COROLLARY 3.13. *Let* I, π, ω_{4n} *and* ω_{8n} *be as in 3.12. If* $\begin{pmatrix} I & \\ \gamma & I \end{pmatrix}$, $\begin{pmatrix} I & \beta \\ & I \end{pmatrix}$, $\begin{pmatrix} I & \\ V & I \end{pmatrix}$, $\begin{pmatrix} I & B \\ & I \end{pmatrix}$, $\begin{pmatrix} I & \\ V' & I \end{pmatrix} \in EQ_{4n}(A,\Lambda)$ *then there are* $E, F \in E_{4n}(A)$ *and*

$$\left(\begin{array}{cc|cc} I & & & \\ & I & & \\ \hline & & I & \\ L & & & I \end{array}\right) \in EQ_{8n}(A,\Lambda) \quad \textit{such that}$$

$$\left(\omega_{4n}\begin{pmatrix} I & \\ \gamma & I \end{pmatrix}\begin{pmatrix} I & \beta \\ & I \end{pmatrix}\right)^{-1} \omega_{4n}\begin{pmatrix} I & \\ V & I \end{pmatrix}\begin{pmatrix} I & B \\ & I \end{pmatrix}\begin{pmatrix} I & \\ V' & I \end{pmatrix} =$$

$$\left(\begin{array}{c|c} E & \\ \hline & \bar{E}^{-1} \end{array}\right) \omega_{8n} \left(\begin{array}{cc|cc} I & & & \\ & I & & \\ \hline (V-\gamma)+(V-\gamma)\beta(V-\gamma) & -\bar{\pi}+(V-\gamma)\beta\bar{\pi} & I & \\ -\bar{\pi}+\bar{\pi}\beta(V-\gamma) & \bar{\pi}\beta\bar{\pi} & & I \end{array}\right)$$

$$\left(\begin{array}{cc|cc} I & & B & \pi \\ & I & \pi & -\pi(V-\gamma)\pi \\ \hline & & I & \\ & & & I \end{array}\right)\left(\begin{array}{cc|cc} I & & & \\ & I & & \\ \hline & & I & \\ L & & & I \end{array}\right)\left(\begin{array}{c|c} F & \\ \hline & \bar{F}^{-1} \end{array}\right).$$

Proof. $\left(\omega_{4n}\begin{pmatrix} I & \\ \gamma & I \end{pmatrix}\begin{pmatrix} I & \beta \\ & I \end{pmatrix}\right)^{-1} \omega_{4n}\begin{pmatrix} I & \\ V & I \end{pmatrix}\begin{pmatrix} I & B \\ & I \end{pmatrix}\begin{pmatrix} I & \\ V' & I \end{pmatrix} =$

$\left(\begin{pmatrix} I & -\pi\gamma\pi \\ & I \end{pmatrix}\omega_{4n}\begin{pmatrix} I & \beta \\ 0 & I \end{pmatrix}\right)^{-1}\begin{pmatrix} I & -\pi V\pi \\ & I \end{pmatrix}\omega_{4n}\begin{pmatrix} I & B \\ & I \end{pmatrix}\begin{pmatrix} I & \\ V' & I \end{pmatrix} =$ (because $\omega_{4n}^{-1} = -\omega_{4n}$) $\begin{pmatrix} -I & \\ & -I \end{pmatrix}\begin{pmatrix} I & -\beta \\ & I \end{pmatrix}\omega_{4n}\begin{pmatrix} I & -\pi(V-\gamma)\pi \\ & I \end{pmatrix}\omega_{4n}\begin{pmatrix} I & B \\ & I \end{pmatrix}$ $\begin{pmatrix} I & \\ V' & I \end{pmatrix} =$ ⓐ. Note that $-I \in E_{4n}(A)$ because I has even rank. If $U = -\pi(V-\gamma)\pi$ and if $E = \begin{pmatrix} -I & \\ & -I \end{pmatrix}\begin{pmatrix} & I \\ -I & \end{pmatrix}\begin{pmatrix} I & U\bar{\pi} \\ & I \end{pmatrix}$ then, applying 3.12 to $\begin{pmatrix} I & -\beta \\ & I \end{pmatrix}\omega_{4n}\begin{pmatrix} I & -\pi(V-\gamma)\pi \\ & I \end{pmatrix}\omega_{4n}\begin{pmatrix} I & B \\ & I \end{pmatrix}$, one can rewrite ⓐ as

$$\left(\begin{array}{c|c} E & \\ \hline & \bar{E}^{-1} \end{array}\right)\left(\begin{array}{cc|cc} I & & -U-U\bar{\pi}\beta\pi\bar{U} & \pi+U\bar{\pi}\beta \\ & I & \pi+\beta\pi\bar{U} & -\beta \\ \hline & & I & \\ & & & I \end{array}\right)$$

$$\omega_{8n}\left(\begin{array}{cc|cc} I & & B & \pi \\ & I & \pi & U \\ \hline & & I & \\ & & & I \end{array}\right)\left(\begin{array}{cc|cc} I & & & \\ & I & & \\ \hline & -\bar{\pi} & I & \\ -\bar{\pi} & \bar{\pi}B\bar{\pi} & & I \end{array}\right)\left(\begin{array}{cc|cc} I & B\bar{\pi} & & \\ & I & & \\ \hline & & I & \\ & & -\bar{\pi}B & I \end{array}\right)$$

$$\left(\begin{array}{cc|cc} I & & & \\ & I & & \\ \hline V' & & I & \\ & & & I \end{array}\right) = ⓑ$$

Furthermore, if $F = \begin{pmatrix} I & B\bar{\pi} \\ & I \end{pmatrix}$ and

$L = \begin{pmatrix} & -\bar{\pi} \\ -\bar{\pi} & \bar{\pi}B\bar{\pi} \end{pmatrix} + \begin{pmatrix} V' & V'B\bar{\pi} \\ -\bar{\pi}BV' & -\bar{\pi}BV'B\bar{\pi} \end{pmatrix}$ then one can replace the last three matrices in ⓑ by $\left(\begin{array}{c|c} I & \\ \hline L & I \end{array}\right)\left(\begin{array}{c|c} F & \\ \hline & \bar{F}^{-1} \end{array}\right)$. This completes the proof of 3.13.

Next we define the group $GH(A,\Lambda,q)$ in analogy with the group $GQ(A,\Lambda,q)$.

Let $a_1,\cdots,a_n \in \Lambda$. Give the free right A-module A^n a basis and give $(A^n)^*$ the dual basis. Let $M_{a_i,\cdots,a_n}(A^n)$ denote the Λ-metabolic module such that with respect to the basis of $A^n \oplus (A^n)^*$ given above, the matrix associated to the Λ-hermitian form on $M_{a_1,\cdots,a_n}(A^n)$ is

$$\left(\begin{array}{cccc|c} a_1 & & & & \\ & \cdot & & & \\ & & \cdot & & \lambda I_n \\ & & & a_n & \\ \hline & & I_n & & 0 \end{array}\right).$$

With respect to this basis, we can identify $\mathrm{Aut}(M_{a_1,\cdots,a_n}(A^n))$ with a subgroup

$$GH_{a_1,\cdots,a_n}(A,\Lambda)$$

of $GL_{2n}(A)$ called the *general hermitian group* of $M_{a_1,\cdots,a_n}(A^n)$. If $\mathfrak{q}$ is an involution invariant ideal, we define

$$GH_{a_1,\cdots,a_n}(A,\Lambda,\mathfrak{q}) = \ker(GH_{a_1,\cdots,a_n}(A,\Lambda) \to GH_{a_1,\cdots,a_n}(A/\mathfrak{q},\Lambda/\mathfrak{q})).$$

$GH_{a_1,\cdots,a_n}(A,\Lambda,\mathfrak{q})$ is called the *relative* or *congruence subgroup of level* $\mathfrak{q}$. If $b_1,\cdots,b_n \in \min$ then $M_{a_1,\cdots,a_n}(A^n) \cong M_{a_1+b_1,\cdots,a_n+b_n}(A^n)$. Hence, up to isomorphism, the group $GH_{a_1,\cdots,a_n}(A,\Lambda,\mathfrak{q})$ depends only on the classes of $a_1,\cdots,a_n \bmod (\min)$. Suppose that $\Lambda/\min$ is finite, say of order N, and that $a_1,\cdots,a_N$ is a set of coset representatives for $\Lambda/\min$. For $n \geq N$, we define

$$GH_{2n}(A,\Lambda,\mathfrak{q}) = GH_{a_1,\cdots,a_N,\underbrace{0,\cdots,0}_{n-N}}(A,\Lambda,\mathfrak{q}).$$

There is a natural embedding $GH_{2n}(A,\Lambda,\mathfrak{q}) \to GH_{2(n+1)}(A,\Lambda,\mathfrak{q})$,

$$\begin{pmatrix} \alpha & \beta \\ \gamma & \delta \end{pmatrix} \mapsto \left(\begin{array}{cc|cc} \alpha & 0 & \beta & 0 \\ 0 & 1 & 0 & 0 \\ \hline \gamma & 0 & \delta & 0 \\ 0 & 0 & 0 & 1 \end{array}\right).$$

We define

$$GH(A,\Lambda,\mathfrak{q}) = \varinjlim GH_{2n}(A,\Lambda,\mathfrak{q}) .$$

A functor $G:$ ((form rings)) $\to$ ((groups)) is called *E-surjective* if given any form ring (A,Λ) and an involution invariant ideal $\mathfrak{q}$ of A, the canonical map (commutator subgroup of $G(A,\Lambda)$) $\to$ (commutator subgroup of $G(A/\mathfrak{q},\Lambda/\mathfrak{q})$) is surjective. E-surjectivity will play an important role in formulating the exact sequences of §7D.

LEMMA 3.14. *The functor* GQ *is E-surjective.*

Proof. By 3.9 the commutator subgroup of $GQ(A,\Lambda)$ (resp. $GQ(A/\mathfrak{q},\Lambda/\mathfrak{q})$) is $EQ(A,\Lambda)$ (resp. $EQ(A/\mathfrak{q},\Lambda/\mathfrak{q})$). But every generator of $EQ(A/\mathfrak{q},\Lambda/\mathfrak{q})$ lifts to a generator of $EQ(A,\Lambda)$.

QUESTION. Is GH E-surjective? Try first fields of characteristic 2.

A question related to the one above is

QUESTION. Find a reasonable set of generators for the commutator subgroup of GH. The fact that the commutator subgroup is perfect follows from Lemma 2.11 and Lemma 3.15 below.

LEMMA 3.15 (Bass). *Let* M *be an object in a category with product. Let* $G_n(M) = \mathrm{Aut}(M \perp \cdots \perp M)$ (n *times*). *There is a natural embedding* $G_n(M) \to G_{n+1}(M)$, $\alpha \mapsto \alpha \perp 1_M$. *If* $G(M) = \varinjlim_n G_n(M)$ *then the commutator subgroup of* $G(M)$ *is perfect.*

Proof. Let $N = M \perp \cdots \perp M$ (n times). If $\alpha, \beta \in \mathrm{Aut}(N)$ then $\alpha^{-1}\beta^{-1}\alpha\beta = (\alpha \perp \alpha^{-1} \perp 1_N)^{-1}(\beta \perp 1_N \perp \beta^{-1})^{-1}(\alpha \perp \alpha^{-1} \perp 1_N)(\beta \perp 1_N \perp \beta^{-1})$. But according to [10, VII 1.8], $\alpha \perp \alpha^{-1} \perp 1_N$ and $\beta \perp 1_N \perp \beta^{-1}$ are commutators in $\mathrm{Aut}(N \perp N \perp N)$.

Let G be a group. A *covering* or *extension* $V \to G$ of G is a group V together with a surjective homomorphism $V \to G$. The homomorphism

$V \to G$ is called *central* if its kernel lies in the center(V). A central covering $f: V \to G$ is called *universal* if given any other central covering $f': V' \to G$, there is a homomorphism $g: V \to V'$ such that $f = f'g$. If V is perfect then it is easy to show that the homomorphism $V \to V'$ is unique (cf. proof of Lemma 3.24 below).

Next, we write down the list of identities promised for $EQ_{2n}(A,\Lambda)$. Then we show that the identities define the perfect, universal, central covering of $EQ(A,\Lambda)$. We let $r_{ij}(a)$, $\ell_{ij}(a)$, and $\varepsilon_{i,j}(a)$ $(1 \le i, j \le n)$ denote respectively an upper right hand, lower left hand, and hyperbolic elementary Λ-quadratic matrix.

LEMMA 3.16.

1. $[r_{ij}(a), \ell_{k\ell}(b)] = 1$ $\quad (\{i,j\} \cap \{k,\ell\} = \emptyset)$
2. $[r_{ij}(a), \ell_{ik}(b)] = H\varepsilon_{jk}(-\lambda\bar{a}b)$ $\quad$ (i, j, *and* k *distinct*)
3. $[r_{ij}(a), \ell_{jj}(b)] = H\varepsilon_{ij}(ab) r_{ii}(-\lambda ab\bar{a})$ $\quad (i \neq j)$
4. $[r_{jj}(b), \ell_{ji}(a)] = H\varepsilon_{ji}(ba)\ell_{ii}(\overline{\lambda\bar{a}}ba)$ $\quad (i \neq j)$

E1. $H\varepsilon_{ij}(a+b) = H\varepsilon_{ij}(a) H\varepsilon_{ij}(b)$ $\quad (i \neq j)$

E2. $[H\varepsilon_{ij}(a), H\varepsilon_{k\ell}(b)] = 1$ $\quad (i \neq \ell$ *and* $j \neq k)$

E3. $[H\varepsilon_{ij}(a), H\varepsilon_{jk}(b)] = H\varepsilon_{ik}(ab)$ $\quad$ (i, j, *and* k *distinct*)

L1. $\ell_{ij}(a+b) = \ell_{ij}(a)\ell_{ij}(b)$

L2. $\ell_{ij}(a)\ell_{rs}(b) = \ell_{rs}(b)\ell_{ij}(a)$

L3. $\ell_{ij}(a) = \ell_{ji}(-\overline{\lambda a})$

L4. $[\ell_{ij}(a), H\varepsilon_{rs}(b)] = 1$ $\quad (r \neq s,\ r \neq i, j)$

L5a. $[\ell_{ij}(a), H\varepsilon_{jk}(b)] = \ell_{ik}(ab)$ $\quad$ (i, j, *and* k *distinct*)

L5b. $[\ell_{ij}(a), H\varepsilon_{ji}(b)] = \ell_{ii}(ab - \overline{\lambda\bar{a}b})$ $\quad (i \neq j)$

L6. $[\ell_{ii}(a), H\varepsilon_{ik}(b)] = \ell_{ik}(ab)\ell_{kk}(\bar{b}ab)$ $\quad (i \neq k)$

R1. $r_{ij}(a+b) = r_{ij}(a) r_{ij}(b)$

R2. $r_{ij}(a) r_{k\ell}(b) = r_{k\ell}(b) r_{ij}(a)$

R3. $r_{ij}(a) = r_{ji}(-\lambda\bar{a})$

R4. $[r_{ij}(a), H\varepsilon_{k\ell}(b)] = 1 \qquad (k \neq \ell,\ \ell \neq i, j)$

R5a. $[r_{ij}(a), H\varepsilon_{kj}(b)] = r_{ik}(-a\bar{b}) \qquad$ (i, j, *and* k *distinct*)

R5b. $[r_{ij}(a), H\varepsilon_{ij}(b)] = r_{ii}(-a\bar{b}+\lambda a\bar{b}) \qquad (i \neq j)$

R6. $[r_{ii}(a), H\varepsilon_{ki}(b)] = r_{ik}(-a\bar{b})r_{kk}(ba\bar{b}) \qquad (i \neq k)$

Let

$$StQ(A,\Lambda)$$

denote the free group generated by the elements $\tilde{r}_{ij}(a)$, $\tilde{\ell}_{ij}(a)$, $H(\tilde{\varepsilon}_{ij}(a))$ modulo the relations given in 3.16. Thus, if $i=j$, it is understood that in the case $\tilde{r}_{ii}(a)$ (resp. $\tilde{\ell}_{ii}(a)$), the element $a \in \Lambda$ (resp. $\bar{\Lambda}$). $StQ(A,\Lambda)$ is called the Λ-*quadratic Steinberg group.*

THEOREM 3.17. *The canonical map*

$$StQ(A,\Lambda) \to EQ(A,\Lambda),\ \tilde{x}_{ij}(a) \mapsto x_{ij}(a)\,,$$

such that $x_{ij} = r_{ij}, \ell_{ij}$, *or* $H\varepsilon_{ij}$, *is a perfect, universal, central covering.*

The proof of 3.17 will be postponed till the end of the section.

A convenient description due to R. Sharpe [26] of $StQ(A,\Lambda)$ is given as follows. Let

$$St(A)$$

denote the free group generated by the symbols $\tilde{\varepsilon}_{ij}(a)$ such that $i \neq j$ and $a \in A$ modulo the relations $\tilde{\varepsilon}_{ij}(a)\tilde{\varepsilon}_{ij}(b) = \tilde{\varepsilon}_{ij}(a+b)$, $[\tilde{\varepsilon}_{ij}(a), \tilde{\varepsilon}_{k\ell}(b)] = 1$ if $i \neq \ell$ and $j \neq k$, and $[\tilde{\varepsilon}_{ij}(a),\tilde{\varepsilon}_{j\ell}(b)] = \tilde{\varepsilon}_{i\ell}(ab)$ if $i \neq \ell$. $St(A)$ is called the *Steinberg group.* By a theorem of M. Kervaire [20], the canonical map

$$St(A) \to E(A),\ \tilde{\varepsilon}_{ij}(a) \mapsto \varepsilon_{ij}(a)\,,$$

is a perfect, universal, central covering. $St(A)$ has an involution $\tilde{\varepsilon} \mapsto \bar{\tilde{\varepsilon}}$ defined by $\bar{\tilde{\varepsilon}}_{ij}(a) = \tilde{\varepsilon}_{ji}(\bar{a})$ which covers that on $E(A)$. Let

$$M_n(\Lambda)$$

denote the subgroup of the ring of $n\times n$ matrices $M_n(A)$ consisting of all σ such that $\sigma = -\lambda\bar{\sigma}$ and the diagonal coefficients of σ lie in Λ. There is an embedding $M_n(\Lambda) \to M_{n+1}(\Lambda)$, $\sigma \mapsto \begin{pmatrix}\sigma & 0\\ 0 & 0\end{pmatrix}$, and we let

$$M(\Lambda) = \varinjlim M_n(\Lambda)\,.$$

THEOREM 3.18. $\mathrm{StQ}(A,\Lambda)$ *is isomorphic to the following group. The generators are the symbols*

$$\begin{pmatrix}I & \\ \gamma & I\end{pmatrix}, \begin{pmatrix}I & \beta\\ & I\end{pmatrix}, \begin{pmatrix}\tilde{\varepsilon} & \\ & \tilde{\bar{\varepsilon}}^{-1}\end{pmatrix}$$

such that $\gamma \in M(\bar{\Lambda})$, $\beta \in M(\Lambda)$, *and* $\tilde{\varepsilon} \in \mathrm{St}(A)$. *The relations are designed to coincide with the matrix identities in 3.6 and are as follows. Write*

$$\beta = \begin{pmatrix}\beta_{11} & \beta_{12}\\ -\lambda\bar{\beta}_{12} & \beta_{22}\end{pmatrix} \quad \textit{and} \quad \gamma = \begin{pmatrix}\gamma_{11} & \gamma_{12}\\ -\bar{\lambda}\bar{\gamma}_{12} & \gamma_{22}\end{pmatrix}.$$

Suppose that both β *and* γ *are* $n\times n$ *matrices,* β_{11} *and* γ_{11} *are* $r\times r$ *matrices and* β_{12} *and* γ_{12} *are* $r\times(n-r)$ *matrices; thus* $\bar{\beta}_{12}$ *and* $\bar{\gamma}_{12}$ *are* $(n-r)\times r$ *matrices and* β_{22} *and* γ_{22} *are* $(n-r)\times(n-r)$ *matrices. If* $\beta_{12} = 0$ *and* $\beta_{22} = 0$ *and if* $\gamma_{11} = 0$ *then in 3.6 c) the matrix* $I+\beta\gamma =$ $I+\beta\gamma+\beta\gamma\beta\gamma = \begin{pmatrix}I & \beta_{11}\gamma_{12}\\ & I\end{pmatrix}$ *has an obvious lifting to* $\mathrm{St}(A)$. *We shall denote this lifting also with the symbol* $\begin{pmatrix}I & \beta_{11}\gamma_{12}\\ & I\end{pmatrix} = I+\beta\gamma$. *If* $\beta_{11} = 0$ *and if* $\gamma_{12} = 0$ *and* $\gamma_{22} = 0$ *then the matrix* $I+\beta\gamma = I+\beta\gamma+\beta\gamma\beta\gamma =$ $\begin{pmatrix}I & -\lambda\bar{\beta}_{12}\gamma_{11}\\ & I\end{pmatrix}$ *also has an obvious lifting to* $\mathrm{St}(A)$. *We shall denote this lifting with the symbol* $\begin{pmatrix} I & -\lambda\bar{\beta}_{12}\gamma_{11} \\ & I \end{pmatrix} = I+\beta\gamma$. *The relations we want to make are the following*:

1. $\left[\begin{pmatrix} I & \beta \\ & I \end{pmatrix}, \begin{pmatrix} I & \\ \gamma & I \end{pmatrix}\right] = \begin{pmatrix} I+\beta\gamma & \\ & \overline{I+\beta\gamma}^{-1} \end{pmatrix}\begin{pmatrix} I & \\ \bar{\gamma}\beta\gamma & I \end{pmatrix}$ *providing*

$$\beta_{12} = 0, \quad \beta_{22} = 0, \quad \textit{and} \quad \gamma_{11} = 0 .$$

2. $\left[\begin{pmatrix} I & \beta \\ & I \end{pmatrix}, \begin{pmatrix} I & \\ \gamma & I \end{pmatrix}\right] = \begin{pmatrix} I+\beta\gamma & \\ & \overline{I+\beta\gamma}^{-1} \end{pmatrix}\begin{pmatrix} I & -\overline{\beta\gamma}\beta \\ & I \end{pmatrix}$ *providing*

$$\beta_{11} = 0, \quad \gamma_{12} = 0, \quad \textit{and} \quad \gamma_{22} = 0 .$$

E. $\begin{pmatrix} \tilde{\varepsilon}\tilde{\tau} & \\ & \overline{\tilde{\varepsilon}\tilde{\tau}}^{-1} \end{pmatrix} = \begin{pmatrix} \tilde{\varepsilon} & \\ & \bar{\tilde{\varepsilon}}^{-1} \end{pmatrix}\begin{pmatrix} \tilde{\tau} & \\ & \bar{\tilde{\tau}}^{-1} \end{pmatrix}$.

L1. $\begin{pmatrix} I & \\ \gamma & I \end{pmatrix}\begin{pmatrix} I & \\ \gamma' & I \end{pmatrix} = \begin{pmatrix} I & \\ \gamma+\gamma' & I \end{pmatrix}$.

L2. *If* ε *denotes the image of* $\tilde{\varepsilon}$ *in* E(A) *then*

$$\left[\begin{pmatrix} \tilde{\varepsilon} & \\ & \bar{\tilde{\varepsilon}}^{-1} \end{pmatrix}, \begin{pmatrix} I & \\ \gamma & I \end{pmatrix}\right] = \begin{pmatrix} I & \\ \gamma - \bar{\varepsilon}\gamma\varepsilon & I \end{pmatrix}.$$

R1. $\begin{pmatrix} I & \beta \\ & I \end{pmatrix}\begin{pmatrix} I & \beta' \\ & I \end{pmatrix} = \begin{pmatrix} I & \beta+\beta' \\ & I \end{pmatrix}$.

R2. *If* ε *denotes the image of* $\tilde{\varepsilon}$ *in* E(A) *then*

$$\left[\begin{pmatrix} \tilde{\varepsilon} & \\ & \bar{\tilde{\varepsilon}}^{-1} \end{pmatrix}, \begin{pmatrix} I & \beta \\ & I \end{pmatrix}\right] = \begin{pmatrix} I & \beta - \varepsilon^{-1}\beta\bar{\varepsilon}^{-1} \\ & I \end{pmatrix}.$$

Let $\tilde{\ell}(\gamma) = \begin{pmatrix} I & \\ \gamma & I \end{pmatrix}$, $\tilde{r}(\beta) = \begin{pmatrix} I & \beta \\ & I \end{pmatrix}$, *and* $H(\tilde{\varepsilon}) = \begin{pmatrix} \tilde{\varepsilon} & \\ & \bar{\tilde{\varepsilon}}^{-1} \end{pmatrix}$.

Let e_{ij} *denote the matrix with* 1 *in the* (i, j)'th *coordinate and* 0 *in all the other coordinates. If* StQ(A, Λ)′ *denotes the group given by the generators and relations above then the map below is an isomorphism*

$$\begin{aligned}
\mathrm{StQ}(A,\Lambda) &\xrightarrow{\cong} \mathrm{StQ}(A,\Lambda)' \\
H(\tilde{\varepsilon}_{ij}(a)) &\longmapsto H(\tilde{\varepsilon}_{ij}(a)) \\
\tilde{\ell}_{ij}(a) &\longmapsto \tilde{\ell}(a\,e_{ij}-\overline{\lambda a}\,e_{ji}) \qquad (i\neq j) \\
\tilde{\ell}_{ii}(a) &\longmapsto \tilde{\ell}(a\,e_{ii}) \\
\tilde{r}_{ij}(a) &\longmapsto \tilde{r}(a\,e_{ij}-\lambda\bar{a}\,e_{ji}) \qquad (i\neq j) \\
\tilde{r}_{ii}(a) &\longmapsto \tilde{r}(a\,e_{ii})\,.
\end{aligned}$$

Proof. One constructs in the obvious way an inverse to the map above. All the necessary verifications are straightforward.

The assertions in 3.10 - 3.13 for $EQ(A,\Lambda)$ are stated so that there are obvious analogous assertions for $StQ(A,\Lambda)$. Furthermore, the proofs of 3.10 - 3.13 are made only with manipulations which are valid in $StQ(A,\Lambda)$. Thus, the analogous assertions for $StQ(A,\Lambda)$ are valid. We record next these assertions.

Let π denote the $2n\times 2n$ matrix $\pi = \begin{pmatrix} 0 & -1 \\ \lambda & 0 \end{pmatrix} \oplus \cdots \oplus \begin{pmatrix} 0 & -1 \\ \lambda & 0 \end{pmatrix}$ (n times) and let

$$\omega_{4n} = \begin{pmatrix} I & \\ -\pi^{-1} & I \end{pmatrix}\begin{pmatrix} I & \pi \\ & I \end{pmatrix}\begin{pmatrix} I & \\ -\pi^{-1} & I \end{pmatrix} \in \mathrm{StQ}(A,\Lambda)\,.$$

THEOREM 3.19. *The analogy of 3.10 for* $StQ(A,\Lambda)$. *Namely, given an element* $x \in StQ(A,\Lambda)$, *there is an integer* n *and matrices* U, B, *and* L *of size* n *and an element* $\tilde{E} \in St_n(A)$ *such that*

$$x = \begin{pmatrix} I & U \\ & I \end{pmatrix}\omega_{4n}\begin{pmatrix} I & B \\ & I \end{pmatrix}\begin{pmatrix} I & \\ L & I \end{pmatrix}\begin{pmatrix} \tilde{E} & \\ & \tilde{\bar{E}}^{-1} \end{pmatrix}.$$

REMARK. Since in $StQ(A,\Lambda)$, $\omega_{4n}^{-1}\begin{pmatrix} I & U \\ & I \end{pmatrix}\omega_{4n} = \begin{pmatrix} I & \\ -\pi^{-1}U\pi & I \end{pmatrix}$, it follows from 3.19 that every element of $StQ(A,\Lambda)$ can be written as a product

$$\omega_{4n}\begin{pmatrix} I & \\ K & I \end{pmatrix}\begin{pmatrix} I & B \\ & I \end{pmatrix}\begin{pmatrix} I & \\ L & I \end{pmatrix}\begin{pmatrix} \tilde{E} & \\ & \bar{\tilde{E}}^{-1} \end{pmatrix} .$$

PROBLEM. For $\Lambda = \min$, R. Sharpe [26] determined when two products as in 3.14 determine the same element of $StQ(A, \Lambda)$. The same problem for an arbitrary form parameter Λ is still open.

REDUCTION LEMMA 3.20. *The analogy of 3.11 for* $StQ(A, \Lambda)$.

COROLLARY 3.21. *The analogy of 3.12 for* $StQ(A, \Lambda)$.

COROLLARY 3.22. *The analogy of 3.13 for* $StQ(A, \Lambda)$.

By definition

$$KQ_2(A, \Lambda) = \ker(StQ(A, \Lambda) \to EQ(A, \Lambda)) .$$

COROLLARY 3.23. *Every element of* $KQ_2(A, \Lambda)$ *can be written as a product below such that* $\bar{\bar{E}}$ *is some lifting of* $-\pi^{-1}B$ *to* $St_n(A)$. *Conversely, any such product lies in* $KQ_2(A, \Lambda)$.

$$\omega_{4n}\begin{pmatrix} I & \\ -B^{-1} & I \end{pmatrix}\begin{pmatrix} I & B \\ & I \end{pmatrix}\begin{pmatrix} I & \\ -B^{-1} & I \end{pmatrix}\begin{pmatrix} \tilde{E} & \\ & \bar{\tilde{E}}^{-1} \end{pmatrix} .$$

Proof. By the remark following 3.19, each element of $KQ_2(A, \Lambda)$ can be written as a product $\omega_{4n}\begin{pmatrix} I & \\ K & I \end{pmatrix}\begin{pmatrix} I & B \\ & I \end{pmatrix}\begin{pmatrix} I & \\ L & I \end{pmatrix}\begin{pmatrix} \tilde{E} & \\ & \bar{\tilde{E}}^{-1} \end{pmatrix}$. Let $\begin{pmatrix} & \pi \\ -\pi^{-1} & \end{pmatrix}$ (resp. E) denote the image of ω_{4n} (resp. $\tilde{E}$) in $EQ(A, \Lambda)$ (resp. $E(A)$). Pushing the product above down to $EQ(A, \Lambda)$ and multiplying out, one obtains that

$$\begin{pmatrix} \pi(K+L+KBL) & \pi(I+KB) \\ -\pi^{-1}(I+BL) & -\pi^{-1}B \end{pmatrix}\begin{pmatrix} E & \\ & \bar{E}^{-1} \end{pmatrix} = \begin{pmatrix} I & \\ & I \end{pmatrix} .$$

Transposing $\begin{pmatrix} E & \\ & \bar{E}^{-1} \end{pmatrix}$ to the right-hand side and equating coefficients, one obtains that $0 = I+KB = I+BL$ and $\bar{E} = -\pi^{-1}B$. Thus, $K = -B^{-1}$, $L = -B^{-1}$, and $\bar{\bar{E}}$ is some lifting of $-\pi^{-1}B$.

We prepare now for the proof of Theorem 3.17.

We shall need a relative version of the expressions perfect and central covering defined above. Let G be a group and let M be a group (not necessarily abelian) with an action $m \mapsto m^g$ of G. If $m \in M$ and $g \in G$ then let $[m, g] = m^{-1}m^g$ and let $[M, G]$ denote the subgroup of M generated by all $[m, g]$. M is called *G-perfect* if $M = [M, G]$. If X is another G-group then a surjective homomorphism $p : X \to M$ of G-groups is called a *G-central covering* or *extension* if G acts trivially on the $\ker(p)$ and the $\ker(p) \subseteq \text{center}(X)$. Call a G-central covering $p : X \to M$ *universal* if given any G-central covering $p' : X' \to M$, there is a G-homomorphism $q : X \to X'$ such that $p = p'q$. If $M = G$ and G acts by conjugation on itself then the expressions G-perfect and G-central covering coincide respectively with the expressions *perfect* and *central* covering defined previously.

LEMMA 3.24. *Let* $M \to N$ *be a G-homomorphism. Let* $X \to M$ *and* $Y \to N$ *be G-central coverings. If* $\alpha, \beta : X \to Y$ *are G-homomorphisms which cover the G-homomorphism* $M \to N$ *then* $\alpha|_{[X,G]} = \beta|_{[X,G]}$.

Proof. If $x \in X$ then there is a $c \in \ker(Y \to N)$ such that $c\,\alpha(x) = \beta(x)$. Thus, $\alpha[x, g] = \alpha(x)^{-1}\alpha(x)^g = (c\,\alpha(x))^{-1}(c\,\alpha(x))^g = \beta[x, g]$.

LEMMA 3.25. *If* M *is G-perfect then* M *has up to unique isomorphism a unique G-perfect, universal, G-central covering.*

Proof. Let $\tilde{M}$ denote the free G-group on the elements of M and let p denote the canonical map $p : \tilde{M} \to M$. If R is the normal subgroup of $\tilde{M}$ generated by $[\ker(p), \tilde{M}]$ and $[\ker(p), G]$ and if $X = \tilde{M}/R$ then the canonical homomorphism $X \to M$ is a universal, G-central covering of M. The canonical map $[X, G] \to M$ is clearly also a universal, G-central covering. Thus, to complete the proof, it suffices to show that $[X, G]$ is G-perfect. Since $[X, G]$ covers M, it follows that if $x \in X$ then there is a $c \in \ker(X \to M)$ and a $y \in [X, G]$ such that $cx = y$. Thus, for any $g \in G$, one has that $[x, g] = [cx, g] = [y, g]$. The uniqueness assertions follow from 3.24.

The next result will be required in the proof of Theorem 3.17. If $\varepsilon \in E(A)$ and if $m \in M(\Lambda)$ (resp. $M(\bar{\Lambda})$) then the rule $m \mapsto \varepsilon^{-1} m \bar{\varepsilon}^{-1}$ (resp. $m \mapsto \bar{\varepsilon} m \varepsilon$) defines an action of $E(A)$ on $M(\Lambda)$ (resp. $M(\bar{\Lambda})$).

THEOREM 3.26. $M(\Lambda)$ *and* $M(\bar{\Lambda})$ *are* $E(A)$-*perfect, universal,* $E(A)$-*central coverings of themselves.*

Theorem 3.26 follows from Theorem 3.28 which is given following the proof below of Theorem 3.17.

The following commutator formulas will be handy to keep in mind for the rest of the section.

LEMMA 3.27. a) *If* x, y, *and* z *are elements of a group let* $[x,y] = x^{-1}y^{-1}xy$. *Then* $[y,x] = [x,y]^{-1}$, $[xy,z] = [x,z]^y[y,z]$, $[x,y]^z = [x^z,y^z] = [x^z,[z,y^{-1}]y]$, *and if* z *commutes with* y *then* $[zx,y] = [x,y]$.

b) *Let* M *and* G *be groups such that* G *acts on* M *via* $m \mapsto m^g$. *If* $m, n \in M$ *and if* $f, g \in G$ *let* $[m,g] = m^{-1}m^g$. *Then* $[mn,f] = [m,f]^n[n,f]$, $[m,fg] = [m,g][m,f]^g$, $[m,f]^g = [m^g, f^g]$, $[m,f]^n = [m^n,[n,f^{-1}]f]$, *and if* n *is fixed by* f *then* $[nm,f] = [m,f]$.

The verification of the formulas in the lemma is straightforward.

Proof of Theorem 3.17. We want to show that $StQ(A,\Lambda)$ is perfect and universal for central coverings of $EQ(A,\Lambda)$.

PERFECTNESS follows directly from relations 3.16 E2, E3, R5a), R6, L5a), and L6.

CENTRALITY is established as follows. Let $\rho: StQ(A,\Lambda) \to EQ(A,\Lambda)$ denote the canonical map. Let $x \in \ker \rho$. Write x as a product $x = \prod x_{k\ell}$ of the generators $x_{k\ell}$ of $StQ(A,\Lambda)$. Let q be an integer such that $q >$ all k and ℓ appearing in the product above. Let L denote the subgroup of $StQ(A,\Lambda)$ generated by all $H\tilde{\varepsilon}_{iq}(a)$ and all $\tilde{\ell}_{qi}(a)$ $(a \in A,\ i \geq 1)$. Pictorially, L looks like

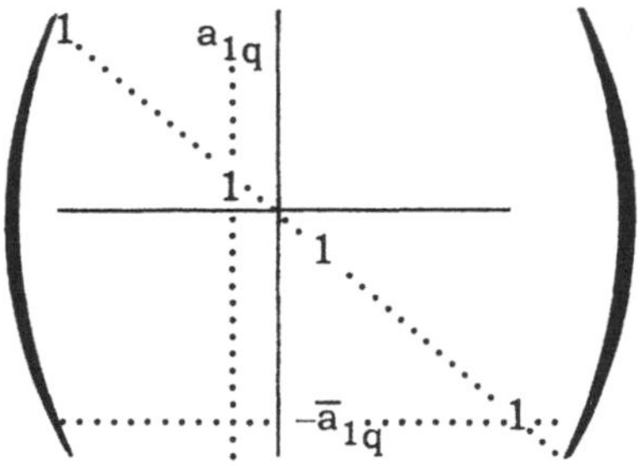

.

Let R denote the subgroup of $StQ(A,\Lambda)$ generated by all $H\tilde{\varepsilon}_{qi}(a)$ and all $\tilde{r}_{iq}(a)$ $(a \in A,\ i \geq 1)$. Pictorially, R looks like

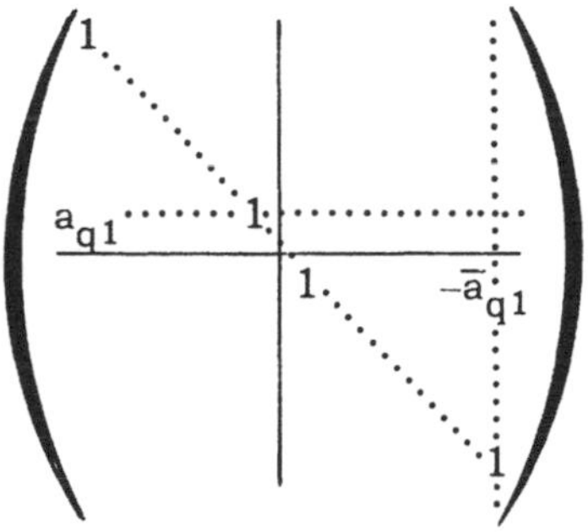

.

Let N denote the subgroup of $StQ(A,\Lambda)$ generated by L and R. We shall show that $N = StQ(A,\Lambda)$. Relation 3.16 (2) shows that N contains all $H\tilde{\varepsilon}_{k\ell}(a)$. Combining this fact with relations 3.16 L5 a) and L6, one deduces easily that N contains all $\tilde{\ell}_{k\ell}(a)$. Similarly, using relations 3.16 R5 a) and R6, one deduces that N contains all $\tilde{r}_{k\ell}(a)$. Thus, $N = StQ(A,\Lambda)$. To show that $x \in \text{center}(StQ(A,\Lambda))$, it suffices to show that conjugation by x fixes each element of L and R. Let us concentrate on the case of L. The case of R is handled similarly. First, we claim that every element of L can be written as a product $\prod_i H\tilde{\varepsilon}_{iq}(a_i)$ $\prod_i \tilde{\ell}_{qi}(b_i)$ such that the indices i are taken in increasing order. Relations E2 and L4 show that all pairs of generators of L commute except for a pair $\tilde{\ell}_{qi}(a)$, $H\tilde{\varepsilon}_{iq}(b)$. But, relation L5b) says that $[\tilde{\ell}_{qi}(a), \tilde{\varepsilon}_{iq}(b)] = \tilde{\ell}_{qq}(ab-\lambda\overline{ab})$. Thus, every element of L can be written as product $\prod_i(\prod_j H\tilde{\varepsilon}_{iq}(a_{ij}))\prod_i(\prod_j \tilde{\ell}_{qi}(b_{ij}))$ such that the indices i are taken in increasing order. But, by E1 (resp. L1) $\prod_j H\tilde{\varepsilon}_{iq}(a_{ij}) = H\tilde{\varepsilon}_{iq}(\sum_j a_{ij})$ (resp. $\prod_j \tilde{\ell}_{qi}(b_{ij}) = \tilde{\ell}_{qi}(\sum_j b_{ij})$). The claim follows. Now, $\rho(\prod_i H\tilde{\varepsilon}_{iq}(a_i)\prod_i \tilde{\ell}_{qi}(a_i)) = 1 \iff a_i = b_i = 0$ for all i. Thus, $\rho|_L$ is injective. Since $\rho(x) = 1$, it follows that conjugation by $\rho(x)$ fixes each element of $\rho(L)$. Thus, conjugation by x fixes each element of $L \iff$ conjugation by x normalizes L (because $\rho|_L$ is injective). If k and $\ell < q$ then it follows from relations E2, E3, L4, L5 that $H\tilde{\varepsilon}_{k\ell}(a)$ normalizes L, from relations L2, L4, L5, that $\tilde{\ell}_{k\ell}(a)$ normalizes L, and from relations 1, 2, 4 that $\tilde{r}_{k\ell}(a)$ normalizes L. Thus, $x = \prod x_{k\ell}$ normalizes L.

UNIVERSALITY. Let $f: F \to EQ(A,\Lambda)$ be a central covering. Using the canonical homomorphism $M(\Lambda) \to EQ(A,\Lambda)$, $\beta \mapsto \begin{pmatrix} I & \beta \\ & I \end{pmatrix}$, and the canonical homomorphism $M(\overline{\Lambda}) \to EQ(A,\Lambda)$, $\gamma \mapsto \begin{pmatrix} I & \\ \gamma & I \end{pmatrix}$, we shall identify $M(\Lambda)$ and $M(\overline{\Lambda})$ with their images in $EQ(A,\Lambda)$. Let $F(\Lambda) = f^{-1}(M(\Lambda))$ and $F(\overline{\Lambda}) = f^{-1}(M(\overline{\Lambda}))$. If $H: E(A) \to EQ(A,\Lambda)$, $\varepsilon \mapsto \begin{pmatrix} \varepsilon & \\ & \bar{\varepsilon}^{-1} \end{pmatrix}$, let $H = f^{-1}(H(E(A)))$. Thus, H is a central covering of $H(E(A))$ and, by universality, there is a homomorphism $h: St(A) \to H$ such the diagram

$$\begin{array}{ccc} St(A) & \xrightarrow{\;h\;} & H \\ \big\downarrow & & \big\downarrow f \\ E(A) & \xrightarrow[\;H\;]{} & EQ(A,\Lambda) \end{array}$$

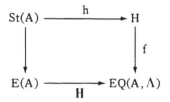

commutes. Since conjugation by elements of $fh(St(A))$ leaves $M(\Lambda)$ and $M(\bar{\Lambda})$ invariant, it is clear that conjugation by elements of $h(St(A))$ leaves $F(\Lambda)$ and $F(\bar{\Lambda})$ invariant. Thus, we have an action of $St(A)$ on $F(\Lambda)$ and $F(\bar{\Lambda})$. Furthermore, since conjugation factors through $H/\ker f$ (because the $\ker f$ is contained in the center), it follows that the action of $St(A)$ on $F(\Lambda)$ and $F(\bar{\Lambda})$ factors through $E(A)$. The action of $E(A)$ induced on $M(\Lambda)$ and $M(\bar{\Lambda})$ is the same as the action in Theorem 3.26 and the coverings $F(\Lambda) \to M(\Lambda)$ and $F(\bar{\Lambda}) \to M(\bar{\Lambda})$ are $E(A)$-central, because $f : F \to EQ(A, \Lambda)$ is central. Thus, by 3.26 there are $E(A)$-splittings $s : M(\Lambda) \to F(\Lambda)$ and $t : M(\bar{\Lambda}) \to F(\bar{\Lambda})$ to the coverings above. If $H\hat{\varepsilon}_{ij}(a) = h\tilde{\varepsilon}_{ij}(a)$, $\hat{r}_{ij}(a) = s\, r_{ij}(a)$, and $\hat{\ell}_{ij}(a) = t\,\ell_{ij}(a)$ then it is clear that $H\hat{\varepsilon}_{ij}(a)$, $\hat{r}_{ij}(a)$, and $\hat{\ell}_{ij}(a)$ satisfy all of the relations of 3.16, except possibly relations 1-4. Next, we show that relations 1-4 are also satisfied. To simplify notation, we shall write ε_{ij}, r_{ij}, ℓ_{ij} in place of $H\hat{\varepsilon}_{ij}$, $\hat{r}_{ij}$, $\hat{\ell}_{ij}$.

1). Choose $p \neq i, j, k$ or ℓ. Then $r_{ij}(a)^{\ell_{k\ell}(b)} =$ (R5a) $[r_{ip}(a), \varepsilon_{jp}(-1)]^{\ell_{k\ell}(b)} =$ (for some $z, z_1 \in \ker \rho$) $[z r_{ip}(a), z_1 \varepsilon_{jp}(-1)] =$ (R5a) $r_{ij}(a)$.

2). Choose $p \neq i, j$ or k. Then $r_{ij}(a)^{\ell_{ik}(b)} =$ (R5a) $[r_{ip}(a), \varepsilon_{jp}(-1)]^{\ell_{ik}(b)} =$ (for some $z, z_1 \in \ker \rho$) $[z r_{ip}(a) \varepsilon_{pk}(-\lambda\bar{a}b), z_1 \varepsilon_{jp}(-1)] =$ $[r_{ip}(a), \varepsilon_{jp}(-1)]^{\varepsilon_{pk}(-\lambda\bar{a}b)} [\varepsilon_{pk}(-\lambda\bar{a}b), \varepsilon_{jp}(-1)] =$ (R5a, E3) $=$ $r_{ij}(a)^{\varepsilon_{pk}(-\lambda\bar{a}b)} \varepsilon_{jk}(-\lambda\bar{a}b) =$ (R4) $r_{ij}(a)\varepsilon_{jk}(-\lambda\bar{a}b)$.

3). Choose $p \neq i$ or j. Then $r_{ij}(a) =$ (R3) $r_{ji}(-\lambda\bar{a}) =$ (R5a) $[r_{jp}(-\lambda\bar{a}), \varepsilon_{ip}(-1)] =$ (R3) $[r_{pj}(a), \varepsilon_{ip}(-1)]$. Hence, $r_{ij}(a)^{\ell_{jj}(b)} = [r_{pj}(a), \varepsilon_{ip}(-1)]^{\ell_{jj}(b)} =$ (for some $z, z_1 \in \ker \rho$) $[z r_{pj}(a) \varepsilon_{pj}(ab) r_{pp}(-\lambda ab\bar{a}), z_1 \varepsilon_{ip}(-1)] = [r_{pj}(a),$ $\varepsilon_{ip}(-1)]^{\varepsilon_{pj}(ab) r_{pp}(-\lambda ab\bar{a})} [\varepsilon_{pj}(ab), \varepsilon_{ip}(-1)]^{r_{pp}(-\lambda ab\bar{a})} [r_{pp}(-\lambda ab\bar{a}, \varepsilon_{ip}(-1)] =$ (R5a, E3, R6) $r_{ij}(a)^{\varepsilon_{pj}(ab) r_{pp}(-\lambda ab\bar{a})} \varepsilon_{ij}(ab)^{r_{pp}(-\lambda ab\bar{a})} r_{pi}(-\lambda ab\bar{a}) r_{ii}(-\lambda ab\bar{a})$.

R2 $\Rightarrow r_{pp}(-\lambda ab\bar{a})$ fixes $r_{ij}(a)^{\varepsilon_{pj}(ab)}$ = (R5a) $r_{ij}(a)r_{ip}(-ab\bar{a})$ and $\varepsilon_{ij}(ab)$. Thus, $r_{ij}(a)^{\ell_{jj}(b)} = r_{ij}(a)r_{ip}(-a\bar{b}\bar{a})\varepsilon_{ij}(ab)r_{pi}(-\lambda ab\bar{a})r_{ii}(-\lambda ab\bar{a})$. R4 $\Rightarrow$ $\varepsilon_{ij}(ab)$ commutes with $r_{pi}(-\lambda ab\bar{a})$ and R3 $\Rightarrow r_{pi}(-\lambda ab\bar{a}) = r_{ip}(\overline{ab\bar{a}})$. Thus, $r_{ij}(a)^{\ell_{jj}(b)} = r_{ij}(a)\varepsilon_{ij}(ab)r_{ii}(-\lambda ab\bar{a})$.

4) is proved similarly to 3).

If $\varepsilon \in E_n(A)$ and $m \in M_n(\Lambda)$ (resp. $M_n(\bar{\Lambda})$) then the rule $m \mapsto \varepsilon^{-1}m\bar{\varepsilon}^{-1}$ (resp. $m \mapsto \bar{\varepsilon}m\varepsilon$) defines an action of $E_n(A)$ on $M_n(\Lambda)$ (resp. $M_n(\bar{\Lambda})$).

THEOREM 3.28. *If* $n \geq 5$ *then* $M_n(\Lambda)$ *and* $M_n(\bar{\Lambda})$ *are* $E_n(A)$-*perfect, universal,* $E_n(A)$-*central coverings of themselves.*

Proof. Suppose that the assertions for $M_n(\bar{\Lambda})$ have been proved. Then the assertions for $M_n(\Lambda)$ can be deduced as follows. If M is any $E_n(A)$-group whose $E_n(A)$-action is given by $m \mapsto m^{\varepsilon}$, let M' denote the $E_n(A)$-group whose underlying group is that of M and whose $E_n(A)$-action is given by $m \mapsto m^{\bar{\varepsilon}^{-1}}$. Clearly $(M')' = M$. Thus, the rule $M \mapsto M'$ defines an equivalence of the category of $E_n(A)$-groups with itself. It follows that $M_n(\bar{\Lambda})'$ is an $E_n(A)$-perfect, universal, $E_n(A)$-central covering of itself. But, the rule $m \mapsto \bar{m}$ defines an $E_n(A)$-isomorphism $M_n(\bar{\Lambda})' \to M_n(\Lambda)$.

We prove now the assertions for $M_n(\bar{\Lambda})$. We begin by defining some notation suggested by the monomorphism $M_n(\bar{\Lambda}) \to EQ_{2n}(A,\Lambda)$, $\gamma \mapsto \begin{pmatrix} I & \\ \gamma & I \end{pmatrix}$. If $1 \leq i,\ j \leq n$, let e_{ij} denote the $n \times n$-matrix with 1 in the (i,j)'th coordinate and 0 in all the other coordinates. For $i \neq j$ and $a \in A$, let $\ell_{ij}(a) = ae_{ij} - \bar{\lambda}\bar{a}e_{ji}$ and for $i = j$ and $a \in \bar{\Lambda}$, let $\ell_{ii}(a) = ae_{ii}$. The elements $\ell_{ij}(a)$ generate $M_n(\bar{\Lambda})$ and satisfy the following relations.

1. $\ell_{ij}(a+b) = \ell_{ij}(a) + \ell_{ij}(b)$.
2. $\ell_{ij}(a) + \ell_{rs}(b) = \ell_{rs}(b) + \ell_{ij}(a)$.
3. $\ell_{ij}(a) = \ell_{ji}(-\bar{\lambda}\bar{a})$.
4. $[\ell_{ij}(a), \varepsilon_{rs}(b)] = 0$ if $r \notin \{i,j\}$.

5a. $[\ell_{ij}(a), \varepsilon_{jk}(b)] = \ell_{ik}(ab)$ if i, j, and k are distinct.

5b. $[\ell_{ij}(a), \varepsilon_{ji}(b)] = \ell_{ji}(ab - \bar{\lambda}\bar{a}\bar{b})$.

6. $[\ell_{ij}(a), \varepsilon_{ik}(b)] = \ell_{jk}(ab) + \ell_{kk}(\bar{b}ab)$.

Relations 4-6 show that $M_n(\Lambda)$ is $E_n(A)$-perfect. Thus, it remains to show that $M_n(\bar{\Lambda})$ is universal for $E_n(A)$-central coverings of $M_n(\bar{\Lambda})$. The symbols $\ell_{ij}(a)$ and the relations 1-3 form a set of generators and relations for the group $M_n(\bar{\Lambda})$. Furthermore, the symbols $\ell_{ij}(a)$ and the relations 1-6 present $M_n(\bar{\Lambda})$ as an $E_n(A)$-group.

We outline the rest of the proof. Let $\rho : Y \to M_n(\bar{\Lambda})$ be an $E_n(A)$-central covering. First, we show that Y is abelian. Then for each generator $\ell_{ij}(a)$, we pick a lifting $y_{ij}(a) \in Y$ and define, for i, j, p distinct, $\ell^p_{ij}(a) = [y_{ip}(1), \varepsilon_{pj}(a)]$. The definition is clearly independent of the lifting $y_{ip}(1)$. We show that for a fixed p, the $\ell^p_{ij}(a)$ satisfy relations 1-5a). Furthermore, it follows from 5a) that if k, i, j, p, q are distinct then $\ell^p_{ij}(a) = [\ell^p_{ik}(a), \varepsilon_{kj}(1)] =$ (for some $z \in \ker \rho$) $[z\ell^q_{ik}(a), \varepsilon_{kj}(1)] = \ell^q_{ij}(a)$. Thus, we can define unambiguously $\tilde{\ell}_{ij}(a) = \ell^p_{ij}(a)$ where p is any integer such that i, j, and p are distinct. Clearly, the $\tilde{\ell}_{ij}(a)$ satisfy relations 1-5a). For $i \neq p$, we define $\tilde{\ell}^p_{ii}(a) = \tilde{\ell}_{pi}(a)\,[y_{pp}(a), \varepsilon_{pi}(1)]$ and show that for a fixed p, the $\tilde{\ell}_{ij}(a)$ and $\tilde{\ell}^p_{ii}(a)$ satisfy relations 1-6. Furthermore, from 6, it follows that if i, k, p, q are distinct then $\tilde{\ell}^p_{ii}(a) = \tilde{\ell}^p_{pi}(-a)$ $[\tilde{\ell}^p_{kk}(a), \varepsilon_{ki}(1)] =$ (for some $z \in \ker \rho$) $\tilde{\ell}^q_{ki}(-a)\,[z\tilde{\ell}^q_{kk}(a), \varepsilon_{ki}(1)] = \tilde{\ell}^q_{ii}(a)$. Thus, we can define unambiguously $\tilde{\ell}_{ii}(a) = \tilde{\ell}^p_{ii}(a)$ where p is any integer such that $p \neq i$. Clearly, the $\tilde{\ell}_{ij}(a)$'s satisfy relations 1-6. Since $M_n(\bar{\Lambda})$ is presented by the relations 1-6, it follows that there is an $E_n(A)$-homomorphism $M_n(\bar{\Lambda}) \to Y$, $\ell_{ij}(a) \mapsto \tilde{\ell}_{ij}(a)$, such that the diagram

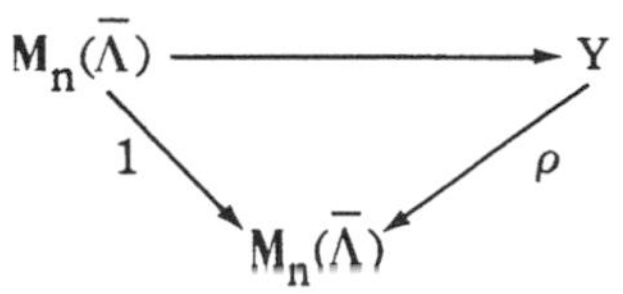

commutes.

We fill in now the details of the outline above.

Y IS ABELIAN. For each $\ell_{ij}(a)$, pick a lifting $y_{ij}(a) \in Y$.

It suffices to show that $y_{ij}(a)^{y_{k\ell}(b)} = y_{ij}(a)$. Suppose $i \neq j$. Choose $p \neq i, j, k$ or ℓ and choose $z \in \ker \rho$ such that $y_{ij}(a) = z[y_{ip}(1), \varepsilon_{pj}(a)]$. Then $y_{ij}(a)^{y_{k\ell}(b)} = z[y_{ip}(1)^{y_{k\ell}(b)}, [y_{k\ell}(b), \varepsilon_{pj}(-a)]\varepsilon_{pj}(a)] =$ (for some $z_1 \in \ker \rho$) $z[z_1 y_{ip}(1), \varepsilon_{pj}(a)] = z[y_{ip}(1), \varepsilon_{pj}(a)] = y_{ij}(a)$. Suppose $i = j$. Choose $p \neq i, k$ or ℓ and choose $z \in \ker \rho$ such that $y_{ii}(a) = z\, y_{pi}(-a)$ $[y_{pp}(a), \varepsilon_{pi}(1)]$. Then $y_{ii}(a)^{y_{k\ell}(b)} = z\, y_{pi}(-a)[y_{pp(a)}{}^{y_{k\ell}(b)}, [y_{k\ell}(b), \varepsilon_{pi}(-1)]\varepsilon_{pi}(1)] =$ (for some $z_1 \in \ker \rho$) $z\, y_{pi}(-a)[z_1 y_{pp}(a), \varepsilon_{pi}(1)] = y_{ii}(a)$.

Next we shall show that if $k \neq i$ or j then $y_{ij}(a)^{\varepsilon_{k\ell}(b)} = y_{ij}(a)$. Suppose $i \neq j$. Choose $p \neq i, j, k$ or ℓ and choose $z \in \ker \rho$ such that $y_{ij}(a) = z[y_{ip}(1), \varepsilon_{pj}(a)]$. Then $y_{ij}(a)^{\varepsilon_{k\ell}(b)} = z[y_{ip}(1)^{\varepsilon_{k\ell}(b)}, [\varepsilon_{k\ell}(b), \varepsilon_{pj}(-a)]\varepsilon_{pj}(a)] =$ (for some $z_1 \in \ker \rho$) $z[z_1 y_{ip}(1), \varepsilon_{pj}(a)] = y_{ij}(a)$. Suppose $i = j$. Choose $p \neq i$, k or ℓ and choose $z \in \ker \rho$ such that $y_{ii}(a) =$ $z\, y_{pi}(-a)[y_{pp}(a), \varepsilon_{pi}(1)]$. Then $y_{ii}(a)^{\varepsilon_{k\ell}(b)} = z\, y_{pi}(-a)[y_{pp}(a)^{\varepsilon_{k\ell}(b)}, [\varepsilon_{k\ell}(b), \varepsilon_{pi}(-1)]\varepsilon_{pi}(1)] =$ (for some $z_1 \in \ker \rho$) $z\, y_{pi}(-a)[z_1 y_{pp}(a), \varepsilon_{pi}(1)] = y_{ii}(a)$.

THE $\ell_{ij}{}^p(a)$ SATISFY RELATIONS 1-5a. For i, j, and p distinct, we define $\ell_{ij}{}^p(a) = [y_{ip}(1), \varepsilon_{pj}(a)]$ and we show that for a fixed p, the $\ell_{ij}{}^p(a)$ satisfy relations 1-5a. Relations 2 and 4 were demonstrated in the proof that Y is abelian. To simplify notation, write h_{ij} in place $\ell_{ij}{}^p$.

1). $h_{ij}(a+b) = [y_{ip}(1), \varepsilon_{pj}(a+b)] = [y_{ip}(1), \varepsilon_{pj}(b)\varepsilon_{pj}(a)] = [y_{ip}(1), \varepsilon_{pj}(a)]$ $[y_{ip}(1), \varepsilon_{pj}(b)]^{\varepsilon_{pj}(a)} = h_{ij}(a) h_{ij}(b)^{\varepsilon_{pj}(a)} = (4)\ h_{ij}(a) h_{ij}(b)$.

5a). $h_{ij}(a)^{\varepsilon_{jk}(b)} = [y_{ip}(1), \varepsilon_{pj}(a)]^{\varepsilon_{jk}(b)} = [y_{ip}(1)^{\varepsilon_{jk}(b)}, [\varepsilon_{jk}(b), \varepsilon_{pj}(-a)]\varepsilon_{pj}(a)] = (4)\ [y_{ip}(1), \varepsilon_{pk}(ab)\varepsilon_{pj}(a)] = [y_{ip}(1), \varepsilon_{pj}(a)][y_{ip}(1), \varepsilon_{pk}(ab)]^{\varepsilon_{pj}(a)} = h_{ij}(a) h_{ik}(ab)^{\varepsilon_{pj}(a)} = (4)\ h_{ij}(a) h_{ik}(ab)$.

5b). If i, j, k, p are distinct then $[h_{ji}(a), \varepsilon_{jk}(b)] = h_{ki}(\bar{b}a)$.

$h_{jk}(a)^{\varepsilon_{jk}(b)} = [y_{jp}(1), \varepsilon_{pi}(a)]^{\varepsilon_{jk}(b)} =$ (for some $z \in \ker \rho$) $[zy_{jp}(1)y_{kp}(\bar{b}), \varepsilon_{pi}(a)] =$ (2) $[y_{jp}(1), \varepsilon_{pi}(a)][y_{kp}(\bar{b}), \varepsilon_{pi}(a)] = h_{ji}(a)[y_{kp}(\bar{b}), \varepsilon_{pi}(a)] = h_{ji}(a)[\ell^q_{kp}(\bar{b}), \varepsilon_{pi}(a)] =$ (if i, k, p, q are distinct) (5a) $h_{ji}(a)\ell^q_{ki}(\bar{b}a)$. However, if i', k, i, p, q are distinct then $\ell^q_{ki}(\bar{b}a) =$ (5a) $[\ell^q_{ki'}(\bar{b}a), \varepsilon_{i'i}(1)] = \ell^p_{ki'}(\bar{b}a), \varepsilon_{i'i}(1)] =$ (5a) $\ell^p_{ki}(\bar{b}a) = h_{ki}(\bar{b}a)$. This completes the proof of 5b).

3). Suppose k, i, j, p are distinct. Since $h_{ik}(a)(h_{ki}(-\lambda\bar{a}))^{-1} \in \ker \rho$, it is fixed by $E_n(A)$. So if $\varepsilon = \varepsilon_{jk}(1)$ then $h_{ik}(a)h_{ki}(-\lambda\bar{a})^{-1} = [h_{ik}(a)h_{ki}(-\lambda\bar{a})^{-1}]^{\varepsilon} =$ (5a, 5b) $h_{ik}(a)h_{ij}(a)(h_{ki}(-\lambda\bar{a})h_{ji}(-\lambda\bar{a}))^{-1} =$ (2) $h_{ik}(a)h_{ki}(-\lambda\bar{a})^{-1}h_{ij}(a)h_{ji}(-\lambda\bar{a})$. Hence, $h_{ij}(a) = h_{ji}(-\lambda\bar{a})$.

THE $\tilde{\ell}_{ij}(a)$ AND $\tilde{\ell}_{ii}{}^p(a)$ SATISFY RELATIONS 1-6. For $i \neq j$, we define $\tilde{\ell}_{ij}(a) = \ell_{ij}{}^p(a)$ where $p \neq i, j$ and, for $i \neq p$, we define $\tilde{\ell}_{ii}{}^p(a) = \tilde{\ell}_{pi}(a)\,[y_{pp}(a), \varepsilon_{pi}(1)]$. We have shown above already that the definition of $\tilde{\ell}_{ij}(a)$ is independent of the choice of p. We shall show that for a fixed p the elements $\tilde{\ell}_{ij}(a)$ and $\tilde{\ell}_{ii}{}^p(a)$ satisfy relations 1-6. The $\tilde{\ell}_{ij}(a)$ satisfy relations 1-5a), because the $\ell_{ij}{}^p(a)$ satisfy these relations. That the $\tilde{\ell}_{ii}{}^p(a)$ satisfy relations 2 and 4 was proved already in the demonstration that Y is abelian. Relation 5a) doesn't apply to the $\tilde{\ell}_{ii}{}^p(a)$. It remains to verify relations 1 and 3 for the $\ell_{ii}{}^p(a)$ and relations 5b and 6 for both the $\tilde{\ell}_{ij}(a)$ and $\ell_{ii}{}^p(a)$. To simplify notation, let $h_{ij} = \tilde{\ell}_{ij}$ or $\tilde{\ell}_{ii}{}^p$.

If $a, b \in \bar{\Lambda}$ then $h_{ii}(a+b) = \tilde{\ell}_{pi}(-(a+b))\,[y_{pp}(a+b), \varepsilon_{pi}(1)] =$ (for some $z \in \ker \rho$) $= \tilde{\ell}_{pi}(-a)\tilde{\ell}_{pi}(-b)\,[z\,y_{pp}(a)y_{pp}(b), \varepsilon_{pi}(1)] = \tilde{\ell}_{pi}(-a)\tilde{\ell}_{pi}(-b)\,[y_{pp}(a), \varepsilon_{pi}(1)]\,[y_{pp}(b), \varepsilon_{pi}(1)] = h_{ii}(a)h_{ii}(b)$.

3). If $a \in \bar{\Lambda}$ then $a = -\lambda\bar{a}$ and $h_{ii}(a) = h_{ii}(-\lambda\bar{a})$.

We prove now an intermediate relation which we denote by 6′.

6′). If i, j and p are distinct and if $a \in \bar{\Lambda}$ then $h_{ii}(a)^{\varepsilon_{ij}(1)} = h_{ii}(a)h_{jj}(a)h_{ij}(a)$. $h_{ii}(a)^{\varepsilon_{ij}(1)} = \tilde{\ell}_{pi}(-a)^{\varepsilon_{ij}(1)}[y_{pp}(a), \varepsilon_{pi}(1)]^{\varepsilon_{ij}(1)} =$ (for some $z \in \ker \rho)\tilde{\ell}_{pi}(-a)\tilde{\ell}_{pj}(-a)[zy_{pp}(a), [\epsilon_{ij}(1), \epsilon_{pi}(-1)]\epsilon_{pi}(1)] = \tilde{\ell}_{pi}(-a)\tilde{\ell}_{pj}(-a)$ $[y_{pp}(a), \epsilon_{pi}(1)]\,[y_{pp}(a), \epsilon_{pj}(1)]^{\epsilon_{pi}(1)} = h_{ii}(a)\tilde{\ell}_{pj}(-a)\,[y_{pp}(a)^{\epsilon_{pi}(1)}, \epsilon_{pj}(1)^{\epsilon_{pi}(1)}]$ $=$ (for some $z \in \ker \rho$) $h_{ii}(a)\tilde{\ell}_{pj}(-a)\,[z\,y_{pp}(a)\,y_{pi}(a)\,y_{ii}(a), \epsilon_{pj}(1) =$ $h_{ii}(a)\tilde{\ell}_{pj}(-a)\,[y_{pp}(a), \epsilon_{pj}(1)]\,[\tilde{\ell}_{ip}(-\lambda\bar{a}), \epsilon_{pj}(1)]\,[y_{ii}(a), \epsilon_{pj}(1)] =$ (5a, 4) $h_{ii}(a)\,h_{jj}(a)\tilde{\ell}_{ij}(-\lambda\bar{a}) =$ (since $a = \lambda\bar{a}$) $h_{ii}(a)\,h_{jj}(a)\,h_{ij}(a)$.

Set $h = h_{kk}(ab-\bar{\lambda}\bar{a}\bar{b})^{\epsilon_{ki}(1)}$, and suppose that i, j, k, p are distinct. $[h_{kj}(a), \epsilon_{jk}(b)]\,h_{kk}(ab-\bar{\lambda}\bar{a}\bar{b})^{-1} \in \ker \rho$ and is fixed by $E_n(A)$. Hence, it $= [h_{jk}(-\bar{\lambda}\bar{a})^{\epsilon_{ki}(1)}, \epsilon_{jk}(b)^{\epsilon_{ki}(1)}]\,h^{-1} = [h_{kj}(a)\,h_{ij}(a), [\epsilon_{ki}(1), \epsilon_{jk}(-b)]\,\epsilon_{jk}(b)]\,h^{-1}$ $= [h_{kj}(a)\,h_{ij}(a), \epsilon_{ji}(b)\,\epsilon_{jk}(b)]\,h^{-1} = [h_{kj}(a), \epsilon_{jk}(b)]\,[h_{ij}(a), \epsilon_{jk}(b)]\,[h_{kj}(a),$ $\epsilon_{ji}(b)]^{\epsilon_{jk}(b)}\,[h_{ij}(a), \epsilon_{ji}(b)]^{\epsilon_{jk}(b)}h^{-1} = [h_{kj}(a), \epsilon_{jk}(b)]\,h_{ik}(ab)\,h_{ki}(ab)\,[h_{ij}(a),$ $\epsilon_{ji}(b)]\,(h_{kk}(ab-\bar{\lambda}\bar{a}\bar{b})\,h_{ki}(ab-\bar{\lambda}\bar{a}\bar{b})\,h_{ii}(ab-\bar{\lambda}\bar{a}\bar{b})^{-1} = [h_{kj}(a), \epsilon_{jk}(b)]\,[h_{ij}(a), \epsilon_{ji}(b)]$ $(h_{kk}(ab-\bar{\lambda}\bar{a}\bar{b})\,h_{ii}(ab-\bar{\lambda}\bar{a}\bar{b}))^{-1}$. Equating the first term with the last, we establish 5b.

6). Define the element $(a, b)_{ik} \in \ker \rho$ by $[h_{ii}(a), \epsilon_{ik}(b)] =$ $h_{ik}(ab)\,h_{kk}(bab)\,(a, b)_{ik}$. We shall show that the symbol $(\ ,\ )_{ik}$ is biadditive and that $(a, b)_{ik} = (a, b)_{i'k'}$. Let $(a, b) = (a, b)_{ik}$. $(a+a_1, b) =$ $[h_{ii}(a+a_1), \epsilon_{ik}(b)] - h_{ik}((a+a_1)b) - h_{kk}(\bar{b}(a+a_1)b) = [h_{ii}(a), \epsilon_{ik}(b)] +$ $[h_{ii}(a_1), \epsilon_{ik}(b)] - h_{ik}(ab) - h_{ik}(a_1 b) - h_{kk}(\bar{b}ab) - h_{kk}(\bar{b}a_1 b) = (a, b) + (a_1, b)$. $(a, b+b_1) = [h_{ii}(a), \epsilon_{ik}(b+b_1)] - h_{ik}(a(b+b_1)) - h_{kk}((\overline{b+b_1})a(b+b_1)) = [h_{ii}(a),$ $\epsilon_{ik}(b_1)] + [h_{ii}(a), \epsilon_{ik}(b)]^{\epsilon_{ik}(b_1)} - h_{ik}(ab) - h_{ik}(ab_1) - h_{kk}(\bar{b}ab) - h_{kk}(\bar{b}_1 ab_1)$ $- h_{kk}(\bar{b}ab_1+\bar{b}_1 ab)$. However, $[h_{ii}(a), \epsilon_{ik}(b)]^{\epsilon_{ik}(b_1)} = [h_{ii}(a)^{\epsilon_{ik}(b_1)}, \epsilon_{ik}(b)]$ $= [h_{ii}(a), \epsilon_{ik}(b)] + [h_{ik}(ab_1), \epsilon_{ik}(b)] + [h_{kk}(\bar{b}_1 ab_1, \epsilon_{ik}(b)]$. The last term is zero (by 4), and $[h_{ik}(ab_1), \epsilon_{ik}(b)] =$ (3) $[h_{ki}(\bar{b}_1 a), \epsilon_{ik}(b)] =$ (5b) $h_{kk}(\bar{b}_1 ab - \lambda\overline{\bar{b}_1 ab}) = h_{kk}(\bar{b}_1 ab+\bar{b}ab_1)$. Making the appropriate substitutions, we see that $(a, b+b_1) = ([h_{ii}(a), \epsilon_{ik}(b_1)] - h_{ik}(ab_1) - h_{kk}(\bar{b}_1 ab_1)) + ([h_{ii}(a),$ $\epsilon_{ik}(b)] - h_{ik}(ab) - h_{kk}(\bar{b}ab)) = (a, b_1) + (a, b_2)$.

Suppose that i, j, k, p are distinct and that $a \in \bar{\Lambda}$. Set $e = \epsilon_{jk}(-1)\,\epsilon_{ij}(b)\,\epsilon_{jk}(1), e_1 = \epsilon_{ij}(b)\,\epsilon_{jk}(1)$ and $e_2 = \epsilon_{jk}(1)$. Then

$h_{ii}(a)^{\varepsilon_{ik}(b)} = h_{ii}(a)^{[\varepsilon_{ij}(\bar{b}),\varepsilon_{jk}(1)]} = h_{ii}(a)^{\varepsilon_{ij}(-b)e} = h_{ii}(a)^{e} + h_{ij}(-ab)^{e} + h_{jj}(\bar{b}ab)^{e} + (a,-b)_{ij} = h_{ii}(a)^{e_1} + h_{ij}(-ab)^{e_1} + h_{ik}(ab)^{e_1} + h_{jj}(\bar{b}ab)^{e_1} + h_{jk}(-\bar{b}ab)^{e_1} + h_{kk}(\bar{b}ab)^{e_1} + (\bar{b}ab,-1)_{jk} + (a,-b)_{ij} = h_{ii}(a)^{e_2} + h_{ij}(ab)^{e_2} + h_{jj}(\bar{b}ab)^{e_2} + (a,b)_{ij} + h_{ij}(-ab)^{e_2} + h_{jj}(-2\bar{b}ab)^{e_2} + h_{ik}(ab)^{e_2} + h_{kj}(\bar{b}ab)^{e_2} + h_{jj}(\bar{b}ab)^{e_2} + h_{jk}(-\bar{b}ab)^{e_2} + h_{kk}(\bar{b}ab)^{e_2} + (\bar{b}ab,-1)_{jk} + (a,-b)_{ij}$ = (after cancelling $h_{ij}(ab)^{e_2}$ with $h_{ij}(-ab)^{e_2}, h_{jj}(\bar{b}ab)^{e_2} + h_{jj}(\bar{b}ab)^{e_2}$ with $h_{jj}(-2\bar{b}ab)^{e_2}$, $(a_1,b)_{ij}$ with $(a,-b)_{ij}$, and $h_{kj}(\bar{b}ab)^{e_2}$ with $h_{jk}(-\bar{b}ab)^{e_2} = h_{kj}(-\bar{b}ab)^{e_2}$) $h_{ii}(a) + h_{ik}(ab) + h_{kk}(\bar{b}ab) + (\bar{b}ab,-1)_{jk}$. However, 6′) $\Longrightarrow$ $(\bar{b}ab,1)_{jk} = 0$, and since $(\ ,\ )_{jk}$ is biadditive, it follows that $(\bar{b}ab,-1)_{jk} = 0$. This completes the proof.

§4. K-THEORY GROUPS OF NONSINGULAR MODULES

In this chapter, we define the functors KQ_i, KH_i, WQ_i, and WH_i for $i = 1, 2$. The corresponding functors for $i = 0$ were defined already in §1 B and C. A formal treatment of the functors K_0, K_1, and K_2 in the setting of a category with product will be given in §6. The definitions given in the current chapter are such that they agree with those in §6, but do not require that the reader has read §6. We take the opportunity also to relate the definitions here with those in Bass [10] and [11].

A. *The* K_1*-functors*

We begin by recalling the functor K_1 : ((rings)) $\to$ ((abelian groups)). Let A be a ring. Let $GL_n(A)$ denote the group of invertible $n \times n$ matrices with coefficients in A and let $E_n(A)$ denote the subgroup of $GL_n(A)$ generated by all elementary matrices (cf. §3). There is a natural embedding $GL_n(A) \to GL_{n+1}(A)$, $\alpha \mapsto \begin{pmatrix} \alpha & 0 \\ 0 & 1 \end{pmatrix}$, and one defines $GL(A) = \varinjlim_n GL_n(A)$ and $E(A) = \varinjlim_n E_n(A)$. By the Whitehead lemma [20, 3.1], $E(A)$ is the commutator subgroup $[GL(A), GL(A)]$ of $GL(A)$. One defines

$$K_1(A) = GL(A)/E(A) .$$

If A has an involution $a \mapsto \bar{a}$ then $GL(A)$ has an involution $\alpha \mapsto \bar{\alpha}$ such that if $\alpha = (a_{ij}) \in GL_n(A)$ then $\bar{\alpha} = \text{transpose } (\bar{a}_{ij})$. The involution on $GL(A)$ induces a $\mathbf{Z}/2\mathbf{Z}$-action on $K_1(A)$.

Let (A, Λ) be a form ring (cf. §1B). Let $\mathfrak{q}$ be an involution invariant ideal of A. Recall the groups $GQ(A, \Lambda)$, $GQ(A, \Lambda, \mathfrak{q})$, and $EQ(A, \Lambda, \mathfrak{q})$ defined in §3. By 3.9, $EQ(A, \Lambda, \mathfrak{q}) = [GQ(A, \Lambda), GQ(A, \Lambda, \mathfrak{q})]$. Define

$$KQ_1(A,\Lambda,\mathfrak{q}) = GQ(A,\Lambda,\mathfrak{q})/EQ(A,\Lambda,\mathfrak{q})$$

$$WQ_1(A,\Lambda,\mathfrak{q}) = KQ_1(A,\Lambda,\mathfrak{q})/\text{hyperbolic matrices.}$$

Suppose that $\Lambda/\min$ is finite and recall the groups $GH(A,\Lambda)$ and $GH(A,\Lambda,\mathfrak{q})$ defined in §3. Call a matrix in $GH(A,\Lambda,\mathfrak{q})$ *metabolic* if it has the form $\begin{pmatrix}\alpha & 0\\ 0 & \bar{\alpha}^{-1}\end{pmatrix}$ for some $\alpha \in GL(A)$. Define

$$KH_1(A,\Lambda,\mathfrak{q}) = GH(A,\Lambda,\mathfrak{q})/[GH(A,\Lambda),GH(A,\Lambda,\mathfrak{q})]$$

$$WH_1(A,\Lambda,\mathfrak{q}) = KH_1(A,\Lambda,\mathfrak{q})/\text{metabolic matrices.}$$

Let

$$KQ_1(A,\Lambda) = KQ_1(A,\Lambda,A)$$

etc.

It is clear that the rules $(A,\Lambda) \mapsto KQ_1(A,\Lambda)$, etc., define functors ((form rings)) $\to$ ((abelian groups)). If we want to emphasize the symmetry λ with respect to which Λ is defined then we write $KQ_1^{\lambda}(A,\Lambda)$ in place of $KQ_1(A,\Lambda)$, etc.

There is an abstract procedure in [10, VII 1.4] which can be used to define the groups $KQ_1(A,\Lambda,\mathfrak{q})$, etc., as K_1-groups of certain categories. The procedure does not require a restriction on Λ and goes as follows. Recall the categories $\mathbf{P}(A)$, $\mathbf{Q}(A,\Lambda)$, $\mathbf{H}(A,\Lambda)$, and $\mathbf{S}(A,\Lambda)$ introduced in §1B and §1C. Let $\mathbf{F}(A,\Lambda)$ denote any one of the categories $\mathbf{Q}(A,\Lambda)$ or $\mathbf{H}(A,\Lambda)$. The canonical homomorphism $A \to A/\mathfrak{q}$ induces functors $G: \mathbf{P}(A) \to \mathbf{P}(A/\mathfrak{q})$, $G': \mathbf{S}(A,\Lambda) \to \mathbf{S}(A/\mathfrak{q},\Lambda/\mathfrak{q})$, and $G'': \mathbf{F}(A,\Lambda) \to \mathbf{F}(A/\mathfrak{q},\Lambda/\mathfrak{q})$. Using the notation of [10, VII 1.4], we define

$$K_1(A,\mathfrak{q}) = K_1(\mathbf{P}(A),G)$$

$$KS_1(A,\Lambda,\mathfrak{q}) = K_1(\mathbf{S}(A,\Lambda),G')$$

$$KF_1(A,\Lambda,\mathfrak{q}) = K_1(\mathbf{F}(A,\Lambda),G'')\,.$$

where $F = Q$ or H. The hyperbolic and metabolic functors induce homomorphisms

$$\mathbf{H}: K_1(A,\mathfrak{q}) \to KQ_1(A,\Lambda,\mathfrak{q}) \quad \text{and} \quad \mathbf{M}: KS_1(A,\Lambda,\mathfrak{q}) \to KH_1(A,\Lambda,\mathfrak{q})$$

and we define

$$WQ_1(A,\Lambda,\mathfrak{q}) = \operatorname{coker} \mathbf{H}$$

$$WH_1(A,\Lambda,\mathfrak{q}) = \operatorname{coker} \mathbf{M}\ .$$

The fact that we get the same definitions of KQ_1, etc., as above follows from the general nonsense theorem [10, VII 2.3] and the fact that the Λ-hyperbolic modules $H(A^n)$ are cofinal 2.10 in $\bar{Q}(A,\Lambda)$ and the Λ-metabolic modules $M_{a_1,\cdots,a_N,0,\cdots,0}(A^{N+m})$ are cofinal 2.10 in $H(A,\Lambda)$.

Let Y be an involution invariant subgroup of $K_1(A)$. For convenience, we shall assume that Y contains the classes of the elements -1 and $-\lambda \in GL_1(A)$. If $F = H$ or Q, define

$$KF_1(A,\Lambda)_{\text{based-}Y} = \ker(KF_1(A,\Lambda) \to K_1(A)/Y)\ .$$

One can show as above that

$$KF_1(A,\Lambda)_{\text{based-}Y} = K_1(F(A,\Lambda)_{\text{based-}Y})\ .$$

B. *The* K_2*-functors*

Recall the Steinberg group $St(A)$ and the quadratic Steinberg group $StQ(A,\Lambda)$ defined in §3. By [20] and 3.17, they are the universal, perfect, central extensions of respectively $E(A)$ and $EQ(A,\Lambda)$. Define

$$K_2(A) = \ker(St(A) \to E(A))$$

$$KQ_2(A,\Lambda) = \ker(StQ(A,\Lambda) \to EQ(A,\Lambda))\ .$$

If $\Lambda/\min$ is finite then by the sentence preceding 3.15, we know that $[GH(A,\Lambda), GH(A,\Lambda)]$ is perfect and thus, by Lemma 3.25 has a universal, perfect, central extension, say $U(A,\Lambda)$. If $\Lambda/\min$ is finite, define

$$KH_2(A,\Lambda) = \ker(U(A,\Lambda) \to [GH(A,\Lambda), GH(A,\Lambda)]).$$

The hyperbolic map $E(A) \to EQ(A,\Lambda)$, $\varepsilon \mapsto \begin{pmatrix} \varepsilon & 0 \\ 0 & \bar{\varepsilon}^{-1} \end{pmatrix}$, induces a map $\mathbf{H}: K_2(A) \to KQ_2(A,\Lambda)$. Define

$$WQ_2(A,\Lambda) = \operatorname{coker}(\mathbf{H} : K_2(A) \to KQ_2(A,\Lambda)) .$$

Next, we shall develop a formal construction which will allow us to define KH_2 without any restriction on $\Lambda/\min$.

If G is a group, let $X(G)$ denote the free group on the symbols $x(g)$ such that $g \in G$. If $p: X(G) \to G$, $x(g) \mapsto g$, let $X_o(G) = X(G)/[\ker p, X(G)]$. It is clear that $X_o(G) \to G$ is a universal, central, covering of G. Let

$$U(G) = [X_0(G), X_0(G)]$$

$$H_2(G) = \ker(U(G) \to G) .$$

If G is perfect then from Lemma 3.25 and from the proof of 3.25, it follows that $U(G)$ is up to isomorphism the unique universal, perfect, central extension of G. The rule $G \mapsto X(G)$ defines in the obvious way a functor ((groups)) $\to$ ((groups)). It follows that the rule $G \mapsto U(G)$ (resp. $G \mapsto H_2(G)$) defines a functor ((groups)) $\to$ ((groups)) (resp. ((groups)) $\to$ ((abelian groups))).

LEMMA 4.1. *The functors* U *and* H_2 *commute with direct limits* [10,I §8].

The proof of 4.1 is left as an easy exercise.

We follow now Bass [11] in making the following definition. Let $\mathbb{C}$ be a category with product $\perp$ in which the isomorphism classes of objects form a set. Assume also that all the morphisms in $\mathbb{C}$ are isomorphisms. If M is an object, define $G_n(M)$ as in Lemma 3.15. Let $G_n'(M)$ and $G'(M)$ denote the commutator subgroup of $G_n(M)$ and $G(M)$ respectively. Define $H_2(M) = H_2(G_1'(M))$. The action of $G_1(M)$ on $G_1'(M)$ by conjugation induces an action of $G_1(M)$ on $H_2(M)$. Let $K_2(M) = H_0(G_1(M), H_2(M)) = H_2(M)/\{\sigma x - x \mid x \in H_2(M), \sigma \in G_1(M)\}$.

LEMMA 4.2. *If* $\alpha, \beta : M \to N$ *are isomorphisms in* $\mathbb{C}$ *then* $K_2(\alpha) = K_2(\beta)$.

Proof. It suffices to consider the case $M = N$, $\beta = 1_M$. But $K_2(\alpha) = 1$ by definition of $K_2(M)$.

For convenience, let us assume now that C has a trivial object O such that $M \perp O \simeq M$ for all objects M of C. Let $\mathrm{Tran}[C]$ denote the category whose objects are the isomorphism classes $[M]$ of objects M of C. If $[M]$, $[N] \in \mathrm{Obj}(\mathrm{Tran}[C])$, let $\mathrm{Morph}([M],[N]) = \{[P] \mid [P] \in \mathrm{Obj}(\mathrm{Tran}[C]),\ [M \perp P] = [N]\}$. Composition of morphisms is just $\perp$. By a result of Bass [10, I §8], $\mathrm{Tran}[C]$ is a directed category and hence if F is a functor $F : \mathrm{Tran}[C] \to$ ((abelian groups)), one can form the direct limit $\varinjlim_{[M]} F[M]$. By Lemma 4.2, the functor $K_2 : C \to$ ((abelian groups)) induces a functor $K_2 : \mathrm{Tran}[C] \to$ ((abelian groups)). We define

$$K_2C = \varinjlim_{[M]} K_2[M] .$$

The next two results together with 2.10, 2.11, and 2.12 show that the definitions of KQ_2 and KH_2 given at the beginning of the section are compatible with the one given just above. Moreover, the first result, by itself, shows that the definition above of K_2 agrees with that given in 6.12.

LEMMA 4.3. *If* $K_{2,n}(M) = H_0(G_n(M),\ H_2(M \perp \cdots \perp M))$ *then* $\varinjlim_n K_{2,n}(M) = H_2(G'(M))$.

Proof. Since H_0 and H_2 commute with direct limits, it follows that $\varinjlim_n K_{2,n}(M) = \varinjlim_n H_0(G_n(M),\ H_2(G_n'(M)) = H_0(G(M),\ H_2(G'(M))$ and $H_2(G'(M)) = \varinjlim_n H_2(G_n'(M))$. The canonical map $H_2(G'(M)) \to H_0(G(M),\ H_2(G'(M))$ is clearly surjective. Thus, it suffices to show that the action of $G(M)$ on $H_2(G'(M))$ is trivial. Let $\alpha \in G_n(M)$ and let $x \in H_2(G_n'(M))$. Let α_o denote the inner automorphism of $G_n(M)$ defined by α. Let $m \geq 2n$. The commutative diagram

$$\begin{array}{ccc} H_2(G'_n(M)) & \xrightarrow{\quad H_2(\alpha_o) \quad} & H_2(G'_n(M)) \\ \downarrow & & \downarrow \\ H_2(G'_m(M)) & \xrightarrow[\quad H_2(\alpha_o \perp \alpha_o^{-1} \perp 1_{M^{m-2n}}) \quad]{} & H_2(G'_m(M)) \end{array}$$

implies that if $H_2(\alpha_o \perp \alpha_o^{-1} \perp 1_{M^{m-2n}})y = y$ (y = image of x in $H_2(G'_m(M))$) then $H_2(\alpha_o \perp 1_{M^{m-n}})y = y$. However, $\alpha \perp \alpha^{-1} \in G'_{2n}(M)$ by [10, VII 1.8] and $\varinjlim G'_n(M)$ is perfect by 3.15. Hence, for m suitably large $\alpha \perp \alpha^{-1} \perp 1_{M^{m-2n}}$ lies in the commutator subgroup of $G'_m(M)$. If V is a universal, central covering of $G'_m(M)$ then $\gamma = \alpha \perp \alpha^{-1} \perp 1_{M^{m-2n}}$ lifts to an element of $[V, V]$. Moreover, conjugation by γ on $H_2(G'_m(M)) \subset [V, V]$ corresponds to $H_2(\alpha_o \perp \alpha_o^{-1} \perp 1_{M^{m-2n}})$. But since $H_2(G'_m(M)) \subset$ center $[V, V]$, it follows that $H_2(\alpha_o \perp \alpha_o^{-1} \perp 1_{M^{m-2n}})y = \gamma y \gamma^{-1} = y$.

COROLLARY 4.4 (Bass). *If* $\mathcal{C}$ *has a cofinal object* A *then the canonical homomorphism below is an isomorphism.*

$$H_2(G'(A)) \xrightarrow{\cong} K_2(\mathcal{C}).$$

Proof. The corollary follows directly from [10, I (8.6)].

Recall the metabolic functor $\mathbf{M}: S(A, \Lambda) \to \mathbf{H}(A, \Lambda)$ defined in §1C. Define

$$KH_2(A, \Lambda) = K_2\mathbf{H}(A, \Lambda)$$

$$WH_2(A, \Lambda) = \operatorname{coker}(K_2(\mathbf{M}): K_2S(A, \Lambda) \to K_2\mathbf{H}(A, \Lambda)).$$

It follows from Corollary 4.4 that the definition just given of KH_2 agrees with the one given previously. Of course, one can also define

$$KQ_2(A, \Lambda) = K_2Q(A, \Lambda)$$

$$WQ_2(A, \Lambda) = \operatorname{coker}(K_2H) : K_2P(A) \to K_2Q(A, \Lambda))$$

and then it follows from Corollary 4.4 and Theorem 3.17 that the definitions just given of KQ_2 and WQ_2 agree with those given previously. It is easy to check that KQ_2, KH_2, etc. define functors ((form rings)) → ((abelian groups)).

If $\mathfrak{q}$ is an involution invariant ideal of A, the congruence or relative groups

$$KQ_2(A, \Lambda, \mathfrak{q})$$

etc. will be defined by the relative procedure described immediately below in §4C. It should be noted that the congruence KQ_2-group defined by Quillen's methods is a quotient of the one defined above. The quotient can be proper.

As usual, if we want to emphasize the symmetry λ with respect to which Λ is defined then we shall write $KQ_2^\lambda(A, \Lambda)$ in place of $KQ_2(A, \Lambda)$, etc.

C. *Relativization*

In this section, we adapt the relativization procedure of Stein [28] to our situation. The significance of the procedure rests in the fact that it can reduce questions about relative groups to questions about absolute groups. We have had already an example of this in the proof of 3.9.

DEFINITION 4.4. Let (A, Λ) be a form ring such that Λ is defined with respect to the element $\lambda \in \operatorname{center}(A)$. Let $\mathfrak{q}$ be an involution invariant ideal of A. A *form ideal* of level $\mathfrak{q}$ of (A, Λ) is a pair

$$(\mathfrak{q}, \Lambda_\mathfrak{q})$$

where $\Lambda_\mathfrak{q}$ is an additive subgroup of A which satisfies the following rules:

1. $\{q-\lambda\bar{q} \mid q \in q\} + \left\{\sum_i q_i \Lambda \bar{q}_i \mid q_i \in q\right\} \subset \Lambda_q \subset q \cap \Lambda$

2. $a \Lambda_q \bar{a} \subset \Lambda_q$ for all $a \in A$.

It is clear that if Ω is either of the extremes in (4.4)(1) then (q, Ω) is a form ideal of level q in (A, Λ). Form ideals occur naturally in the classification of normal subgroups of general quadratic groups, cf. [1] and [2].

DEFINITION 4.5. a) Let A be a ring. If q is a two-sided ideal of A, define the *smash product ring*

$$A \ltimes q$$

such that $A \ltimes q = \{(a, q) \mid a \in A, q \in q\}$ and addition is defined by $(a, q) + (a', q') = (a+a', q+q')$ and multiplication by $(a, q)(a', q') = (aa', qa'+aq'+qq')$.

b) Let (A, Λ) be a form ring with involution $a \mapsto \bar{a}$. If (q, Λ_q) is a form ideal of (A, Λ), define the *smash product form ring*

$$(A, \Lambda) \ltimes (q, \Lambda_q)$$

by $(A, \Lambda) \ltimes (q, \Lambda_q) = (A \ltimes q, \Lambda \ltimes \Lambda_q)$ where the involution on $A \ltimes q$ is given by $(a, q) \mapsto (\bar{a}, \bar{q})$ and where $\Lambda \ltimes \Lambda_q = \{(a, q) \mid a \in \Lambda, q \in \Lambda_q\}$. If $\Lambda_q = q \cap \Lambda$ then we shall often write

$$(A, \Lambda) \ltimes q \quad \text{in place of} \quad (A, \Lambda) \ltimes (q, q \cap \Lambda).$$

There is a split ring homomorphism

$$f : A \ltimes q \to A, \quad (a, q) \mapsto a$$

with splitting

$$i : A \to (A \ltimes q), a \mapsto (a, 0).$$

Similarly, there is a split form ring homomorphism

$$g : (A, \Lambda) \ltimes (q, \Lambda_q) \to (A, \Lambda)$$

induced by f above, with splitting

$$j : (A, \Lambda) \to (A, \Lambda) \ltimes (q, \Lambda_q)$$

induced by i above.

DEFINITION 4.6. Let G be a group. If H is a normal subgroup of G, define the *smash product* (or *wreath product*)

$$G \ltimes H$$

such that $G \ltimes H = \{(\sigma, \rho) \mid \sigma \in G, \rho \in H\}$ and multiplication is defined by $(\sigma', \rho')(\sigma, \rho) = (\sigma'\sigma, (\sigma^{-1}\rho'\sigma)\rho)$.

DEFINITION 4.7. a) If G is a functor $G : ((\text{rings})) \to ((\text{groups}))$ and if A is a ring and $q \subset A$ is a two-sided ideal, we define the *relative group* of level q

$$G(A, q) = \ker(Gf : G(A \ltimes q) \to G(A)) .$$

Using the splitting Gi for Gf, we can canonically identify

$$G(A \ltimes q) = G(A) \ltimes G(A, q) .$$

b) If G is a functor $G : ((\text{form rings})) \to ((\text{groups}))$ and if (A, Λ) is a form ring and $(q, \Lambda_q) \subset (A, \Lambda)$ a form ideal, we define the relative group of level (q, Λ_q)

$$G((A, \Lambda), (q, \Lambda_q)) = \ker(Gg : G((A, \Lambda) \ltimes (\Lambda, \Lambda_q)) \to G(A, \Lambda)) .$$

Using the splitting Gj for Gg, we can canonically identify

$$G((A, \Lambda) \ltimes (q, \Lambda_q)) = G((A, \Lambda)) \ltimes G((A, \Lambda), (q, \Lambda_q)) .$$

If $\Lambda_q = q \cap \Lambda$ then we shall often write

$$G(A, \Lambda, q) \quad \text{in place of} \quad G((A, \Lambda), (q, q \cap \Lambda)) .$$

In this case, the equation above becomes

$$G((A, \Lambda) \ltimes q) = G(A, \Lambda) \ltimes G(A, \Lambda, q) .$$

Let (A, Λ) be a form ring and let $\mathfrak{q}$ be an involution invariant ideal of A. Consider the commutative diagram of form rings

$$\begin{array}{ccccc} & & (a, \mathfrak{q}) & \longmapsto & a \\ (a, \mathfrak{q}) & & (A, \Lambda) \ltimes \mathfrak{q} & \xrightarrow{\ g\ } & (A, \Lambda) \\ \Big\downarrow & & \Big\downarrow h & & \Big\downarrow \\ a + \mathfrak{q} & & (A, \Lambda) & \longrightarrow & (A/\mathfrak{q}, \Lambda/\mathfrak{q} \cap \Lambda) . \end{array}$$

g induces a canonical homomorphism $g_* : G(A, \Lambda, \mathfrak{q}) \to G(A, \Lambda)$. It is easy to check that if G is one of the functors GQ_{2n}, EQ_{2n}, or GH_{2n} defined in §A then Gg_* identifies the relative groups defined in this section with the relative groups defined in §A. Hence, the relative groups for the functors KQ_1, WQ_1, KH_1, WH_1 defined by the procedures in this section agree with their counterparts defined in §A. Moreover, the identification for the case $\mathfrak{q} = A$ eliminates a possible ambiguity in notation. Similar remarks apply, of course, to the situation in which the square of form rings above is replaced by the analogous square of rings.

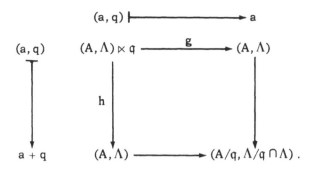

§5. HOMOLOGY EXACT SEQUENCES

A. *Homology groups and central extensions*

Throughout this section, ((groups)) denotes the category of all groups and group homomorphisms and G is an object of ((groups)).

DEFINITION 5.1. The *commutator subgroup*

$$[G, G]$$

of G is the subgroup of G generated by all elements $[\sigma, \rho] = \sigma\rho\sigma^{-1}\rho^{-1}$ such that $\sigma, \rho \in G$. G is called *perfect* if $G = [G, G]$. If N is a subgroup of G then the *mixed commutator subgroup*

$$[G, N]$$

of G is the subgroup of G generated by all $[\sigma, n] = \sigma n \sigma^{-1} n^{-1}$ such that $\sigma \in G$ and $n \in N$.

DEFINITION 5.2. An *extension* is a surjective homomorphism $\pi : G \to H$ of groups. An extension $\pi : G \to H$ is called *central* if the kernel$(\pi) \subset$ center(G). When the homomorphism π is a priori clear, the group G itself is often referred to as a central extension of H.

Let X(G) denote the free group on the set of symbols $\{x(\sigma) \mid \sigma \in G\}$. Let $Y(G) = \ker(X(G) \to G, x(\sigma) \mapsto \sigma)$ and let $X_0(G) = X(G)/[X(G), Y(G)]$. The homomorphism $X_0(G) \to G, [x(\sigma)] \mapsto \sigma$, is a central extension. If $f : G \to H$ is a group homomorphism then $X(f) : X(G) \to X(H), x(\sigma) \mapsto x(f(\sigma))$, and $X_0(f) : X_0(G) \to X_0(H), [x(\sigma)] \mapsto [x(f(\sigma))]$, define group homomorphisms. It follows that the constructions X(G) and $X_0(G)$ define functors ((groups)) $\to$ ((groups)).

DEFINITION - LEMMA 5.3. Define

$$U(G) = [X_0(G), X_0(G)] .$$

It is clear that U defines a functor ((groups)) → ((groups)). Furthermore, *if* G *is perfect then the following holds*:

a) U(G) *is perfect.*

b) *The homomorphism* $U(G) \to G$, $[x(\sigma)] \mapsto \sigma$, *is a central extension.*

c) *The extension* $U(G) \to G$ *solves the following*

UNIVERSAL PROBLEM: Given a group homomorphism $G \to H$ and a central extension $V \to H$, there is a homomorphism $U \to V$ such that the square below commutes

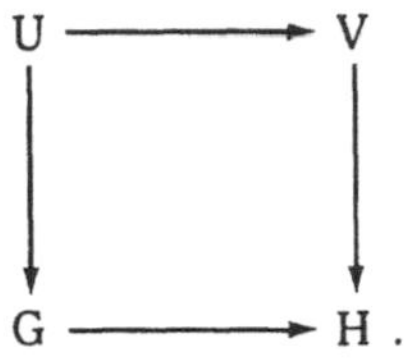

A central extension of G which solves the universal problem will be called *universal. Furthermore, if* $U \to G$ *and* $U' \to G$ *are two perfect, universal extensions then there are unique homomorphisms* $f : U \to U'$ *and* $g : U' \to U$ *such that the diagram*

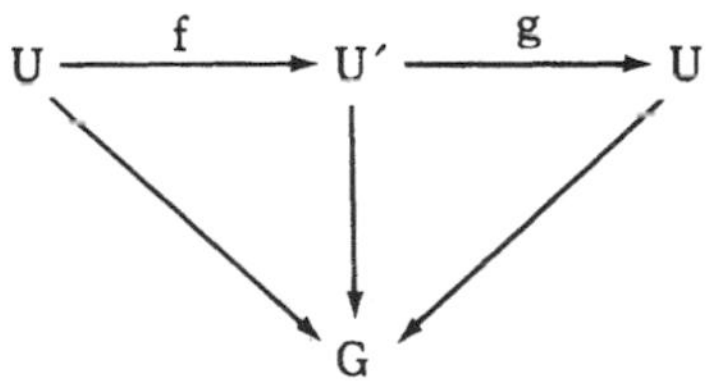

commutes. Again by uniqueness, $gf = 1$ and $fg = 1$. Thus, a perfect, universal extension is unique up to a unique isomorphism.

Proof. b) Since $G = [G, G]$ and since the homomorphism $\pi : X_0(G) \to G$, $[x(\sigma)] \mapsto \sigma$, is surjective, it follows that the restriction of π to $U(G) = [X_0(G), X_0(G)]$ is surjective. Thus, $U(G) \to G$ is an extension. It is clearly central, because $X_0(G) \to G$ is central.

a) Let $x, y \in X_0(G)$. Choose u (resp. v) $\in U(G)$ such that u and x (resp. v and y) have the same image in G. Since $X_0(G) \to G$ is central, there are elements $c, d \in \text{center}(X_0(G))$ such that $x = uc$ and $y = vd$. Thus, $[x,y] = [uc, vd] = [u,v]$. Since $U(G)$ is generated by $\{[x,y] \mid x, y \in X_0(G)\}$, it follows that $U(G)$ is perfect.

c) If f denotes the homomorphism $f: G \to H$, let f_* denote the homomorphism $f_*: X(G) \to V$, $x(\sigma) \mapsto \tilde{f}(\sigma)$, where $\tilde{f}(\sigma)$ is some lifting of the element $f(\sigma)$ to V. If $y \in Y(G)$ then $f_*(y) \in \ker(V \to H)$. Thus, $f_*(y) \in \text{center}(V)$. Thus, if $x \in X(G)$ then $f_*[x,y] = [f_*(x), f_*(y)] = 1$. Thus, f_* factors through $[X(G), Y(G)]$ and induces a homomorphism $f_*: U(G) \to V$. It is clear that the square

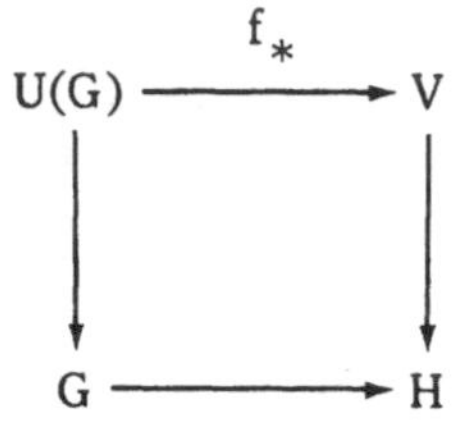

commutes. Suppose that $g: U(G) \to V$ is a homomorphism such that the corresponding square for g commutes. Let $x, y \in U(G)$. Since $f_*(x)$ and $g(x)$ (resp. $f_*(y)$ and $g(y)$) lie over the same element of H, it follows that there is an element c (resp. d) $\in \text{center}(V)$ such that $f_*(x) = g(x)c$ (resp. $f_*(y) = g(y)d$). Thus, $f_*[x,y] = [f_*(x), f_*(y)] = [g(x)c, g(y)d] = [g(x), g(y)] = g[x,y]$. Since $U(G)$ is perfect, it follows that $f_* = g$.

LEMMA 5.4. *Suppose that* G *is perfect. To check whether a central extension* $U \to G$ *is a universal extension in the sense of (5.3), it suffices to check whether it solves the following problem: Given a central extension* $V \to G$, *there is a unique homomorphism* $U \to V$ *such that the triangle below commutes*

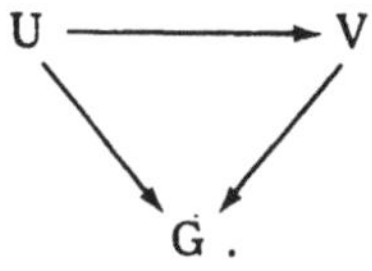

Proof. It suffices to show that a universal extension solves the problem in the lemma. But clearly, $U(G)$ solves the problem in the lemma.

The following lemma will be handy in several circumstances.

LEMMA 5.5. *Let* $V \to G$ *be a central extension. Let* H *be a subgroup of* G *and let* W *be its inverse image in* V. *Let* f *and* g *be automorphisms of* V *which induce automorphisms* $\bar{f}$ *and* $\bar{g}$ *of* G.

a) *If* $\bar{f}$ *and* $\bar{g}$ *agree on* H *then* f *and* g *agree on* $[W,W]$.

b) *If* $\bar{f}$ *fixes each element of* H *then* f *fixes each element of* $[W,W]$.

c) *If* $\bar{f}$ *and* $\bar{g}$ *leave* H *invariant then* f *and* g *leave* $[W,W]$ *invariant. Moreover, if* $\bar{f}$ *and* $\bar{g}$ *commute on* H *then* f *and* g *commute on* $[W,W]$.

Proof. a) If $w \in W$ then $h(w) = f(w)g(w)^{-1} \in \ker(V \to G)$. Thus, $h(w) \in \text{center}(V)$. If $v \in W$ then $h(v)h(w) = (f(v)g(v)^{-1})(f(w)g(w)^{-1}) = f(v)(f(w)g(w)^{-1})g(v)^{-1} = f(vw)g(vw)^{-1} = h(vw)$. Thus, the rule $W \to \text{center}(V)$, $w \mapsto h(w)$, defines a homomorphism which clearly vanishes on $[W,W]$. Thus, if $x \in [W,W]$ then $f(x) = g(x)$.

b) If $g = 1_V$ then $\bar{f}$ and $\bar{g}$ agree on H. Thus, by part a), f and g agree on $[W,W]$. Thus, f fixes each element of $[W,W]$.

c) It is clear that f and g leave W invariant. Thus, they leave $[W,W]$ also invariant. If $\bar{f}$ and $\bar{g}$ commute on H then $\overline{fg}$ and $\overline{gf}$ agree on H. Thus, by part a), fg and gf agree on $[W,W]$.

DEFINITION 5.6. Define the functors

$$H_i : ((\text{groups})) \to ((\text{abelian groups})) \quad (i = 1, 2)$$

$$H_1(G) = G/[G,G]$$

$$H_2(G) = \ker(U(G) \to G,\ [x(\sigma)] \mapsto \sigma).$$

The next definition is due to H. Bass [11, A.1].

DEFINITION 5.7. A functor $H : ((\text{groups})) \to ((\text{groups}))$ is called *central* if the action of G on $H(G)$ induced by inner automorphisms is trivial.

LEMMA 5.8. *The functors* H_i $(i=1,2)$ *are central.*

Proof. It is clear that H_1 is a central functor. Suppose that f is the inner automorphism of G corresponding to conjugation by the element $\rho \in G$. If $f_*: X_o(G) \to X_o(G)$, $[x(\sigma)] \mapsto [x(\rho\sigma\rho^{-1})]$, then by definition $H_2(f) = f_*|_{H_2(G)}$. If $f'_*: X_o(G) \to X_o(G)$, $[x(\sigma)] \mapsto [x(\rho)][x(\sigma)][x(\rho)]^{-1}$, then f_* and f'_* induce the same automorphism, namely f, of G. Thus, by Lemma 5.5a), f_* and f'_* agree on U(G). Thus, they agree on $H_2(G)$. But, f'_* leaves fixed each element of $Y(G)/[X(G), Y(G)]$ and $H_2(G) \subset Y(G)/[X(G), Y(G)]$. Thus, $H_2(f) = 1$.

LEMMA 5.9. *The functors* U *and* H_i $(i=1,2)$ *commute with direct limits.*

Proof. It is clear that H_1 commutes with direct limits. The functor X clearly commutes with direct limits. Since direct limits preserve exact sequences, it follows that the functor Y commutes with direct limits. Thus, the functor X_o commutes with direct limits. Thus, the functors U and H_2 commute with direct limits.

B. *The relative sequence of a homomorphism*

We begin by recalling the definition of a fibre product of sets with possibly extra structure such as a group multiplication or a topology.

DEFINITION 5.10. Given a diagram

$$\begin{array}{ccc} & & A_2 \\ & & \downarrow{\scriptstyle f_2} \\ A_1 & \xrightarrow[f_1]{} & \hat{A} \end{array}$$

of sets, groups, rings, topological groups or topological rings, we define the *fibre product*

$$A = A_1 \times_{\hat{A}} A_2 = \{(a_1, a_2) \mid a_i \in A_i\ (i=1,2),\ f_1(a_1) = f_2(a_2)\}.$$

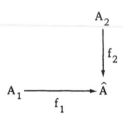

Addition or multiplication is defined componentwise. There are homomorphisms

$$f'_i : A \to A_i, \quad (a_1, a_2) \mapsto a_i$$

$(i = 1, 2)$ and in the case of topological groups or rings, we give A the topology with the least number of open sets such that the maps $f'_i\,(i = 1, 2)$ are continuous. The square

$$\begin{array}{ccc} A & \xrightarrow{f'_2} & A_2 \\ {\scriptstyle f'_1}\downarrow & & \downarrow{\scriptstyle f_2} \\ A_1 & \xrightarrow[f_1]{} & \hat{A} \end{array}$$

is clearly commutative. The fibre product construction solves the following **universal problem**: Given a commutative diagram

$$\begin{array}{ccc} B & \xrightarrow{g_2} & A_2 \\ {\scriptstyle g_1}\downarrow & & \downarrow{\scriptstyle f_2} \\ A_1 & \xrightarrow[f_1]{} & \hat{A} \end{array}$$

there is a *unique* map $g : B \to A$ such that $g_i = f'_i g\,(i = 1, 2)$. In fact, if $b \in B$ then $g(b) = (g_1(b), g_2(b))$ is the required map. On the other hand, if the diagram above itself solves the universal problem then there is a unique map $f' : A \to B$ such that $f'_i = g_i f'\ (i = 1, 2)$. Since $f'g$ and gf' are respectively the unique maps such that $g_i = g_i f' g$ and $f'_i = f'_i g f'$ $(i = 1, 2)$, it follows that $f' = g^{-1}$. Thus, a solution to the universal problem is unique up to a 'unique' isomorphism. A commutative square which solves the universal problem will be called a *fibre product square*.

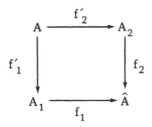

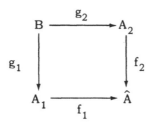

For the rest of the section, we fix the following notation:

$$G \text{ and } \hat{G}$$

denote groups such that their commutator subgroups

$$E \text{ and } \hat{E}$$

are perfect.

$$S \text{ and } \hat{S}$$

denote universal, perfect, central extensions of respectively E and $\hat{E}$.

$$f : G \to \hat{G}$$

denotes a group homomorphism.

$$G \times_{\hat{G}} \hat{S}$$

denotes the fibre product of the diagram

$$\begin{array}{ccc} & & \hat{S} \\ & & \downarrow \\ G & \xrightarrow[f]{} & \hat{G} . \end{array}$$

LEMMA 5.11. *The image of the canonical homomorphism* $S \to G \times_{\hat{G}} \hat{S}$ *is a normal subgroup of* $G \times_{\hat{G}} \hat{S}$.

Proof. If $(\sigma, \hat{s}) \in G \times_{\hat{G}} \hat{S}$ then we must show that the image $(S \to G \times_{\hat{G}} \hat{S})$ is stable under conjugation by $(\sigma, \hat{s})$. If U is the functor defined in (5.3) then we can canonically identify $S = U(E)$ and $\hat{S} = U(\hat{E})$. Let $\hat{\varepsilon}$ denote the image of $\hat{s}$ in $\hat{E}$. If σ' and $\hat{\varepsilon}'$ denote conjugation by respectively σ and $\hat{\varepsilon}$ then applying the functor U, we get automorphisms $U(\sigma') \in \mathrm{Aut}(S)$ and $U(\hat{\varepsilon}') \in \mathrm{Aut}(\hat{S})$ which cover respectively σ' and $\hat{\varepsilon}'$. Furthermore, by (1.3)c), $U(\hat{\varepsilon}')$ is conjugation by $\hat{s}$.

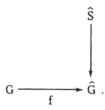

Let $t \in S$. If ε is the image of t in E then the image of t in $G \times_{\hat{G}} \hat{S}$ is $(\varepsilon, U(f)t)$ where $f: G \to \hat{G}$ is our fixed homomorphism. Conjugating by $(\sigma, \hat{s})$, we obtain $(\sigma, \hat{s})(\varepsilon, U(f)t)(\sigma, \hat{s})^{-1} = (\sigma\varepsilon\sigma^{-1}, \hat{s}(U(f)t)\hat{s}^{-1}) = (\sigma'(\varepsilon), U(\hat{\varepsilon}')U(f)t)$. Applying U to the commutative diagram

$$\begin{array}{ccc} E & \xrightarrow{f} & \hat{E} \\ {\scriptstyle \sigma'}\downarrow & & \downarrow{\scriptstyle \hat{\varepsilon}'} \\ E & \xrightarrow[f]{} & \hat{E} \end{array}$$

we obtain that $U(f)U(\sigma') = U(\hat{\varepsilon}')U(f)$. Thus, $(\sigma'(\varepsilon), U(\hat{\varepsilon}')U(f)t) = (\sigma'(\varepsilon), U(f)(U(\sigma')t))$ which is the image of the element $U(\sigma')t \in S$.

Using the previous lemma, we make the following definition:

DEFINITION 5.12. Define the *relative group*

$$H_1(f) = \operatorname{coker}(S \to G \times_{\hat{G}} \hat{S}) .$$

The next result will be a principal tool to construct relative K-Theory exact sequences in §6B. The result extends a related result [11, A.10] of H. Bass where it is assumed that the homomorphism $E \to \hat{E}$ is surjective.

THEOREM 5.13. *The following sequence is exact*

$$\begin{array}{c} [\hat{s}] \mapsto [1, \hat{s}] \\ H_2(E) \to H_2(\hat{E}) \to H_1(f) \to H_1(G) \to H_1(\hat{G}) . \\ [\sigma, \hat{s}] \mapsto [\sigma] \end{array}$$

Furthermore, the sequence is natural with respect to morphisms $(G \xrightarrow{f} \hat{G}) \to (G_1 \xrightarrow{f_1} \hat{G}_1)$ *between homomorphisms of groups.*

Proof. Let $(G, f) = \ker(f: G \to H_1(\hat{G}))$. Consider the following commutative diagram

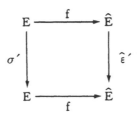

$$\begin{array}{ccccccc}
1 & & 1 & & & & \\
\downarrow & & \downarrow & & & & \\
H_2(E) & \longrightarrow & H_2(\hat{E}) & & \hat{s} & & \\
\downarrow & & \downarrow & & \downarrow & & \\
S & \longrightarrow & G \times_{\hat{G}} \hat{S} & & (1, \hat{s}) & & (\sigma, \hat{s}) \\
\downarrow & & \downarrow & & & & \downarrow \\
E & \longrightarrow & (G, f) & & & & \sigma \\
\downarrow & & \downarrow & & & & \\
1 & \longrightarrow & 1\,. & & & &
\end{array}$$

The first column is exact by definition and it is trivial to check that the second column is also exact. It is clear that the map $E \to (G, f)$ is injective. Thus, by the snake lemma, there is an exact sequence $1 \to \operatorname{coker}(H_2(E) \to H_2(\hat{E})) \to \operatorname{coker}(S \to G \times_{\hat{G}} \hat{S}) \to \operatorname{coker}(E \to (G, f)) \to 1$. Thus, there is an exact sequence $H_2(E) \to H_2(\hat{E}) \to H_1(f) \to (G, f)/E \to 1$. By the definition of (G, f), the sequence $1 \to (G, f)/E \to H_1(G) \to H_1(\hat{G})$ is exact. Splicing the last two exact sequences together, we get the exact sequence in the lemma. The naturality assertion is routine to verify. The details will be left to the reader.

C. *The Mayer-Vietoris sequence of a fibre square*

In this section, we fix the following notation: Let

$$(5.14)\qquad \begin{array}{ccc}
G & \xrightarrow{f_2'} & G_2 \\
{\scriptstyle f_1'}\downarrow & & \downarrow{\scriptstyle f_2} \\
G_1 & \xrightarrow[f_1]{} & \hat{G}
\end{array}$$

be a commutative square of groups whose commutator subgroups are perfect.

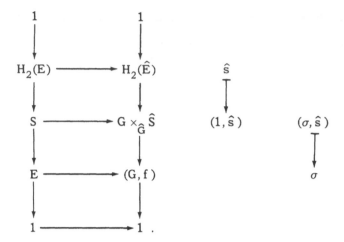

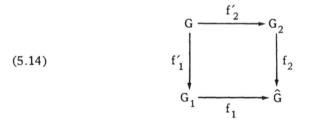

Let

$$(5.15)\qquad \begin{array}{ccc} E & \longrightarrow & E_2 \\ \downarrow & & \downarrow \\ E_1 & \longrightarrow & \hat{E} \end{array}$$

be the commutative square of commutator subgroups.

Let

$$(5.16)\qquad \begin{array}{ccc} S & \longrightarrow & S_2 \\ \downarrow & & \downarrow \\ S_1 & \longrightarrow & \hat{S} \end{array}$$

be the corresponding commutative square of universal, perfect, central extensions of the groups in (5.15). Factoring $H_2(E)$ and $H_2(E_1)$ out of the square above, one obtains the commutative square

$$(5.17)\qquad \begin{array}{ccc} E & \longrightarrow & S_2/H_2(E) \\ \downarrow & & \downarrow \\ E_1 & \longrightarrow & \hat{S}/H_2(E_1) \end{array}$$

and factoring $H_2(E_2)$ out of the square above, one obtains the commutative square

$$(5.18)\qquad \begin{array}{ccc} E & \longrightarrow & E_2 \\ \downarrow & & \downarrow \\ E_1 & \longrightarrow & \hat{S}/H_2(E_1)+H_2(E_2)\,. \end{array}$$

DEFINITION 5.19. The commutative square (5.14) will be called:

(i) E-*surjective*, if given $x_i \in G_i$ $(i=1,2)$ such that $f_1(x_1)f_2(x_2) \in \hat{E}$, there are elements $e_i \in E_i$ $(i=1,2)$ such that $f_1(x_1)f_2(x_2)=f_1(e_1)f_2(e_2)$.

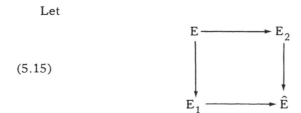

(ii) E-*fibred*, if (5.17) is a fibre square.

(iii) *weak* E-*fibred*, if (5.18) is a fibre square.

(iv) S-*surjective*, if it is E-surjective and if given $c \in H_2(\hat{E})$, there are elements $s_i \in S_i$ $(i = 1, 2)$ such that $c = f_1(s_1)f_2(s_2)$.

(v) S-*exact*, if given $s_i \in S_i$ $(i = 1, 2)$ such that $f_1(s_1) = f_2(s_2) \in \hat{S}$, there is an element $s \in S$ such that $f_i'(s) = s_i$ $(i = 1, 2)$.

REMARK 5.20. The definition of E-surjective above is weaker than that given by H. Bass [10, VII 3.3]. Nevertheless, all of the results in [10, VII] which require E-surjectivity as a hypothesis remain valid when the stronger notion of E-surjectivity is replaced by the weaker one above. Bass' definition of E-surjective is as follows. The square (5.14) is called E-surjective if given $\hat{e} \in \hat{E}$, there are elements $e_i \in E_i$ $(i = 1, 2)$ such that $\hat{e} = f_1(e_1)f_2(e_2)$. An example which demonstrates the difference between the two notions is as follows. If A is a commutative ring, let GL(A), SL(A), and E(A) denote respectively the infinite general linear group, special linear group, and elementary group. Let

$$\begin{array}{ccc} A & \longrightarrow & B \\ \downarrow & & \downarrow f_2 \\ \hat{A} & \underset{f_1}{\longrightarrow} & \hat{B} \end{array}$$

be a commutative square of commutative rings such that the $f_1\hat{A} \subset f_2B$ and such that $SL(f_2B) = E(f_2B)$. If $f_2B \neq \hat{B}$ then the square

$$\begin{array}{ccc} GL(A) & \longrightarrow & GL(B) \\ \downarrow & & \downarrow \\ GL(\hat{A}) & \longrightarrow & GL(\hat{B}) \end{array}$$

is E-surjective in the weaker sense, but not in the stronger sense. It is E-surjective in the stronger sense $\Longleftrightarrow f_2B = \hat{B}$.

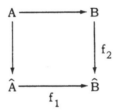

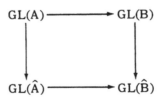

LEMMA 5.21. *If each element of* $\hat{S}$ *can be written as a product* $s_1 s_2$ *such that* $s_i \in$ *image* $(S_i \to \hat{S})$ $(i=1,2)$ *then (5.14) is* S*-surjective.*

Proof. Clear.

LEMMA 5.22. S*-surjective* $\Rightarrow$ E*-surjective.*

Proof. Clear.

LEMMA 5.23. *If (5.14) is a fibre square then the following are equivalent:*

(i) *(5.14) is* E*-surjective.*

(ii) *The sequence below is exact*

$$H_1(G) \to H_1(G_1) \oplus H_1(G_2) \to H_1(\hat{G}).$$
$$[\sigma] \mapsto [f'_1(\sigma), f'_2(\sigma^{-1})]$$

Proof. a) *(5.14) is* E*-surjective*: It is clear that the composite mapping at $H_1(G_1) \oplus H_1(G_2)$ is zero. Thus, it suffices to show that the $\ker(H_1(G_1) \oplus H_2(G_2) \to H_1(\hat{G})) \subset \text{image } (H_1(G) \to H_1(G_1) \oplus H_1(G_2))$. Let $\sigma_i \in G_i$ $(i=1,2)$ such that $[f_1(\sigma_1) f_2(\sigma_2)] = 1$. By E-surjectivity, there are $\varepsilon_i \in E_i$ such that $f_1(\varepsilon_1 \sigma_1) f_2(\sigma_2 \varepsilon_2) = 1$. Since (5.14) fibred, there is a $\sigma \in G$ such that $f'_1(\sigma) = \varepsilon_1 \sigma_1$ and $f'_2(\sigma) = (\sigma_2 \varepsilon_2)^{-1}$. Thus, $[\sigma_1, \sigma_2] = [\varepsilon_1 \sigma_1, \sigma_2 \varepsilon_2] = [f'_1(\sigma), f'_2(\sigma^{-1})]$.

b) *The sequence (ii) is exact*: Let $\sigma_i \in G_i$ $(i=1,2)$ such that $f_1(\sigma_1) f_2(\sigma_2) \in \hat{E}$. By the exactness of (ii), there are $\varepsilon_i \in E_i$ and $\sigma \in G$ such that $f'_1(\sigma) = \varepsilon_1 \sigma_1$ and $f'_2(\sigma^{-1}) = \sigma_2 \varepsilon_2$. Thus, $f_1(\varepsilon_1 \sigma_1) f_2(\sigma_2 \varepsilon_2) = f_1 f'_1(\sigma) f_2 f'_2(\sigma^{-1}) = 1$, because (5.14) is commutative. Thus, $f_1(\sigma_1) f_2(\sigma_2) = f_1(\varepsilon_1^{-1}) f_2(\varepsilon_2^{-1})$.

LEMMA 5.24. *If (5.14) is a fibre square then the condition* S*-exact* $\Rightarrow$ *the condition* E*-fibre.*

Proof. Since (5.14) is a fibre square, it follows that the map $E \to E_1 \times E_2$, $\varepsilon \mapsto (f'_1(\varepsilon), f'_2(\varepsilon))$, is injective. Thus, to prove (5.14) is an E-fibre square, it suffices to show that given $\varepsilon_1 \in E_1$ and $\bar{s}_2 \in S_2/H_2(E)$ such that $f_1(\varepsilon_1) = f_2(\bar{s}_2) \in \hat{S}/H_2(E_1)$, there is an $\varepsilon \in E$ such that $f'_1(\varepsilon) = \varepsilon_1$ and

$f'_2(\varepsilon) = \bar{s}_2$. Let $s_i \in S_i$ $(i=1;2)$ such that s_1 covers ε_1 and s_2 covers $\bar{s}_2$. Since the images of the s_i in $\hat{S}/H_2(E_1)$ coincide, it is clear that we can adjust s_1 so that their images in $\hat{S}$ coincide. Now, by S-exactness, the s_i have a common preimage s in S. But, then the image of s in E is a preimage for ε_1 and $\bar{s}_2$.

LEMMA 5.25. *If (5.14) is a fibre square then the condition* E*-fibre* $\Longrightarrow$ *the condition weak* E*-fibre.*

The proof of Lemma 5.25 is similar to that of Lemma 5.24. The details will be omitted.

Let

$$f'\colon G \to G_1 \times G_2, \quad \sigma \mapsto (f'_1(\sigma), f'_2(\sigma))$$

$$(G, f') = \ker(f'\colon G \to H_1(G_1) \oplus H_1(G_2)).$$

It is clear that also

$$(G, f') = \ker(G \to H_1(G_1) \oplus H_1(G_2))$$
$$\sigma \mapsto [f'_1(\sigma), f'_2(\sigma^{-1})].$$

LEMMA 5.26. *If (5.14) is a fibre square the commutative square below is a fibre square*

$$\begin{array}{ccc} (G, f') & \xrightarrow{f'_2} & E_2 \\ {\scriptstyle f'_1}\downarrow & & \downarrow{\scriptstyle f_2} \\ E_1 & \xrightarrow[f_1]{} & \hat{E} \end{array}$$

Proof. Let $\varepsilon_i \in E_i$ $(i=1,2)$ such that $f_1(\varepsilon_1) = f_2(\varepsilon_2)$. Since (5.14) is a fibre square, it follows that the ε_i $(i=1,2)$ have a unique common preimage $\sigma \in G$. Since $f'_i(\sigma) = \varepsilon_i \in E_i$ $(i=1,2)$, it follows by definition that $\sigma \in (G, f')$.

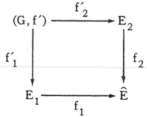

DEFINITION-LEMMA 5.27. *Consider the not necessarily commutative square*

$$\begin{array}{ccc} (G, f') & \xrightarrow{f'_2} & E_2 \\ {\scriptstyle f'_1}\downarrow & & \downarrow{\scriptstyle f_2} \\ E_1 & \xrightarrow[f_1]{} & \hat{S}/H_2(E_1)+H_2(E_2)\,. \end{array}$$

Since (5.14) is commutative, it follows that if $\sigma \in (G, f')$ *then* $f_1 f'_1(\sigma^{-1}) f_2 f'_2(\sigma) \in H_2(\hat{E})/H_2(E_1) + H_2(E_2)$. *We assert that the function*

$$\psi : (G, f') \to H_2(\hat{E})/H_2(E_1) + H_2(E_2)$$
$$\sigma \mapsto f_1 f'_1(\sigma^{-1}) f_2 f'_2(\sigma)$$

defines a homomorphism

$$\psi : (G, f')/E \to H_2(\hat{E})/H_2(E_1) + H_2(E_2)\,.$$

Proof. If $\sigma, \rho \in (G, f')$ then $\psi(\sigma)\psi(\rho) = (f_1 f'_1(\sigma^{-1}) f_2 f'_2(\sigma))(f_1 f'_1(\rho^{-1}) f_2 f'_2(\rho))$ $= f_1 f'_1(\rho^{-1})(f_1 f'_1(\sigma^{-1}) f_2 f'_2(\sigma)) f_2 f'_2(\rho) = f_1 f'_1((\sigma\rho)^{-1}) f_2 f'_2(\sigma\rho) = \psi(\sigma\rho)$. Thus, $\psi : (G, f') \to H_2(\hat{E})/H_2(E_1) + H_2(E_2)$ is a homomorphism. ψ factors through E, because the square (5.14) is commutative.

LEMMA 5.28. *Suppose that (5.14) is a fibre square. Define the group* J *by the fibre square*

$$\begin{array}{ccc} J & \longrightarrow & E_2 \\ \downarrow & & \downarrow \\ E_1 & \longrightarrow & \hat{S}/H_2(E_1)+H_2(E_2)\,. \end{array}$$

Since (5.14) is a fibre square, we can canonically identify $E \subset J \subset (G, f')$. *We assert that the sequence below is exact*

$$1 \to J/E \to (G, f')/E \xrightarrow{\psi} H_2(\hat{E})/H_2(E_1) + H_2(E_2)\,.$$

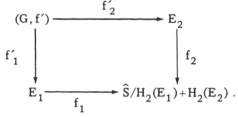

Proof. Since the square in the lemma is commutative, it is clear that $\psi(J/E) = 1$. On the other hand, suppose that $\sigma \in (G, f')$ such that $\psi(\sigma) = 1$. Since the square in the lemma is fibred, there is an element $\tau \in J$ such that $f'_i(\tau) = f'_i(\sigma)$ $(i = 1, 2)$. But, since (5.14) is fibred, it follows that $\tau = \sigma$.

LEMMA 5.29. *Suppose that (5.14) is a fibre square.*

a) *(5.14) is weak* E-*fibred* $\Longleftrightarrow \psi$ *is injective.*

b) *(5.14) is* S-*surjective* $\Longleftrightarrow$ *it is* E-*surjective and* ψ *is surjective.*

Proof. a) follows directly from Lemma 5.28.

b) We must show that under the assumption that (5.14) is E-surjective, the following are equivalent: (i) (5.14) is surjective. (ii) ψ is surjective. It is clear that (ii) $\Rightarrow$ (i). Conversely, suppose that (i) holds. Thus, given $c \in H_2(\hat{E})$, there are elements $s_i \in S_i$ $(i = 1, 2)$ such that $c = f_1(s_1)f_2(s_2)$. Thus, given $c \in H_2(\hat{E})/H_2(E_1) + H_2(E_2)$, there are elements $\varepsilon_i \in E_i$ $(i = 1, 2)$ such that $c = f_1(\varepsilon_1)f_2(\varepsilon_2)$. Since (5.14) is commutative and fibred, there is an element $\sigma \in G$ such that $f'_1(\sigma) = \varepsilon_1^{-1}$ and $f'_2(\sigma) = \varepsilon_2$. From the definition of (G, f'), it is clear that $\sigma \in (G, f')$. Moreover, $\psi(\sigma) = f_1 f'_1(\sigma^{-1}) f_2 f'_2(\sigma) = f_1(\varepsilon_1) f_2(\varepsilon_2) = c$.

DEFINITION 5.30. Suppose that (5.14) is a weak E-fibre, S-surjective, fibre square. Using Lemma 5.29, we define

$$\partial : H_2(\hat{E}) \to H_1(G)$$

as the composite of the following homomorphisms

$$H_2(\hat{E}) \longrightarrow H_2(\hat{E})/H_2(E_1) + H_2(E_2) \xrightarrow{\psi^{-1}} (G, f')/E \longrightarrow H_1(G) .$$

The next result will be a principal tool to construct Mayer-Vietoris, K-theory, exact sequences in §6B. The result extends a similar result [11, A.14] of H. Bass where it is assumed that the homomorphisms $E_i \to \hat{E}$ $(i = 1, 2)$ are surjective.

THEOREM 5.31. *If (5.14) is an S-exact, S-surjective, fibre square then there is an exact Mayer-Vietoris sequence*

$$H_2(E) \to H_2(E_1) \oplus H_2(E_2) \to H_2(\hat{E}) \xrightarrow{\partial} H_1(G) \to H_1(G_1) \oplus H_1(G_2) \to H_1(\hat{G})$$

$$c \mapsto (H_2(f_1')(c), H_2(f_2')(c^{-1})) \qquad [\sigma] \mapsto [f_1'(\sigma), f_2'(\sigma^{-1})]$$

where ∂ *is as in (5.30). Furthermore, the sequence is natural with respect to morphisms between commutative squares of groups.*

Proof. a) *Exactness at* $H_1(G_1) \oplus H_2(G_2)$ follows from Lemma 5.23.

b) *Exactness at* $H_1(G)$: From the definition of (G, f'), it follows that the sequence $1 \mapsto (G, f')/E \to H_1(G) \to H_1(G_1) \oplus H_2(G_2)$ is exact and from the definition of ψ, it follows that the image $\partial = (G, f')/E$.

c) *Exactness at* $H_2(\hat{E})$ follows directly from the definition of ψ.

d) *Exactness at* $H_2(E_1) \oplus H_2(E_2)$: It is clear the composite homomorphism is trivial. Suppose now that $c_i \in H_2(E_i)$ $(i = 1, 2)$ such that the image of (c_1, c_2) in $H_2(\hat{E})$ is trivial. By S-exactness, there is a $c \in S$ which is a preimage for both c_1 and c_2. We will be finished if the image of c in E is trivial. If ε denotes the image of c in E then we know that $f_i'(\varepsilon) = 1 \in G_i$ $(i = 1, 2)$. Since (5.14) is a fibre square, it follows that $\varepsilon = 1$.

e) *The naturality assertion* is straightforward to verify. The required checking will be left to the reader.

COROLLARY 5.32. *If in Theorem 5.29 the assumption that the square (5.14) is S-exact is weakened to the assumption that it is weak E-fibred then the conclusion of the theorem holds except for exactness at* $H_2(E_1) \oplus H_2(E_2)$.

Proof. The corollary is established already in the proof of Theorem 5.31.

D. *Excision*

We adopt the notational conventions (5.14 - 5.18) of the previous section.

DEFINITION 5.33. The commutative square (5.14) is called:

(i) *weak* E-*surjective*, if given $x_1 \in G_1$ such that $f_1(x_1) \in E_1$, there are elements $e_i \in E_i$ $(i=1,2)$ such that $f_1(x_1) = f_1(e_1)f_2(e_2)$.

(ii) *weak* S-*surjective*, if it is weak E-surjective and if given $c \in H_2(\hat{E})$, there are elements $s_i \in S_i$ $(i=1,2)$ such that $c = f_1(s_1)f_2(s_2)$.

LEMMA 5.34. a) E-*surjective* $\Rightarrow$ *weak* E-*surjective.*

b) S-*surjective* $\Rightarrow$ *weak* S-*surjective.*

Proof. Clear.

LEMMA 5.35. *(5.14) is weak* S-*surjective* $\Longleftrightarrow$ *it is weak* E-*surjective and the map* $\psi : (G, f')/E \to H_2(\hat{E})/H_2(E_1) + H_2(E_2)$ *in (5.27) is surjective.*

Lemma 5.35 is proved similarly to Lemma 5.29b). Details will be left to the reader.

DEFINITION 5.36. The commutative square (5.14) is said to satisfy *excision* for H_1 if the canonical homomorphism $H_1(f_2') \to H_1(f_1)$ is an isomorphism where $H_1(f_2')$ and $H_1(f_1)$ are the relative groups constructed in (5.12).

THEOREM 5.37. *A fibre square of groups satisfies excision for* H_1 *if and only if it is* E-*fibred and weak* S-*surjective. In fact, if (5.14) is a fibre square then the canonical homomorphism* $H_1(f_2') \to H_1(f_1)$ *is injective (resp. surjective) if and only if the square is* E-*fibred (resp. weak* S-*surjective).*

Proof. To simply notation, we let $f = f_2$ and $\hat{f} = f_1$.

a) Assume the square is E-fibred: Let $(\sigma, s_2) \in G \times_{G_2} S_2$ and let $(\sigma_1, \hat{s}_2)$ be its image in $G_1 \times_{\hat{G}} \hat{S}$. If $(\sigma_1, \hat{s}_2)$ vanishes in $H_1(\hat{f})$ then there is an element $s_1 \in S_1$ such that if ε_1 and $\hat{s}_1$ denote respectively its image in G and $\hat{S}$ then $(\sigma_1, \hat{s}_2) = (\varepsilon_1, \hat{s}_1)$. Consider the following commutative diagram

$$\begin{array}{ccccc} E & \longrightarrow & S_2/H_2(E) & & [s_2] \\ \downarrow & & \downarrow & & \downarrow \\ E_1 & \longrightarrow & \hat{S}/H_2(E_1) & & \\ \sigma_1 = \varepsilon_1 & \longmapsto & & & \hat{f}(\sigma_1) = \hat{f}(\varepsilon_1) = [\hat{s}_1] = [\hat{s}_2] \end{array}$$

By the E-fibre condition, there is an element $\varepsilon \in E$ which is a preimage for σ_1 and $[s_2]$. Since σ is a preimage for σ_1 and for the image of $[s_2]$ in E_2 and since (5.14) is a fibre square, it follows that $\sigma = \varepsilon$. If s is a preimage in S for ε, then from the commutative diagram

$$\begin{array}{cccccc} s & S & \longrightarrow & S_2 & & s_2 \\ \downarrow & \downarrow & & \downarrow & & \downarrow \\ & E & \longrightarrow & S_2/H_2(E) & & \\ \varepsilon & & \longmapsto & & & [s_2] \end{array}$$

it follows that we can adjust s by an element of $H_2(E)$ so that s becomes also a preimage for s_2. Thus, $(\sigma, s_2) = (\varepsilon, s_2)$ vanishes in $H_1(f)$.

b) *Assume the homomorphism* $H_1(f) \to H_1(\hat{f})$ *is injective*: Let $\varepsilon_1 \in E_1$ and $s_2 \in S_2$ which coincide in $\hat{S}/H_2(E_1)$. We shall show there is an element $\varepsilon \in E$ which is a preimage for ε_1 and for the image $[s_2]$ of s_2 in $S_2/H_2(E)$. It is clear that ε must be unique, because (5.14) is a fibre square. Let ε_2 denote the image of s_2 in E_2. Since (5.14) is a fibre square, there is a $\sigma \in G$ which is a preimage for both ε_1 and ε_2. Clearly $(\sigma, s_2) \in G \times_{G_2} S_2$. If s_1 is a lifting of ε_1 to S_1 then after adjusting s_1 by an element of $H_2(E_1)$, we can assume that the image $\hat{s}_1$ of s_1 in $\hat{S}$ coincides with the image of s_2 in $\hat{S}$. Thus, the image of (σ, s_2) in $G_1 \times_{\hat{G}} \hat{S}$ is the element $(\varepsilon_1, \hat{s}_1)$. Since $(\varepsilon_1, \hat{s}_1)$ vanishes in $H_1(\hat{f})$ and since the map $H_1(f) \to H_1(\hat{f})$ is injective, it follows that

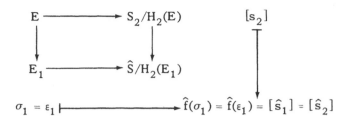

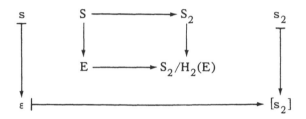

$(\sigma, s_2) = (\varepsilon, s_2)$ for some $\varepsilon \in E$. It is clear that ε is a preimage for ε_1 and $[s_2]$.

c) *Assume the square is weak* S-*surjective*: Let $(\sigma_1, \hat{s}) \in G_1 \times_{\hat{G}} \hat{S}$. Let $\hat{\varepsilon}$ denote the image of $\hat{s}$ in $\hat{E}$. By weak S-surjectivity, there are elements $\varepsilon_i \in E_i$ $(i = 1, 2)$ such that if $\hat{\varepsilon}_i$ denotes the image of ε_i in $\hat{E}$ then $\hat{\varepsilon} = \hat{\varepsilon}_1 \hat{\varepsilon}_2$. Let s_i $(i = 1, 2)$ be a preimage for ε_i in S_i and let $\hat{s}_i$ denote the image of s_i in $\hat{S}$. If $(\sigma_1, \hat{s})$ is multiplied on the left by $(\varepsilon_1, \hat{s}_1)^{-1}$ then its class in $H_1(\hat{f})$ is unchanged. After performing the multiplication, we get the element $(\varepsilon_1^{-1}\sigma_1, \hat{s}_1^{-1}\hat{s})$. Consider the commutative diagram

$$
\begin{array}{ccccc}
G & \longrightarrow & G_2 & & \varepsilon_2 \\
\downarrow & & \downarrow & & \big\downarrow \\
G_1 & \longrightarrow & \hat{G} & & \\
 & & & & \\
\varepsilon_1^{-1}\sigma_1 & \longmapsto & & & \hat{\varepsilon}_1^{-1}\hat{\varepsilon} \, .
\end{array}
$$

Since the square is fibred, there is an element $\rho \in G$ which is a preimage for both $\varepsilon_1^{-1}\sigma_1$ and ε_2. It is clear that $(\rho, s_2) \in G \times_{G_2} S_2$. Let $(\rho_1, \hat{s}_2)$ denote the image of (ρ, s_2) in $G_1 \times_{\hat{G}} \hat{S}$. After multiplying $(\varepsilon_1^{-1}\sigma_1, \hat{s}_1^{-1}\hat{s})$ on the left by $(\rho_1, \hat{s}_2)^{-1}$, we get the element $(1, \hat{s}_2^{-1}\hat{s}_1^{-1}\hat{s})$. Let $x = \hat{s}_2^{-1}\hat{s}_1^{-1}\hat{s}$. By Lemma 5.35, the map $\psi : (G, f')/E \to H_2(\hat{E})/H_2(E_1) + H_2(E_2)$ is surjective. Thus, there is an element $\tau \in (G, f')$ and elements $c_i \in H_2(E_i)$ $(i = 1, 2)$ such that $x = \hat{t}_1^{-1}\hat{t}_2\hat{c}_1\hat{c}_2$ where $\hat{c}_i$ $(i = 1, 2)$ is the image of c_i in $\hat{S}$ and where if τ_i $(i = 1, 2)$ is the image of τ in E_i and t_i is a lifting of τ_i to S_i then $\hat{t}_i$ is the image of t_i in $\hat{S}$. Thus, $(1, x) = (\tau_1, \hat{t}_1)^{-1}(\tau_2, \hat{t}_2)(1, \hat{c}_1)(1, \hat{c}_2)$. But, $(\tau_1, \hat{t}_1)$ and $(1, \hat{c}_1)$ vanish in $H_1(\hat{f})$ and $(\tau_2, \hat{t}_2)$ and $(1, \hat{c}_2)$ have respectively the preimages (τ, t_2) and $(1, c_2) \in G \times_{G_2} S_2$.

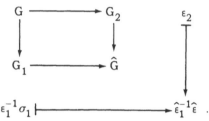

d) *Assume the homomorphism* $H_1(f) \to H_1(\hat{f})$ *is surjective*: First, we shall show that (5.14) is weak E-surjective. Let $\sigma_1 \in G_1$ such that its image $\hat{\sigma}_1$ in $\hat{G}$ lies in $\hat{E}$. If $\hat{s}$ is a lifting of $\hat{\sigma}_1$ to $\hat{S}$ then $(\sigma_1, \hat{s}) \in G_1 \times_{\hat{G}} \hat{S}$. Since $H_1(f) \to H_1(\hat{f})$ is surjective, there is an element $(\rho, s_2) \in G \times_{G_2} S_2$ and an element $s_1 \in S_1$ such that if $(\rho_1, \hat{s}_2)$ (resp. $(\varepsilon_1, \hat{s}_1)$) denotes the image of (ρ, s_2) (resp. s_1) in $G_1 \times_{\hat{G}} \hat{S}$ (resp. $G_1 \times_{\hat{G}} \hat{S}$) then $(\sigma_1, \hat{s}) = (\varepsilon_1, \hat{s}_1)(\rho_1, \hat{s}_2)$. Thus, if $\hat{\varepsilon}, \hat{\varepsilon}_1$, and $\hat{\varepsilon}_2$ denote respectively the images of $\hat{s}, \hat{s}_1$, and $\hat{s}_2$ in $\hat{E}$ then $\hat{\sigma}_1 = \hat{\varepsilon} = \hat{\varepsilon}_1 \hat{\varepsilon}_2$ which shows that (5.14) is weak E-surjective. To show that (5.14) is weak S-surjective, it suffices now by Lemma 5.35 to show that the map $\psi : (G, f')/E \to H_2(\hat{E})/H_2(E_1) + H_2(E_2)$ is surjective. Let $\hat{s} \in H_2(\hat{E})$ and let $[\hat{s}]$ denote its class in $H_2(\hat{E})/H_2(E_1) + H_2(E_2)$. Clearly $(1, \hat{s}) \in G_1 \times_{\hat{G}} \hat{S}$. Applying the work above with $\sigma_1 = 1$, we can write $(1, \hat{s}) = (\varepsilon_1, \hat{s}_1)(\rho_1, \hat{s}_2)$. Since $1 = \hat{\sigma}_1 = \hat{\varepsilon} = \hat{\varepsilon}_1 \hat{\varepsilon}_2$, it follows that $\hat{\varepsilon}_1^{-1} = \hat{\varepsilon}_2$. If ε_i denotes the image of s_i in E_i $(i = 1, 2)$ (this is already the case for $i = 1$) then from the fact (5.26) that the square

$$\begin{array}{ccc} (G, f') & \longrightarrow & E_2 \\ \downarrow & & \downarrow \\ E_1 & \longrightarrow & \hat{E} \end{array} \qquad \begin{array}{ccc} & & \varepsilon_2 \\ & & \downarrow \\ \varepsilon_1^{-1} & \longmapsto & \hat{\varepsilon}_1^{-1} = \hat{\varepsilon}_2 \end{array}$$

is fibred, it follows that ε_1^{-1} and ε_2 have a common preimage $\sigma \in (G, f')$. If $f_i : E_i \to \hat{S}/H_2(E_i)$ $(i - 1, 2)$ then clearly $\psi(\sigma) = f_1(\varepsilon_1) f_2(\varepsilon_2) = [\hat{s}_1][\hat{s}_2] = [\hat{s}]$.

DEFINITION 5.38. A homomorphism $f_1 : G_1 \to \hat{G}$ is called E-*surjective* (resp. S-*surjective*) if the square (5.14) with the restrictions that $G_1 = G_2$ and $f_1 = f_2$ is E-surjective (resp. S-surjective).

LEMMA 5.39. a) *If* $f_1 : G_1 \to \hat{G}$ *is* S-*surjective then it is* E-*surjective.*

b) *If the homomorphism* $f_1 : E_1 \to \hat{E}$ *is surjective then* $f_1 : G_1 \to \hat{G}$ *is* S-*surjective.*

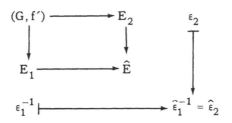

Proof. a) follows directly from the definitions.

b) We shall show that the canonical homomorphism $S_1 \to \hat{S}$ is surjective, from which it will follow that $f_1 : G_1 \to \hat{G}$ is S-surjective. Let $\hat{s}, \hat{t} \in \hat{S}$. Since $E_1 \to \hat{E}$ is surjective, there are elements $\hat{s}_1$ and $\hat{t}_1 \in \operatorname{image}(S_1 \to \hat{S})$ and $\hat{c}$ and $\hat{d} \in H_2(\hat{E})$ such that $\hat{s}_1 = \hat{s}\hat{c}$ and $\hat{t}_1 = \hat{t}\hat{d}$. Thus, $[\hat{s}_1, \hat{t}_1] = [\hat{s}\hat{c}, \hat{t}\hat{d}] = [\hat{s}, \hat{t}]$. But, since $\hat{S}$ is perfect, it follows that the map $S_1 \to \hat{S}$ is surjective.

DEFINITION 5.40. If N is a normal subgroup of G then we define the *smash product group*

$$G \ltimes N = \{(\rho, n) \mid \rho \in G,\ n \in N\}$$

such that multiplication is defined by the rule $(\sigma, m)(\rho, n) = (\sigma\rho, (\rho^{-1}m\rho)n)$.

LEMMA 5.41. *Let* $N = \ker(f_1 : G_1 \to \hat{G})$.

a) *The square below is a fibre square*

$$(\sigma_1, n) \longmapsto \sigma_1$$

$$\begin{array}{ccccc} (\sigma_1, n) & & G_1 \ltimes N & \longrightarrow & G_1 \\ \big\downarrow & & \big\downarrow & & \big\downarrow f_1 \\ \sigma_1 n & & G_1 & \xrightarrow[f_1]{} & \hat{G}\,. \end{array}$$

b) *The commutator subgroup* $E_1 \ltimes [G_1, N]$ *of* $G_1 \ltimes N$ *is perfect* $\Longleftrightarrow$ $[G_1, N] = [E_1, [G_1, N]]$.

Assume now that $[G_1, N] = [E_1, [G_1, N]]$.

c) *If* $f_1 : G_1 \to \hat{G}$ *is* E*-surjective (resp.* S*-surjective) then by definition the square* a) *is* E*-surjective (resp.* S*-surjective).*

d) *If* $f_1 : E_1 \to \hat{E}$ *is surjective then the square* a) *is an* E*-fibre,* S*-surjective, fibre square.*

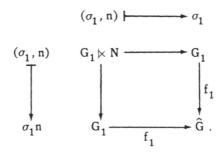

Proof. The proofs of a) and b) are very easy. Details will be left to the reader. c) is clear. We prove now d). By a), the square is a fibre square. By Lemma 5.39, the square is S-surjective. It remains, therefore, to show that the square is an E-fibre square. Thus, we must show that the square below is a fibre square

$$\begin{array}{ccc} E_1 \ltimes [G_1, N] & \longrightarrow & E_1 = S_1/H_2(E_1) \\ \downarrow & & \downarrow \\ E_1 & \longrightarrow & \hat{S}/H_2(E_1)\,. \end{array}$$

This is equivalent to showing that the homomorphism $E_1/[G_1, N] \to \hat{S}/H_2(E_1)$ is injective. Since $[G_1, N] = [E_1, [G_1, N]]$ and $[G_1, N] \subset E_1 \cap N \subset N$, it follows that $[G_1, N] = [E_1, E_1 \cap N]$. It is clear that the covering $E_1/[E_1, E_1 \cap N] \to \hat{E}$ is central. Thus, by the universality of the covering $\pi : \hat{S} \to \hat{E}$ there is a (unique) homomorphism $\gamma : \hat{S} \to E_1/[E_1, E_1 \cap N]$ such that the diagram below commutes

$$\begin{array}{ccc} E_1/[E_1, E_1 \cap N] & \xleftarrow{\ \gamma\ } & \hat{S} \\ & {\scriptstyle\delta}\searrow \quad \swarrow{\scriptstyle\pi} & \\ & \hat{E} & \end{array}$$

To show that the homomorphism $E_1/[E_1, E_1 \cap N] \to \hat{S}/H_2(E_1)$ is injective, it suffices to show that it has a retract. To show the latter, it suffices to show that the canonical diagram below commutes

$$\begin{array}{ccc} & S_1 & \\ & {\scriptstyle\alpha}\swarrow \quad \searrow{\scriptstyle\beta} & \\ E_1/[E_1, E_1 \cap N] & \xleftarrow[\ \gamma\ ]{} & \hat{S}\,. \end{array}$$

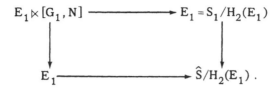

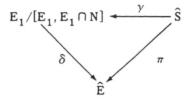

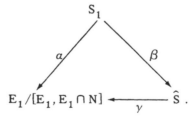

Let $s, t \in S_1$. Because $\delta\alpha = \pi\beta = \delta\gamma\beta$, it follows that $\alpha(s)$ and $\gamma\beta(s)$ differ by an element in the $\ker \delta = E_1 \cap N/[E_1, E_1 \cap N]$. Thus, there are elements $c, d \in E_1 \cap N/[E_1, E_1 \cap N]$ such that $\alpha(s) = (\gamma\beta(s))c$ and $\alpha(t) = (\gamma\beta(t))d$. Thus, $\alpha[s,t] = [\alpha(s), \alpha(t)] = [(\gamma\beta(s))c, (\gamma\beta(t))d] = [\gamma\beta(s), \gamma\beta(t)] = \gamma\beta[s,t]$. Since S_1 is perfect, it follows that $\alpha = \gamma\beta$.

THEOREM 5.42. *Let* $N = \ker(f_1 : G_1 \to \hat{G})$ *and suppose that* $[G_1, N] = [E_1, [G_1, N]]$. *Define the homomorphism*

$$\varphi : N/[G_1, N] \to H_1(f_1)$$
$$[n] \mapsto [n, 1] .$$

a) *If* $f_1 : E_1 \to \hat{E}$ *is surjective then* φ *is bijective.*

b) φ *is surjective* $\Longleftrightarrow$ $f_1 : G_1 \to \hat{G}$ *is S-surjective.*

Proof. Consider the fibre square in Lemma 5.41 a). Let $f : G_1 \ltimes N \to G_1$, $(\sigma_1, n) \mapsto \sigma_1$, denote the top horizontal map in this square. If $\phi : N/[G_1, N] \to H_1(f)$, $[n] \mapsto [(1, n), 1]$, then it is straightforward to check that ϕ is an isomorphism. Furthermore, it is clear that φ is the composite of ϕ with the canonical map $H_1(f) \to H_1(f_1)$.

a) It suffices to show that the map $H_1(f) \to H_1(f_1)$ is bijective. But, this follows directly from Lemma 5.41 d) and Theorem 5.37.

b) It suffices to show that the map $H_1(f) \to H_1(f_1)$ is surjective $\Longleftrightarrow$ $f_1 : G_1 \to \hat{G}$ is S-surjective. It is clear that the square in Lemma 5.41 a) is S-surjective $\Longleftrightarrow$ it is weak S-surjective. Part b) follows directly now from Lemma 5.41c) and Theorem 5.37.

§6. K-THEORY IN CATEGORIES WITH PRODUCT

A. *Fibre product categories*

DEFINITION 6.1. Let $\mathbf{A}$ be a category with product $\perp$. A *trivial object* for $\mathbf{A}$ is an object O in $\mathbf{A}$ and a collection $\{i_M : M \perp O \to M \mid M \in \mathrm{Obj}(\mathbf{A})\}$ of isomorphisms in $\mathbf{A}$ such that for each morphism $\sigma : M \to N$ in $\mathbf{A}$, the diagram below commutes

$$\begin{array}{ccc} M \perp O & \xrightarrow{\ i_M\ } & M \\ {\scriptstyle \sigma \perp 1_O}\downarrow & & \downarrow{\scriptstyle \sigma} \\ N \perp O & \xrightarrow[\ i_N\ ]{} & N\,. \end{array}$$

For example, if $\mathbf{A}$ is the category $\mathbf{P}(A)$ of finitely generated, projective modules over a ring A, with product $\oplus$, then we can choose O to be the trivial A-module.

DEFINITION 6.2. Let $\mathbf{A}$ and $\mathbf{A}'$ be categories with product and with trivial objects O and O' respectively. A product preserving functor $F : \mathbf{A} \to \mathbf{A}'$ is called *trivial object preserving* if there is an isomorphism $\gamma : F(O) \to O'$ in $\mathbf{A}'$ such that for all $M \in \mathrm{Obj}(\mathbf{A})$, the diagram below commutes

$$\begin{array}{ccc} F(M) \perp F(O) & \xrightarrow{\ F(i_M)\ } & F(M) \\ {\scriptstyle 1_{F(M)} \perp \gamma}\downarrow & & \downarrow{\scriptstyle 1_{F(M)}} \\ F(M) \perp O' & \xrightarrow[\ i_{F(M)}\ ]{} & F(M)\,. \end{array}$$

DEFINITION 6.3. Given two categories $\mathbf{A}_1$ and $\mathbf{A}_2$, we define the *product category*

$$\mathbf{A}_1 \times \mathbf{A}_2$$

as follows: Its objects are pairs (A_1, A_2) such that $A_i \in \operatorname{Obj} \mathbf{A}_i$. A morphism $(A_1, A_2) \to (B_1, B_2)$ is a pair (f_1, f_2) of morphisms $f_i : A_i \to B_i$ in $\mathbf{A}_i$ $(i = 1, 2)$. If $\mathbf{A}_1$ and $\mathbf{A}_2$ are categories with product then $\mathbf{A}_1 \times \mathbf{A}_2$ has a product defined by $(A_1, A_2) \perp (B_1, B_2) = (A_1 \perp B_1, A_2 \perp B_2)$, $(f_1, f_2) \perp (g_1, g_2) = (f_1 \perp g_1, f_2 \perp g_2)$. If $\mathbf{A}_i$ $(i = 1, 2)$ have trivial objects O_i $(i = 1, 2)$ then $\mathbf{A} \times \mathbf{B}$ has a trivial object $O = (O_1, O_2)$.

DEFINITION 6.4. A diagram

$$\begin{array}{ccc} \mathbf{A} & \xrightarrow{F_2'} & \mathbf{A}_2 \\ {\scriptstyle F_1'}\downarrow & & \downarrow{\scriptstyle F_2} \\ \mathbf{A}_1 & \xrightarrow[F_1]{} & \hat{\mathbf{A}} \end{array}$$

of categories and functors is called *commutative* if there is an isomorphism $\alpha : F_1F_1' \cdot F_2F_2'$ of functors.

The next definition is taken directly from Bass [10, VII(3.1)].

DEFINITION 6.5. Given a diagram of functors

$$\text{(1)} \qquad \begin{array}{ccc} & & \mathbf{A}_2 \\ & & \downarrow{\scriptstyle F_2} \\ \mathbf{A}_1 & \xrightarrow[F_1]{} & \hat{\mathbf{A}} \end{array}$$

we define the *fibre product category*,

$$\mathbf{A} = \mathbf{A}_1 \times_{\mathbf{A}} \mathbf{A}_2 = \mathbf{co}(F_1, F_2)$$

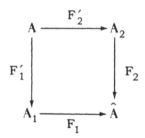

as follows: Its objects are triples (A_1, α, A_2) such that $A_i \in \mathbf{A}_i$ and $\alpha : F_1A_1 \to F_2A_2$ is an isomorphism in $\hat{\mathbf{A}}$. A morphism $(A_1, \alpha, A_2) \to (B_1, \beta, B_2)$ in $\mathbf{A}$ is a pair of morphisms $f_i : A_i \to B_i$ in $\mathbf{A}_i$ $(i = 1, 2)$ such that

$$\begin{array}{ccc} F_1A_1 & \xrightarrow{\alpha} & F_2A_2 \\ {\scriptstyle F_1f_1}\downarrow & & \downarrow{\scriptstyle F_2f_2} \\ F_1B_1 & \xrightarrow[\beta]{} & F_2B_2 \end{array}$$

commutes. There are canonical functors

$$F_i' : \mathbf{A} \longrightarrow \mathbf{A}_i ; \qquad \begin{array}{c} (A_1, \alpha, A_2) \longmapsto A_i \\ \\ (f_1, f_2) \longmapsto f_i \end{array} \qquad (i = 1, 2) ,$$

and the square

$$(2) \qquad \begin{array}{ccc} \mathbf{A} & \xrightarrow{F_2'} & \mathbf{A}_2 \\ {\scriptstyle F_1'}\downarrow & & \downarrow{\scriptstyle F_2} \\ \mathbf{A}_1 & \xrightarrow[F_1]{} & \hat{\mathbf{A}} \end{array}$$

is clearly commutative up to the natural isomorphism

$$\alpha : F_1F_1' \to F_2F_2'$$

which maps $F_1F_1'(A_1, \alpha, A_2) = F_1A_1$ to $F_2F_2'(A_1, \alpha, A_2) = F_2A_2$.

The fibre product construction solves the following

UNIVERSAL PROBLEM: Given a square

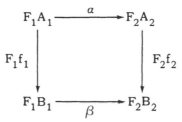

$$(3)\qquad \begin{array}{ccc} B & \xrightarrow{H_2} & A_2 \\ {\scriptstyle H_1}\downarrow & & \downarrow{\scriptstyle F_2} \\ A_1 & \xrightarrow[F_1]{} & \hat{A} \end{array}, \quad \text{and} \quad \beta : F_1H_1 \xrightarrow{\simeq} F_2H_2 ,$$

there is a *unique* (not just up to isomorphism) functor $T : B \to A$ such that $F'_iT = H_i$ (equality, not isomorphism) $(i = 1, 2)$ and such that

$$\beta = a \cdot T : F_1H_1 = F_1F'_1T \to F_2H_2 = F_2F'_2T .$$

Namely, we must have

$$\begin{aligned} T(B) &= (H_1B, \beta_B, H_2B) \\ T(f) &= (H_1f, H_2f), \end{aligned}$$

and this T clearly works. On the other hand, if the diagram above itself solves the universal problem then there is a unique functor $U : A_1 \times_{\hat{A}} A_2 \to B$ such that $F'_i = H_iU$ $(i = 1, 2)$. Since UT and TU are respectively the unique functors such that $F'_i = F'_iUT$ and $H_i = H_iTU$, it follows that $UT = 1_B$ and $TU = 1_{A_1 \times_{\hat{A}} A_2}$, i.e. $\hat{B} \cong A_1 \times_{\hat{A}} A_2$. A commutative square which solves the universal problem will be called a *fibre product square*.

If A_i $(i = 1, 2)$ and $\hat{A}$ are categories with product and the F_i $(i = 1, 2)$ are product preserving functors then $A_1 \times_{\hat{A}} A_2$ has a product defined by

$$(A_1, a, A_2) \perp (B_1, \beta, B_2) = (A_1 \perp B_1, a \perp \beta, A_2 \perp B_2),$$

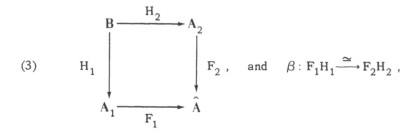

$(f_1, f_2) \perp (g_1, g_2) = (f_1 \perp g_1, f_2 \perp g_2)$ and the F'_i $(i = 1, 2)$ are product preserving functors. If A_i $(i = 1, 2)$ and $\hat{A}$ have trivial objects O_i $(i = 1, 2)$ and $\hat{O}$ respectively and the F_i $(i = 1, 2)$ are trivial object preserving with associated isomorphisms $\gamma_i : F_i O_i \to \hat{O}$ then $A_1 \times_{\hat{A}} A_2$ has a trivial object $O = (O_1, \gamma_2^{-1}\gamma_1, O_2)$ and the F'_i are trivial object preserving.

B. *The relative sequence of a product preserving functor*

Throughout this section, we fix the following data:

> $\mathbf{A}$ and $\mathbf{B}$ are categories with product and trivial object,
> $F : \mathbf{A} \to \mathbf{B}$ is a functor preserving products and trivial objects.

We shall assume that all morphisms in $\mathbf{A}$ and $\mathbf{B}$ are isomorphisms and that the isomorphism classes of $\mathbf{A}$ and $\mathbf{B}$ form a set.

The purpose of the section is to associate to F a relative exact sequence $K_2\mathbf{A} \to K_2\mathbf{B} \to K_1F \to K_1\mathbf{A} \to K_1\mathbf{B} \to K_0F \to K_0\mathbf{A} \to K_0\mathbf{B}$ which is natural with respect to F. No cofinal assumptions on F will be made. This will be particularly important for us since many functors in the K-theory of forms, for example $\mathbf{H}(A, \min) \to \mathbf{H}(A, \max)$, are not cofinal.

DEFINITION 6.6. Define the *relative category* of F

$$\mathbf{Rel}\ F \quad \text{or}$$
$$\mathbf{Rel}\,(\mathbf{A}, \mathbf{B})$$

as follows: Its objects are triples (M, f_Q, N) such that $M, N \in \mathrm{Obj}\,(\mathbf{A})$, $Q \in \mathrm{Obj}\,(\mathbf{B})$, and f_Q is an isomorphism $f_Q : FM \perp Q \to FM \perp Q$ in $\mathbf{B}$. A morphism $(\sigma, \rho, \tau) : (M, f_Q, N) \to (M', f_{Q'}, N')$ in $\mathbf{Rel}\ F$ is a triple (σ, ρ, τ) such that $\sigma : M \to M'$, $\rho : N \to N' \in \mathrm{Morph}(\mathbf{A})$, $\tau : Q \to Q' \in \mathrm{Morph}\,(\mathbf{B})$, and the diagram below commutes

$$\begin{array}{ccc} FM \perp Q & \xrightarrow{\ f_Q\ } & FN \perp Q \\ \Big\downarrow{\scriptstyle F\sigma \perp \tau} & & \Big\downarrow{\scriptstyle F\rho \perp \tau} \\ FM' \perp Q' & \xrightarrow[\ f_{Q'}\ ]{} & FN' \perp Q' . \end{array}$$

We shall find it convenient sometimes to indicate the morphism (σ, ρ, τ) by the pair $(F\sigma \perp \tau, F\rho \perp \tau)$.

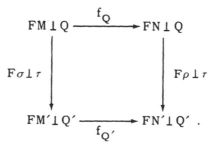

Rel F has a product defined by $(M, f_Q, N) \perp (M', f_{Q'}, N') = (M \perp M', f_Q \perp f_{Q'}, N \perp N')$, $(\sigma, \rho, \tau) \perp (\sigma', \rho', \tau') = (\sigma \perp \sigma', \rho \perp \rho', \tau \perp \tau')$.

DEFINITION 6.7. Define the *relative Grothendieck group* of F

$$K_0F \quad \text{or}$$

$$K_0(A, B)$$

as follows:

$$K_0F = K_0 \operatorname{Rel} F/[M, f_Q, N] + [N, g_Q, P] = [M, g_Q f_Q, P].$$

REMARK 6.8. *In 6.5, we defined the category* **co** F. *For* F *cofinal, Bass [10, VII 5.1] defines the relative Grothendieck group* $K_0'F = K_0 \,\mathbf{co}\, F/\{[M, f, N] + [N, g, P] = [M, gf, N]\}$. *In general, there is a product preserving functor* $\mathbf{co}\, F \to \operatorname{Rel} F$, $(M, f, N) \mapsto (M \perp O, f \perp 1_{FO}, N \perp O)$, *where* O *is the trivial object of* A *and for* F *cofinal, it is shown in 6.22 that the induced homomorphism* $K_0'F \to K_0F$ *is an isomorphism.*

DEFINITION 6.9. A functor $L : A \to$ ((groups)) is called *central* if for each $M \in \operatorname{Obj}(A)$, the action of $\operatorname{Aut}_A(M)$ on FM induced by F is trivial.

DEFINITION 6.10. If $M \in \operatorname{Obj}(A)$, let $M^n = M \perp \cdots \perp M$ (n times). If $G_n(M) = \operatorname{Aut}_A(M^n)$ then there is a natural embedding $G_n(M) \to G_{n+1}(M)$, $a \mapsto a \perp 1_M$. Let U be as in 5.3. We define

$$G(M) = \varinjlim_n G_n(M)$$

$$E(M) = [G(M), G(M)]$$

$$S(M) = U(E(M)).$$

One can check easily that G, E, and S define functors $A \to$ ((groups)).

LEMMA 6.11. E(M) *is perfect and* S(M) *is the universal, perfect, central extension of* E(M).

Proof. The assertion for $E(M)$ follows from 3.15 and the assertion for $S(M)$ from 5.3.

DEFINITION 6.12. Let $H_i : ((\text{groups})) \to ((\text{abelian groups}))$ $(i = 1, 2)$ denote the homology functors defined in 5.6. If $M \in \text{Obj}(A)$, we define

$$K_1(M) = H_1 G(M)$$
$$K_2(M) = H_2 E(M) .$$

It is clear that the K_i $(i = 1, 2)$ define functors $A \to ((\text{abelian groups}))$.

LEMMA 6.13. $K_i : A \to ((\text{abelian groups}))$ $(i = 1, 2)$ *are central.*

Proof. The case $i = 1$ follows from 5.8 and the case $i = 2$ follows from 4.2 and 4.3.

DEFINITION 6.14. If $(M, Q) \in \text{Obj}(A \times B)$, we define

$$S(M, Q)$$

by the fibre product diagram

$$\begin{array}{ccc} S(M, Q) & \longrightarrow & S(FM \perp Q) \\ \downarrow & & \downarrow \\ G(M) & \longrightarrow & G(FM \perp Q) . \end{array}$$

It is easy to check that S defines a functor $A \times B \to ((\text{groups}))$.

Let $S(M) \to G(M)$ denote the canonical homomorphism and let $S(M) \to S(FM \perp Q)$ denote the composite of the canonical homomorphisms $S(M) \to S(FM)$ and $S(FM) \to S(FM \perp Q)$. Since the diagram

$$\begin{array}{ccc} S(M) & \longrightarrow & S(FM \perp Q) \\ \downarrow & & \downarrow \\ G(M) & \longrightarrow & G(FM \perp Q) \end{array}$$

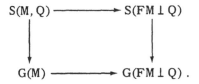

commutes, it follows from the universality of the fibre product that there is a homomorphism

$$S(M) \to S(M,Q) .$$

LEMMA 6.15. *Image* $(S(M) \to S(M,Q)) = [S(M,Q), S(M,Q)]$.

Proof. Since $S(M)$ is perfect, it suffices to show that the image $(S(M) \to S(M,Q)) \supset [S(M,Q), S(M,Q)]$. Recall the construction of $S(M,Q)$ given in 5.10. Thus, the elements of $S(M,Q)$ consist of all pairs (σ, s) such that $\sigma \in G(M)$, $s \in S(FM \perp Q)$, and the images of σ and s in $G(FM \perp Q)$ agree. Let $(\sigma, s), (\rho, t) \in S(M,Q)$. We must show that $[(\sigma, s), (\rho, t)] \in$ image $(S(M) \to S(M,Q))$. Let ε (resp. τ) denote the image of s (resp. t) in $E(FM \perp Q) \subset G(FM \perp Q)$. Let G_n (resp. $G_{n,1}$) (resp. $G_{n,2}$) denote the image of $G_n(M)$ (resp. $1_{M^n} \perp G_n(M)$) (resp. $1_{M^n} \perp 1_{M^n} \perp G_n(M)$) in $G(M)$. Let $G''_n(FM \perp Q)$ denote the double commutator subgroup of $G_n(FM \perp Q)$. Let E_n (resp. $E_{n,1}$) (resp. $E_{n,2}$) denote the image of $G''_n(FM \perp Q)$ (resp. $1_{(FM\perp Q)^n} \perp G''_n(FM \perp Q)$) (resp. $1_{(FM\perp Q)^n} \perp 1_{(FM\perp Q)^n} \perp G''_n(FM \perp Q)$) in $E(FM \perp Q)$. Let S_n (resp. $S_{n,1}$) (resp. $S_{n,2}$) denote the inverse image of E_n (resp. $E_{n,1}$) (resp. $E_{n,2}$) in $S(FM \perp Q)$. Since $E(FM \perp Q)$ is perfect, it follows that $E(FM \perp Q) = \varinjlim_n E_n$. Choose n so that $\sigma, \rho \in G_n$ and $\varepsilon, \tau \in E_n$. Choose $s_1 \in S_{n,1}$ so that s_1 covers $1_{(FM\perp Q)^n} \perp \varepsilon^{-1} \in E_{n,1}$ and choose $t_2 \in S_{n,2}$ such that t_2 covers $1_{(FM\perp Q)^n} \perp 1_{(FM\perp Q)^n} \perp \tau^{-1} \in E_{n,2}$. By the Whitehead lemma 3.15, $\sigma \perp \sigma^{-1} \perp 1_{M^n}$ and $\rho \perp 1_{M^n} \perp \rho^{-1} \in E(M)$. Let $u, v \in S(M)$ such that u covers $\sigma \perp \sigma^{-1} \perp 1_{M^n}$ and v covers $\rho \perp 1_{M^n} \perp \rho^{-1}$. If u' and v' denote respectively the images of u and v in $S(FM \perp Q)$ then u' and v' cover respectively $\varepsilon \perp \varepsilon^{-1} \perp 1_{(FM\perp Q)^n}$ and $\tau \perp 1_{(FM\perp Q)^n} \perp \tau^{-1}$. Since ss_1 and tt_2 cover respectively also $\varepsilon \perp \varepsilon^{-1} \perp 1_{(FM\perp Q)^n}$ and $\tau \perp 1_{(FM\perp Q)^n} \perp \tau^{-1}$, there are elements $c, d \in$ center $(S(FM \perp Q))$ such that $cu' = ss_1$ and $dv' = tt_2$. From the commutator formulas 3.27 a) and from Lemma 5.5 b), it follows that $[s,t] = [ss_1, tt_2]$. Thus, $[(\sigma, s), (\rho, t)] = ([\sigma, \rho], [s,t]) = ([\sigma, \rho], [ss_1, tt_2]) =$ (because $ss_1 = cu', tt_2 = dv'$) $([\sigma, \rho], [u', v']) = ([\sigma \perp \sigma^{-1} \perp 1_{M^n}, \rho \perp 1_{M^n} \perp \rho^{-1}], [u', v']) = [(\sigma \perp \sigma^{-1} \perp 1_{M^n}, u'), (\rho \perp 1_{M^n} \perp \rho^{-1}, v')]$. But the last commutator is the image of the element $[u,v] \in S(M)$.

Using the previous lemma, we make the following definition.

DEFINITION 6.16. If $(M, Q) \in \mathrm{Obj}(\mathbf{A} \times \mathbf{B})$, we define

$$K_1(M, Q) = \operatorname{coker}(S(M) \to S(M, Q)) .$$

It is easy to check that K_1 defines a functor $\mathbf{A} \times \mathbf{B} \to$ ((abelian groups)).

LEMMA 6.17. *If* $(M, Q) \in \mathrm{Obj}(\mathbf{A} \times \mathbf{B})$ *then there is an exact sequence*

$$K_2(M) \longrightarrow K_2(FM \perp Q) \xrightarrow{\partial_1} K_1(M, Q) \xrightarrow{\rho_1} K_1(M) \longrightarrow K_1(FM \perp Q)$$

where ∂_1 *and* ρ_1 *are as in Theorem 5.13. The sequence is natural with respect to morphisms in* $\mathbf{A} \times \mathbf{B}$.

Proof. The lemma follows directly from Theorem 5.13.

COROLLARY 6.18. $K_1 : \mathbf{A} \times \mathbf{B} \to$ *((abelian groups)) is central.*

Proof. Let $(M, Q) \in \mathrm{Obj}(\mathbf{A} \times \mathbf{B})$. If $(\alpha, \beta) \in \mathrm{Aut}_{\mathbf{A} \times \mathbf{B}}(M, Q)$ then we must show that $K_1(\alpha, \beta) = 1$. Consider the diagram

$$\begin{array}{ccccccccc}
K_2(M) & \longrightarrow & K_2(FM) & \longrightarrow & K_1(M, Q) & \longrightarrow & K_1(M) & \longrightarrow & K_1(FM) \\
\Big\downarrow{\scriptstyle K_2(\alpha)} & & \Big\downarrow{\scriptstyle K_2(F\alpha)} & & \Big\downarrow{\scriptstyle K_1(\alpha, \beta)} & & \Big\downarrow{\scriptstyle K_1(\alpha)} & & \Big\downarrow{\scriptstyle K_1(F\alpha)} \\
K_2(M) & \longrightarrow & K_2(FM) & \longrightarrow & K_1(M, Q) & \longrightarrow & K_1(M) & \longrightarrow & K_1(FM) .
\end{array}$$

The diagram is commutative and its rows are exact by Lemma 6.17 and the maps $K_i(\alpha)$ and $K_i(F\alpha)$ $(i=1,2)$ are the identity maps by Lemma 6.13. Thus, $K_2(\alpha,\beta)$ must be the identity map.

Recall the directed category $\mathrm{Tran}[A]$ defined in §4B. Since $\mathrm{Tran}[A]$ is directed, given any functor $L : \mathrm{Tran}[A] \to$ ((abelian groups)), one can form its direct limit $\varinjlim_{[M]} L[M]$ as in [10, I §8]. Using the centrality of the functors $K_i : A \to$ ((abelian groups)) $(i=1,2)$ and $K_1 : A \times B \to$ ((abelian groups)), one can show that they induce respectively functors $K_i : \mathrm{Tran}[A] \to$ ((abelian groups)) $(i=1,2)$ and $K_1 : \mathrm{Tran}[A \times B] \to$ ((abelian groups)).

DEFINITION 6.19. $K_iA = \varinjlim_{[M]} K_i[M] \quad (i=1,2)$

$$K_1F = \varinjlim_{[M,Q]} K_1[M,Q] .$$

THEOREM 6.20. *There is an exact sequence* $K(F)$:

$$K_2A \longrightarrow K_2B \xrightarrow{\partial_1} K_1F \xrightarrow{\rho_1} K_1A \longrightarrow K_1B \xrightarrow{\partial_0} K_0F \xrightarrow{\rho_0} K_0A \longrightarrow K_0B$$

where ∂_1 *and* ρ_1 *are induced by the corresponding maps in 6.17,* $\rho_0[M,f_Q,N] = [M]-[N]$, *and if* $\sigma \in \mathrm{Aut}_B(Q)$, O *is the trivial object of* A, *and* $f_Q = 1_{FO} \perp \sigma$ *then* $\partial_0[\sigma] = [O, f_Q, O]$. *Furthermore,* $K(F)$ *is natural in* F, *i.e. if*

$$\begin{array}{ccc} A & \xrightarrow{F} & B \\ \big\downarrow & & \big\downarrow \\ \hat{A} & \xrightarrow{\hat{F}} & \hat{B} \end{array}$$

is a commutative square of categories and functors then there is induced in the obvious way a homomorphism $K(F) \to K(\hat{F})$ *of exact sequences.*

Proof. The proof of the naturality assertions is easy and will be left to the reader. The exactness of the sequence $K_2A \to K_2B \to K_1F \to K_1A \to K_1B$

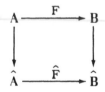

follows from applying $\varinjlim$ to the exact sequence of 6.17. To prove exactness at the remaining terms, we need the following lemma. The proof of the theorem will be continued after the proof of the lemma.

LEMMA 6.21. a) *Each element of* K_0F *is the class* $[M, f_Q, N]$ *of some object* (M, f_Q, N) *of* Rel F.

b) *Let* $(M, f_Q, M) \in \mathrm{Obj}(\mathrm{Rel}\ F)$. *If there is a* $Q_1 \in \mathrm{Obj}(B)$ *such that* $f_Q \perp 1_{Q_1}$ *lies in the commutator subgroup of* $\mathrm{Aut}_B(FM \perp Q \perp Q_1)$ *then* $[M, f_Q, M] = 0$.

c) $[M \perp P, f_Q, N \perp P] = [M, f_{FP \perp Q}, N]$ *where* $f_{FP\perp Q} = f_Q : FM \perp FP \perp Q \to FN \perp FP \perp Q$.

d) *If* $\sigma : M \perp P \to M \perp P$ *is an isomorphism in* A *then* $[M, F(\sigma), N] = 0$.

Proof. a) Each element $x \in K_0F$ can be written as difference $x = [M, f_Q, N] - [M_1, f_{Q_1}, N_1]$. The defining relations for K_0F imply that $-[M_1, f_{Q_1}, N_1] = [N_1, f_{Q_1}^{-1}, M_1]$. Thus, $x = [M, f_Q, N] + [N_1, f_{Q_1}^{-1}, M_1] = [M \perp N_1, f_Q \perp f_{Q_1}^{-1}, N \perp M_1]$.

b) From the defining relations for K_0F, we know that $[M, 1_{FM\perp Q_1}, M] = [M, 1_{FM\perp Q_1}, M] + [M, 1_{FM\perp Q_1}, M]$. Thus, $[M, 1_{FM\perp Q_1}, M] = 0$. Thus, $[M, f_Q, M] = [M, f_Q, M] + [M, 1_{FM\perp Q_1}, M] = [M \perp M, f_Q \perp 1_{FM\perp Q_1}, M \perp M]$. Since $f_Q \perp 1_{FM\perp Q_1}$ is a product of commutators, it follows directly from the defining relations (cf. proof of a)) that $[M \perp M, f_Q \perp 1_{FM\perp Q_1}, M \perp M] = 0$.

c) Let $g = f_{FP\perp Q}$. The defining relations for K_0F imply that $[M \perp P, f_Q, N \perp P] - [M, g, N] = [M \perp P \perp N, f_Q \perp g^{-1}, N \perp P \perp M]$. If $\pi : N \perp P \perp M \to M \perp P \perp N$ is the isomorphism given by the associativity of the product in A and if $\tau = F(\pi) \perp 1_{Q\perp FP\perp Q}$ then the pair $(1_{F(M\perp P\perp N)\perp Q\perp FP\perp Q}, \tau)$ defines an isomorphism $(M \perp P \perp N, f_Q \perp g^{-1}, N \perp P \perp M) \xrightarrow{\cong} (M \perp P \perp N, \tau(f_Q \perp g^{-1}), M \perp P \perp N)$. One checks easily that $\tau(f_Q \perp g^{-1})$ vanishes in $K_1(F(M \perp P \perp N) \perp Q \perp FP \perp Q)$. Thus, by part b), $[M \perp P \perp N, \tau(f_Q \perp g^{-1}), M \perp P \perp N] = 0$.

d) $[M, F(\sigma), M] =$ (cf. proof of b)) $[M, F(\sigma), M] + [M, 1_{FM} \perp 1_{FM}, M] = [M \perp M, F(\sigma) \perp 1_{FM} \perp 1_{FM}, M \perp M] =$ (part c)) $[M \perp P \perp M, F(\sigma) \perp 1_{FM} \perp 1_{FM}, M \perp P \perp M]$. But the pair $(F(\sigma) \perp 1_{FM} \perp 1_{FM}, 1_{F(M \perp P)} \perp 1_{FM} \perp 1_{FM})$ defines an isomorphism $(M \perp P \perp M, F(\sigma) \perp 1_{FM} \perp 1_{FM}, M \perp P \perp M) \xrightarrow{\cong} (M \perp P \perp M, 1_{F(M \perp P)} \perp 1_{FM} \perp 1_{FM}, M \perp P \perp M)$ and the latter object vanishes in K_0F (cf. proof of part b)).

Proof of Theorem 6.20 (continued).

Exactness at K_0A: It is clear that the composite mapping at K_0A is trivial. So, suppose that $[M] - [N] \in K_0A$ vanishes in K_0B. Thus, there is an object Q in B and an isomorphism $f_Q : FM \perp Q \to FN \perp Q$ in B. By definition, $\rho_0[M, f_Q, N] = [M] - [N]$.

Exactness at K_0F: It is clear that the composite mapping at K_0F is trivial. By 6.21 a), each element of K_0F is represented by a class $[M, f_Q, N]$. If $\rho_0[M, f_Q, N] = [M] - [N] = 0$ then there is an object P in A and an isomorphism $\alpha : N \perp P \to M \perp P$ in A. By 6.21 d), $[P, 1_{FP \perp Q}, P] = 0$. Thus, $[M, f_Q, N] = [M \perp P, f_Q \perp 1_{FP \perp Q}, P \perp N]$. From the commutative diagram

$$\begin{array}{ccc}
FM \perp Q \perp FP \perp Q & \xrightarrow{\;f_Q \perp 1_{FP \perp Q}\;} & FN \perp Q \perp FP \perp Q \\
\Big\downarrow 1 & & \Big\downarrow F\alpha \perp 1_{Q \perp Q} \\
FM \perp Q \perp FP \perp Q & \xrightarrow[\;f = (F\alpha \perp 1_{Q \perp Q})(f_Q \perp 1_{FP \perp Q})\;]{} & FM \perp Q \perp FP \perp Q
\end{array}$$

it follows that $(M \perp P, f_Q \perp 1_{FP \perp Q}, N \perp P) \cong (M \perp P, f, M \perp P)$. Thus, $[M, f_Q, N] = [M \perp P, f, M \perp P]$. Let O be the trivial object of A. By 6.21 d), $[O, 1_{FO}, O] = 0$. Thus, $[M, f_Q, N] = [O, 1_{FO}, O] + [M \perp P, f, M \perp P] = [O \perp M \perp P, 1_{FO} \perp f, O \perp M \perp P] =$ (6.21 c)) $[O, 1_{FO} \perp f, O] =$ (definition) $\partial_0[f]$.

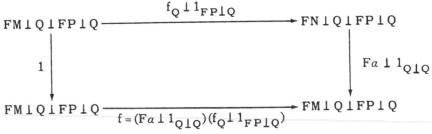

Exactness at K_1B: It should be noted that 6.21 b) implies that ∂_0 vanishes on the commutator subgroup of $\mathrm{Aut}_B(Q)$. By 6.21 d), the composite mapping at K_1B is trivial. Let $Q \in \mathrm{Obj}(B)$ and let $\sigma \in \mathrm{Aut}_B(Q)$ such that $\partial[\sigma] = [O, 1_{FO} \perp \sigma, O] = 0$. From the defining relations for K_0F, it follows that there are objects $(M_1, f_{Q_1}, N_1), (N_1, g_{Q_1}, P_1), (M_2, f_{Q_2}, N_2), (N_2, g_{Q_2}, P_2) \in \mathrm{Rel}\, F$ such that the following equation in $K_0 \mathrm{Rel}\, F$ holds $[O, 1_{FO} \perp \sigma, O] + [M_1, g_{Q_1} f_{Q_1}, P_1] + [M_2, f_{Q_2}, N_2] + [N_2, g_{Q_2}, P_2] = [M_1, f_{Q_1}, N_1] + [N_1, g_{Q_1}, P_1] + [M_2, g_{Q_2} f_{Q_2}, P_2]$. Since $K_0 \mathrm{Rel}\, F$ is an ordinary Grothendieck group, it follows that there is an object (M_3, f_{Q_3}, N_3) in $\mathrm{Rel}\, F$ such that $(O \perp M_1 \perp M_2 \perp N_2 \perp M_3, (1_{FO} \perp \sigma) \perp g_{Q_1} f_{Q_1} \perp f_{Q_2} \perp g_{Q_2} \perp f_{Q_3}, O \perp P_1 \perp N_2 \perp P_2 \perp N_3) \cong (M_1 \perp N_1 \perp M_2 \perp M_3, f_{Q_1} \perp g_{Q_1} \perp g_{Q_2} f_{Q_2} \perp f_{Q_3}, N_1 \perp P_1 \perp P_2 \perp N_3)$. The isomorphism above is given by a 4-tuple $((\alpha, \tau), (\beta, \tau))$ where $\alpha : I = O \perp M_1 \perp M_2 \perp N_2 \perp M_3 \to J = M_1 \perp N_1 \perp M_2 \perp M_3$, $\beta : K = O \perp P_1 \perp N_2 \perp P_2 \perp N_3 \to L = N_1 \perp P_1 \perp P_2 \perp N_3$, and $\tau : R = Q \perp Q_1 \perp Q_2 \perp Q_2 \perp Q_3 \to S = Q_1 \perp Q_1 \perp Q_2 \perp Q_3$. Thus, $(1_{FO} \perp \sigma \perp 1_{FM_1 \perp FM_2 \perp FN_2 \perp FM_3 \perp Q \perp Q_1 \perp Q_2 \perp Q_3}) = (1_{FO} \perp 1_Q \perp g_{Q_1} f_{Q_1} \perp f_{Q_2} \perp g_{Q_2} \perp f_{Q_3})^{-1} (F\beta \perp \tau)^{-1} (f_{Q_1} \perp g_{Q_1} \perp g_{Q_2} f_{Q_2} \perp f_{Q_3}) (F\alpha \perp \tau)$. Now, if one identifies in the obvious way the isomorphisms σ, f_{Q_i}, g_{Q_i}, $F\alpha$, $F\beta$, τ with isomorphisms of $FI \perp FJ \perp FK \perp FL \perp R \perp S$ then it will follow that $[\sigma] = [F\beta]^{-1} [F\alpha] \in K_1B$.

COROLLARY 6.22. *Let O denote the trivial object of Λ. Recall the category* $\mathrm{co}\, F$ *defined in 6.5. Let* $K_0'F = \mathrm{co}\, F / \{[M, f, N] + [N, g, P] = [M, gf, N]\}$. *If* F *is cofinal then the homomorphism below is an isomorphism*

$$K_0'F \xrightarrow{\cong} K_0F$$

$$[M, f, N] \mapsto [M \perp O, f \perp 1_{FO}, N \perp O].$$

Proof. By 6.20, there is an exact sequence $K_1A \to K_1B \to K_0F \to K_0A \to K_0B$ and by [10, VII 5.3] there is an exact sequence $K_1A \to K_1B \to K_0'F \to K_0A \to K_0B$.

It is easy to check that the diagram

$$\begin{array}{ccccc} & & K_0F & & \\ & \nearrow & \uparrow & \searrow & \\ K_1\mathcal{A} \to K_1\mathcal{B} & & & & K_0\mathcal{A} \to K_0\mathcal{B} \\ & \searrow & & \nearrow & \\ & & K_0'F & & \end{array}$$

commutes. The lemma follows.

The next lemma is useful for applications of material in the chapter.

LEMMA 6.23. a) *If $\mathcal{C}$ is a full, faithful, cofinal subcategory of $\mathcal{A}$ which is closed under products and contains the trivial object of $\mathcal{A}$ then the canonical homomorphism below is an isomorphism*

$$K_i\mathcal{C} \xrightarrow{\cong} K_i\mathcal{A} \quad (i=1,2) .$$

b) *If A is a cofinal object for $\mathcal{A}$ then the canonical homomorphism below is an isomorphism*

$$K_i(A) \xrightarrow{\cong} K_i\mathcal{A} \quad (i=1,2) .$$

Proof. a) follows from [10, I 8.5] and b) follows from [10, I 8.6].

COROLLARY 6.24. *Suppose that $F : \mathcal{A} \to \mathcal{B}$ is cofinal. If A is a cofinal object for $\mathcal{A}$ and $B = F(A)$ then the canonical homomorphism below is an isomorphism*

$$K_1(A, B) \longrightarrow K_1(\mathcal{A}, \mathcal{B}) .$$

Proof. Consider the commutative diagram

$$\begin{array}{ccccccccc} K_2\mathcal{A} & \longrightarrow & K_2\mathcal{B} & \longrightarrow & K_1(\mathcal{A},\mathcal{B}) & \longrightarrow & K_1\mathcal{A} & \longrightarrow & K_1\mathcal{B} \\ \uparrow & & \uparrow & & \uparrow & & \uparrow & & \uparrow \\ K_2(A) & \longrightarrow & K_2(B \perp B) & \longrightarrow & K_1(A,B) & \longrightarrow & K_1(A) & \longrightarrow & K_1(B \perp B) . \end{array}$$

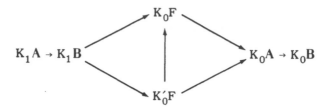

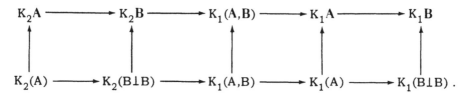

By 6.20, the top row is exact, by 6.17 the bottom row is exact, and by 6.23 b), the vertical arrows, except possibly $K_1(A, B) \to K_1(\mathbf{A}, \mathbf{B})$, are isomorphisms. It follows that $K_1(A, B) \to K_1(\mathbf{A}, \mathbf{B})$ must be an isomorphism also.

C. *The Mayer-Vietoris sequence of a fibre square*

Throughout the section, we fix the following data:

$$(6.25) \qquad \begin{array}{ccc} \mathbf{A} & \xrightarrow{F_2'} & \mathbf{A}_2 \\ {\scriptstyle F_1'}\downarrow & & \downarrow{\scriptstyle F_2} \\ \mathbf{A}_1 & \xrightarrow[F_1]{} & \hat{\mathbf{A}} \end{array} \qquad \alpha : F_1F_1' \to F_2F_2'$$

is a commutative square of categories with product and trivial object and product preserving, trivial object preserving functors. Let

$$\begin{aligned} A &\in \mathrm{Obj}(\mathbf{A}) \\ A_i &= F_i(A) \qquad (i=1,2) \\ \hat{A} &= F_1F_1'(A)\,. \end{aligned}$$

Recall the functors G, E, and S defined in §6B. Using the natural equivalence α above, we shall identify

$$\begin{aligned} G(\hat{A}) &= G(F_2F_2'(A)) \\ E(\hat{A}) &= E(F_2F_2'(A)) \\ S(\hat{A}) &= S(F_2F_2'(A))\,. \end{aligned}$$

Consider the following commutative diagrams:

$$(6.26) \qquad \begin{array}{ccc} G(A) & \longrightarrow & G(A_2) \\ \downarrow & & \downarrow \\ G(A_1) & \longrightarrow & G(\hat{A}) \end{array}$$

(6.27)
$$\begin{array}{ccc} E(A) & \longrightarrow & E(A_2) \\ \downarrow & & \downarrow \\ E(A_1) & \longrightarrow & E(\hat{A}) \end{array}$$

(6.28)
$$\begin{array}{ccc} S(A) & \longrightarrow & S(A_2) \\ \downarrow & & \downarrow \\ S(A_2) & \longrightarrow & S(\hat{A}) \end{array}$$

(6.29)
$$\begin{array}{ccc} E(A) & \longrightarrow & S(A_2)/H_2(E(A)) \\ \downarrow & & \downarrow \\ E(A_1) & \longrightarrow & S(\hat{A})/H_2(E(A_1)) \end{array}$$

(6.30)
$$\begin{array}{ccc} E(A) & \longrightarrow & E(A_2) \\ \downarrow & & \downarrow \\ E(A_1) & \longrightarrow & S(\hat{A})/H_2(E(A_1)) + H_2(E(A_2)) \end{array}$$

The squares (6.26)-(6.30) correspond to the squares (5.14)-(5.18) in §5C. Moreover, if 6.25 is a fibre product square of categories then 6.26 is a fibre square of groups.

METATHEOREM 6.31. *All of the definitions and assertions of §5C have obvious, analogous counterparts in* K*-theory.*

We shall give now some applications of the theorem.

THEOREM 6.32. *Suppose the following*:

(i) *6.25 is a fibre product square.*

(ii) F_i *and* F'_i $(i=1,2)$ *are cofinal.*

(iii) A *has a cofinal object* A *such that 6.26 is* S*-exact and* S*-surjective.*

Then for $n = 0,1,2$, *there is an exact Mayer-Vietoris sequence*

$$\cdots \longrightarrow K_n A \longrightarrow K_n A_1 \oplus K_n A_2 \longrightarrow K_n \hat{A} \xrightarrow{\partial_{n-1}} \cdots$$

such that ∂_1 *is induced from* ∂ *in 5.31 and such that if* $\sigma \in \mathrm{Aut}_{\hat{A}}(\hat{A}^n)$ *and* α *is as in 6.24 then* $\partial_0[\sigma] = [A_1, \alpha\sigma, A_2] \in K_0(A_1 \times_{\hat{A}} A_2) = K_0 A$. *Furthermore, the sequence is natural with respect to morphisms of commutative squares of categories with product and trivial object.*

Proof. Since A is cofinal in $\mathcal{A}$ and since F_i and F'_i $(i = 1, 2)$ are cofinal functors, it follows that A_i $(i = 1,2)$ and $\hat{A}$ are cofinal in respectively $\mathcal{A}_i$ $(i = 1, 2)$ and $\hat{\mathcal{A}}$. By Theorem 5.31, there is an exact Mayer-Vietoris sequence $K_2(A) \to K_2(A_1) \oplus K_2(A_2) \to K_2(\hat{A}) \xrightarrow{\partial} K_1(A) \to K_1(A_1) \oplus K_1(A_2) \to K_1(\hat{A})$. Thus, by 6.23 b), there is an exact Mayer-Vietoris sequence $K_2\mathcal{A} \to K_2\mathcal{A}_1 \oplus K_2\mathcal{A}_2 \to K_2\hat{\mathcal{A}} \xrightarrow{\partial_1} K_1\mathcal{A} \to K_1\mathcal{A}_1 \oplus K_1\mathcal{A}_2 \to K_1\hat{\mathcal{A}}$. By 5.29 b), S-surjective implies E-surjective. Thus, by Bass [10, VII 4.3], there is an exact Mayer-Vietoris sequence $K_1\mathcal{A} \to K_1\mathcal{A}_1 \oplus K_1\mathcal{A}_2 \to K_1\hat{\mathcal{A}} \xrightarrow{\partial_0} K_0\mathcal{A} \to K_0\mathcal{A}_1 \oplus K_0\mathcal{A}_2 \to K_0\hat{\mathcal{A}}$. Splicing the two exact sequences above together, one obtains the exact sequence of the theorem. The naturality assertions follow also from Theorem 5.31 and [10, VII 4.3]. It should be noted that the notion of E-surjectivity we use is weaker than that in Bass [10, VII] (cf. remark 5.20) but that the proof of [10, VII 4.3] requires only the weaker notion.

THEOREM 6.33. *The assertions of 6.32 remain valid when assumption 6.32 (iii) is replaced by the following assumption*:

(iii)′ *For each* $M \in \mathrm{Obj}(\mathcal{A})$, *there is an* $N \in \mathrm{Obj}(\mathcal{A})$ *such that for* $A = M \perp N$, *the square 6.26 is* S*-exact and* S*-surjective.*

The proof of 6.33 is similar to that of 6.32. Details will be left to the reader.

THEOREM 6.34. *If one replaces in the assumptions of 6.32 and 6.33, the expression* S*-exact by the expression* E*-fibred then the assertions of 6.32 and 6.33 remain valid, except possibly the assertion of exactness at* $K_2A_1 \oplus K_2A_2$.

6.34 is proved similar to 6.32 and 6.33, except the reference to Theorem 5.31 is replaced by a reference to Corollary 5.32.

D. *Excision*

We adopt the data (6.25)-(6.30) of the previous section.

DEFINITION 6.35. The commutative square 6.25 is said to satisfy *excision* for K_1 if the canonical homomorphism $K_1F'_2 \xrightarrow{\cong} K_1F_1$ is an isomorphism.

The definition above corresponds to definition 5.35 in §5D.

METATHEOREM 6.36. *All of the definitions and assertions of §5D have obvious, analogous counterparts in* K*-theory.*

We shall give now some applications of the theorem.

THEOREM 6.37. *Suppose the following*:

(i) *6.25 is a fibre product square.*
(ii) F_i *and* F'_i $(i=1,2)$ *are cofinal.*
(iii) $\mathbf{A}$ *has a cofinal object* A.

Then 6.25 satisfies excision for $K_1 \Longleftrightarrow$ *6.26 is* E*-fibred and weak* S*-surjective for* A. *In fact, the canonical homomorphism* $K_1F'_2 \to K_1F_1$ *is injective (resp. surjective)* $\Longleftrightarrow$ *6.26 is* E*-fibred (resp. weak* S*-surjective) for* A.

Proof. Since A is cofinal in $\mathbf{A}$, it follows that A_1 is cofinal in $\mathbf{A}_1$. The theorem follows now directly from Corollary 6.24 and Theorem 5.37.

COROLLARY 6.38. *Suppose the following*:

(i) *6.25 is a fibre product square.*
(ii) F_i *and* F'_i $(i=1,2)$ *are cofinal.*

(iii) A *has a cofinal object* A *such that 6.26 is* S*-surjective.*

(iv) *6.25 satisfies excision.*

Then there is a $K_0 - K_1 - K_2$ *Mayer-Vietoris sequence satisfying all of the assertions of 6.32, except possibly the assertion of exactness at* $K_2A_1 \oplus K_2A_2$.

Proof. By 6.37, the square 6.26 is an E-fibre square. One establishes now 6.38 exactly as one established 6.32, except the reference to Theorem 5.31 is replaced by a reference to Corollary 5.32.

THEOREM 6.39. *Suppose the following*:

(i) *6.25 is a fibre product square.*

(ii) F_i *and* F'_i $(i = 1, 2)$ *are cofinal.*

If for each $M \in \mathrm{Obj}(A)$, *there is an* $N \in \mathrm{Obj}(A)$ *such that 6.26 is* E*-fibred (resp. weak* S*-surjective) for* $A = M \perp N$ *then the canonical map* $K_1F'_2 \to K_1F_1$ *is injective (resp. surjective).*

The proof of 6.39 is similar to that of 6.37. Details will be left to the reader.

THEOREM 6.40. *Suppose the following*:

(i) $F_1 : A_1 \to \hat{A}$ *is cofinal.*

(ii) A_1 *has a cofinal object* C.

(iii) $G(C, F_1) = \ker(G(C) \to G(F_1C))$.

Define

$$\varphi : G(C, F_1)/[G(C), G(C, F_1)] \to K_1F_1 .$$
$$[\sigma] \mapsto [\sigma, 1]$$

a) *If* $E(C) \to E(F_1C)$ *is surjective then* φ *is bijective.*

b) φ *is surjective* $\Longleftrightarrow$ $G(C) \to G(F_1C)$ *is* S*-surjective.*

Proof. We want to apply Theorem 5.43. Let $f_1 : G(C) \to G(F_1C)$. By definition, $K_1(C, F_1C) = H_1f_1$ and by 6.24, we can identify $K_1(C, F_1C) = K_1F_1$. Thus, 6.40 will follow from 5.43, once we can show that the hypothesis of 5.43 is satisfied, namely that $[G(C), G(C, F_1)] = [E(C), [G(C), G(C, F_1)]]$. By Lemma 5.42 b), it suffices to show that the commutator subgroup of $G(C) \ltimes G(C, F_1)$ is perfect. But this follows from Lemma 5.42 a) and the proof of the Whitehead lemma 3.15.

§7. K-THEORY OF NONSINGULAR AND PROJECTIVE MODULES

In this chapter, it is shown that the hypotheses in §6 are satisfied for commutative squares of categories

$$
(1)\qquad \begin{array}{ccc} X(A) & \longrightarrow & X(B) \\ \downarrow & & \downarrow \\ X(\hat{A}) & \longrightarrow & X(\hat{B}) \end{array},
$$

such that

$$
(2)\qquad \begin{array}{ccc} A & \longrightarrow & B \\ \downarrow & & \downarrow \\ \hat{A} & \longrightarrow & \hat{B} \end{array}
$$

is a commutative square of rings or form rings satisfying suitable hypotheses and such that $X(A) = P(A)$ if (2) is a square of rings and $X(A) = Q(A)$ or $H(A)$ or a suitable basing of these categories if (2) is a square of form rings. The plan of the chapter is as follows. In §A, we say that (2) is an approximation square of rings or form rings if it is a fibre product square of topological rings or topological form rings such that the horizontal maps are open and the vertical maps are dense. If $G = GL$ or GQ, it is shown that if (2) is an approximation square then the square

$$
(3)\qquad \begin{array}{ccc} G(A) & \longrightarrow & G(B) \\ \downarrow & & \downarrow \\ G(\hat{A}) & \longrightarrow & G(\hat{B}) \end{array}
$$

is an S-exact, S-surjective, fibre product square of groups. In §B, we give some examples of approximation squares. In §C, we describe some squares which are closely related to approximation squares. In §D, we show that (1) is a fibre product square of categories for squares of rings and form rings as in §B and §C. In §E, we show that in certain circumstances K-theory commutes with infinite and restricted direct products. In §F, we apply all of the above to the case of an order A on a semisimple algebra B.

The results of the chapter are drawn essentially from my preprint [3]. The exact sequences in §F are the backbone of many computations of obstruction groups in algebraic topology, cf. [4]-[7], and of the solution to the congruence subgroup problem for nonexceptional, simply connected, simple algebraic groups of rank > 1 over a global field, cf. [8.3]-[8.4]. In Wall [34, IV], there are some exact sequences related to those in §F. One important difference, however, for the applications above is that the sequences in [34, IV] are somewhat courser than those in §F and as a result, certain fine arithmetical information arising in the sequences in §F vanishes in the sequences in [34, IV].

Finally, we should like to point out that the method of fibre squares was first developed by Milnor [20] for the K-theory of projective modules and then expanded and abstracted by Bass [10].

A. *Approximation squares*

A *topological group* is a group with a topology such that the operations of multiplication and inversion are continuous. A *topological ring* is a ring with a topology such that the operations of addition, multiplication, and additive inversion are continuous. A *topological form ring* is a form ring whose ring is a topological ring on which the involution acts continuously and whose form parameter has the subspace topology with respect to the ring.

Throughout the section, we fix the following data: The square

(7.1)
$$\begin{array}{ccc} A & \xrightarrow{g'} & B \\ {\scriptstyle f'}\downarrow & & \downarrow{\scriptstyle g} \\ \hat{A} & \xrightarrow[f]{} & \hat{B} \end{array}$$

is a fibre product square of topological rings and continuous homomorphisms such that the following three conditions hold:

(i) f and g' are open, i.e. take open sets to open sets.

(ii) f' and g are dense, i.e. their images are dense respectively in $\hat{A}$ and $\hat{B}$, and $f': A \to \text{image}(f')$ and $g: B \to \text{image}(g)$ are open.

(iii) Given an open neighbourhood U of zero in B (resp. $\hat{B}$), there is an open neighbourhood U' of zero in B (resp. $\hat{B}$) such that each element of U' can be written as a sum $\sum_{j=1}^{k} c_j d_j$ such that $c_j, d_j \in U \, (1 \leq j \leq k)$.

The square

(7.2)
$$\begin{array}{ccc} (A,\Lambda) & \longrightarrow & (B,\Gamma) \\ \downarrow & & \downarrow \\ (\hat{A},\hat{\Lambda}) & \longrightarrow & (\hat{B},\hat{\Gamma}) \end{array}$$

is a fibre product square of topological form rings and continuous homomorphisms such that the corresponding square of rings satisfies (7.1)(i)-(iii) and such that the following three conditions hold:

(i) $f: \hat{\Lambda} \to \hat{\Gamma}$ and $g': \Lambda \to \Gamma$ are open.

(ii) $f': \Lambda \to \hat{\Lambda}$ and $g: \Gamma \to \hat{\Gamma}$ are dense.

(iii) Given an open neighbourhood U of zero in B (resp. $\hat{B}$) and an open neighbourhood V of zero in Γ (resp. $\hat{\Gamma}$), there is a neighbourhood V' of zero in Γ (resp. $\hat{\Gamma}$) such that each element of V' can be written as a sum $\sum_{j-1} c_j \delta_j \bar{c}_j$ such that $c_j \in U$ and $\delta_j \in V$ $(1 \leq j \leq k)$.

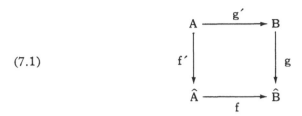

DEFINITION 7.3. a) A square as in (7.1) is called an *approximation square of rings.*

b) A square as in (7.2) is called an *approximation square of form rings.*

REMARK 7.4. (i) *Given a fibre product square of topological rings and continuous homomorphisms, the condition that the top homomorphism is open and injective follows from the analogous condition for the bottom homomorphism.*

(ii) *Given a fibre product square of topological rings and continuous homomorphisms, the condition that the left hand map is either dense or an open map onto its image follows from the analogous condition for the right hand map.*

REMARK 7.5. *Condition 7.1 (iii) is satisfied for a ring* C *in the following circumstances. Let* I *be a family of additive subgroups of* C *such that if* $\mathfrak{p}$ *and* $\mathfrak{q} \in I$ *then* $\mathfrak{p} \cap \mathfrak{q} \in I$ *and such that if* $\mathfrak{q} \in I$ *and* $c \in C$ *then there is a* $\mathfrak{q}_c \in I$ *such that* $c\mathfrak{q}_c \subset \mathfrak{q}$ *and* $\mathfrak{q}_c c \subset \mathfrak{q}$. *If one gives* C *the topology such that an open set is an arbitrary union of additive cosets of members of* I *then* C *is a topological ring. Furthermore, if* I *is closed under the operation*

$$\mathfrak{q} \mapsto \mathfrak{q}^2 = \{\sum_{j=1}^{k} p_j q_j \mid p_j, q_j \in \mathfrak{q}\ (1 \le j \le k)\}$$ *then* C *satisfies condition 7.1 (iii).*

The main result for approximation squares is the following:

THEOREM 7.6. a) *The square*

$$\begin{array}{ccc} GL(A) & \longrightarrow & GL(B) \\ \downarrow & & \downarrow \\ GL(\hat{A}) & \longrightarrow & GL(\hat{B}) \end{array}$$

is an S*-exact,* S*-surjective, fibre product square of groups.*

b) *The square*

$$\begin{array}{ccc} GQ(A,\Lambda) & \longrightarrow & GQ(B,\Gamma) \\ \downarrow & & \downarrow \\ GQ(\hat{A},\hat{\Lambda}) & \longrightarrow & GQ(\hat{B},\hat{\Gamma}) \end{array}$$

is an S-*exact,* S-*surjective, fibre product square of groups.*

c) *Suppose that we are in the situation of (7.2). Let*

$$\begin{array}{ccc} X & \longrightarrow & Y \\ \downarrow & & \downarrow \\ \hat{X} & \longrightarrow & \hat{Y} \end{array} \qquad \begin{array}{ccc} E(A) & \longrightarrow & E(B) \\ \downarrow & & \downarrow \\ E(\hat{A}) & \longrightarrow & E(\hat{B}) \end{array}$$

be a fibred, involution invariant subsquare of a) containing the corresponding subsquare of elementary groups. If

$$GQ(A,\Lambda)_{based\text{-}X} = \{\sigma \mid f(\sigma) \in X, f : GQ(A,\Lambda) \to GL(A)\},$$

etc., then the square

$$\begin{array}{ccc} GQ(A,\Lambda)_{based\text{-}X} & \longrightarrow & GQ(B,\Gamma)_{based\text{-}Y} \\ \downarrow & & \downarrow \\ GQ(\hat{A},\hat{\Lambda})_{based\text{-}\hat{X}} & \longrightarrow & GQ(\hat{B},\hat{\Gamma})_{based\text{-}\hat{Y}} \end{array}$$

is an S-*exact,* S-*surjective, fibre product square of groups.*

Proof. The proof of b) is similar to that of a) and will be omitted. The presentation of the quadratic Steinberg group which is required in the proof of b) is given in Theorem 3.17. Part c) is deduced very easily from part b). Details will be left to the reader.

Part a) is proved as follows. Since the square of rings is fibred, it follows that the corresponding square of general linear groups is fibred.

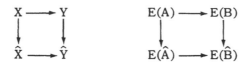

Let St(A) denote the Steinberg group of A. The definition of St(A) is given after Theorem 3.17 (cf. also Milnor [20, §5]). Let

$$S(A) = \operatorname{image}(St(A) \to St(B))$$
$$S(\hat{A}) = \operatorname{image}(St(\hat{A}) \to St(\hat{B}))\,.$$

Let $St(B)/S(A)$ and $St(\hat{B})/S(\hat{A})$ denote respectively the space of all right cosets of $S(A)$ and $S(\hat{A})$ in respectively $St(B)$ and $St(\hat{B})$. Consider the canonical map

$$\psi : St(B)/S(A) \to St(\hat{B})/S(\hat{A})\,.$$

By definition, the square in a) is S-exact (resp. S-surjective) if ψ is injective (resp. surjective). To prove that ψ is bijective will require the preparation of several technical lemmas. The fact that ψ is bijective will be achieved in Theorem 7.16 below.

Before starting the preparation for Theorem 7.16, we want to list a useful extension property of approximation squares.

Recall from §4C, the definition of the smash product $\ltimes$ and the definition of a form ideal.

RELATIVE LEMMA 7.7. a) *If*

$$\begin{array}{ccc} \mathfrak{a} & \longrightarrow & \mathfrak{b} \\ {\scriptstyle f'}\big\downarrow & & \big\downarrow{\scriptstyle g} \\ \hat{\mathfrak{a}} & \longrightarrow & \hat{\mathfrak{b}} \end{array}$$

is a fibred subsquare of (7.1) consisting of open two-sided ideals such that f′ *and* g *are dense then the square*

$$\begin{array}{ccc} A \ltimes \mathfrak{a} & \longrightarrow & B \ltimes \mathfrak{b} \\ \big\downarrow & & \big\downarrow \\ \hat{A} \ltimes \hat{\mathfrak{a}} & \longrightarrow & \hat{B} \ltimes \hat{\mathfrak{b}} \end{array}$$

is an approximation square of rings.

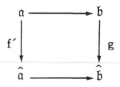

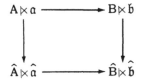

b) *If*

$$\begin{array}{ccc} (\mathfrak{a},\Lambda_{\mathfrak{a}}) & \longrightarrow & (\mathfrak{b},\Gamma_{\mathfrak{b}}) \\ {\scriptstyle f'}\downarrow & & \downarrow{\scriptstyle g} \\ (\hat{\mathfrak{a}},\hat{\Lambda}_{\hat{\mathfrak{a}}}) & \longrightarrow & (\hat{\mathfrak{b}},\hat{\Gamma}_{\hat{\mathfrak{b}}}) \end{array}$$

is a fibred subsquare of (7.2) consisting of open forms ideals such that f' *and* g *are dense then the square*

$$\begin{array}{ccc} (A,\Lambda)\ltimes(\mathfrak{a},\Lambda_{\mathfrak{a}}) & \longrightarrow & (B,\Gamma)\ltimes(\mathfrak{b},\Gamma_{\mathfrak{b}}) \\ \downarrow & & \downarrow \\ (\hat{A},\hat{\Lambda})\ltimes(\hat{\mathfrak{a}},\hat{\Lambda}_{\hat{\mathfrak{a}}}) & \longrightarrow & (\hat{B},\hat{\Gamma})\ltimes(\hat{\mathfrak{b}},\hat{\Gamma}_{\hat{\mathfrak{b}}}) \end{array}$$

is an approximation square of form rings.

The proof of 7.7 is easy and will be left to the reader.

We begin now our preparation for the proof of Theorem 7.16. We recall that if C is a ring then the *Steinberg group* $St(C)$ is the free group generated by all symbols $s_{\alpha\beta}(c)$ such that $\alpha \neq \beta$ are natural numbers and $c \in C$, modulo the *Steinberg relations*:

(1) $s_{\alpha\beta}(c+d) = s_{\alpha\beta}(c)s_{\alpha\beta}(d)$

(2) $[s_{\alpha\beta}(c), s_{\gamma\delta}(d)] = 1$ if $\alpha \neq \delta$, $\beta \neq \gamma$

(3) $[s_{\alpha\beta}(c), s_{\beta\delta}(d)] = s_{\alpha\delta}(cd)$ if $\alpha \neq \delta$.

LEMMA 7.8. *Let* C *be any topological ring which satisfies condition (7.1)(iii). If* $c_1, \cdots, c_n \in C$ *and if* U *is an open neighbourhood of zero then there are open neighbourhoods*

$$U = U(c_1) \supset \cdots \supset U(c_1, \cdots, c_n)$$

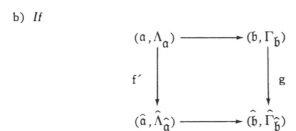

of zero such that for each m $(2 \leq m \leq n)$, *each element of* $U(c_1, \cdots, c_m)$ *can be written as a sum* $\sum_{j=1}^{k} c'_j d_j$ *such that* $c'_j d_j c_{m-1} c'_j, c_{m-1} d_j, c'_j c_{m-1}, d_j c_{m-1} \in U(c_1, \cdots, c_{m-1})$ $(1 \leq j \leq k)$.

Proof. The proof is by induction on n. The case $n = 1$ is clear. Suppose that $c_1, \cdots, c_n \in C$ and that $U(c_1) \supset \cdots \supset U(c_1, \cdots, c_n)$ satisfy the conclusions of the lemma. We must show that there is a neighbourhood $U(c_1, \cdots, c_{n+1}) \subset U(c_1, \cdots, c_n)$ of zero such that $U(c_1, \cdots, c_{n+1})$ satisfies the conclusion of the lemma with respect to c_n. But, clearly such a neighbourhood exists by condition (7.1)(iii).

LEMMA 7.9. *Let* C *be a ring. Let* $c_1, \cdots, c_n \in C$ *and let* $U(c_1) \supset \cdots \supset U(c_1, \cdots, c_n)$ *be a sequence of sets as in (7.8). Let* $S(U(c_1))$ *denote the subgroup of* $St(C)$ *generated by all Steinberg generators* $s_{\alpha\beta}(u)$ *such that* $u \in U(c_1)$. *If* $s = \prod_{i=1}^{n-1} s_{\alpha(i)\beta(i)}(c_i) \in St(C)$ *and if* $a \in U(c_1, \cdots, c_n)$ *then* ${}^{s}s_{\gamma\delta}(a) = s\, s_{\gamma\delta}(a) s^{-1} \in S(U(c_1))$.

Proof. For the case $n = 1$, there is nothing to prove. For the case $n > 1$, it suffices by induction to show that ${}^{s_{\alpha(n-1)\beta(n-1)}(c_{n-1})}s_{\gamma\delta}(a) \in S(U_1(c_1, \cdots, c_{n-1}))$. Clearly, it suffices to consider the case $n = 2$. Let $\alpha = \alpha(1)$, $\beta = \beta(1)$, and $c = c_1$. Write $a = \sum_{j=1}^{k} c'_j d_j$ where c'_j and d_j are as in (7.8) $(1 \leq j \leq k)$. Since $s_{\gamma\delta}(a) = \prod_{j=1}^{k} s_{\gamma\delta}(c'_j d_j)$, we can assume that $k = 1$. Let $c' = c_1$, $d = d_1$. Choose $\eta \neq \alpha, \beta, \gamma, \delta$. Since $s_{\gamma\delta}(a) = [s_{\gamma\eta}(c'), s_{\eta\delta}(d)]$, it suffices to show that ${}^{s_{\alpha\beta}(c)}s_{\gamma\eta}(c')$ and ${}^{s_{\alpha\beta}(c)}s_{\eta\gamma}(d) \in S(U(c))$. By the Steinberg relations, if $\beta \neq \gamma$ then ${}^{s_{\alpha\beta}(c)}s_{\gamma\eta}(c') = s_{\gamma\eta}(c') \in S(U(c))$ and if $\beta = \gamma$ then ${}^{s_{\alpha\beta}(c)}s_{\gamma\eta}(c') = s_{\alpha\eta}(cc') s_{\gamma\eta}(c') \in S(U(c))$. Similarly, one shows that ${}^{s_{\alpha\beta}(c)}s_{\eta\gamma}(d) \in S(U(c))$.

DEFINITION 7.10. Let $\hat{b}_1, \cdots, \hat{b}_n \in \hat{B}$. A *system of neighbourhoods* for $\hat{b}_1, \cdots, \hat{b}_n$ is a sequence of open neighbourhoods of zero

$$g'(A) \supset U(\hat{b}_1) \supset \cdots \supset U(\hat{b}_1, \cdots, \hat{b}_n)$$

in B and sequence of open neighbourhoods of zero

$$f(\hat{A}) \supset \hat{V}(\hat{b}_1) \supset \cdots \supset \hat{V}(\hat{b}_1, \cdots, \hat{b}_n)$$

in $\hat{B}$ which have the following properties:

(i) If $b, c \in B$ such that $\hat{b}_i - g(b)$ and $\hat{b}_i - g(c) \in \hat{V}(\hat{b}_1, \cdots, \hat{b}_i)$ then $b - c \in U(\hat{b}_1, \cdots, \hat{b}_i)$.

(ii) If $b \in B$ such that $\hat{b}_i - g(b) \in \hat{V}(\hat{b}_1, \cdots, \hat{b}_i)$ then each element of $U(\hat{b}_1, \cdots, \hat{b}_{i+1})$ can be written as a sum $\sum_{j=1}^{k} c_j d_j$ such that $c_j, d_j, bc_j, bd_j, c_j b, d_j b \in U(\hat{b}_1, \cdots, \hat{b}_i)$.

LEMMA 7.11. *Let* $J \subseteq I$ *be finite nonempty sets of natural numbers. If* $\{U_i \mid i \in I, U_i \subset B\}$ *and* $\{\hat{V}_i \mid i \in I, \hat{V}_i \subset \hat{B}\}$ *is a system of sets for* $\{\hat{b}_i \mid i \in I, \hat{b}_i \in \hat{B}\}$ *then* $\{U_i \mid i \in J\}$ *and* $\{\hat{V}_i \mid i \in J\}$ *is a system of sets for* $\{\hat{b}_i \mid i \in J\}$.

Proof. Clear.

LEMMA 7.12. *If* $\hat{b}_1, \cdots, \hat{b}_n \in \hat{B}$ *then* $\hat{b}_1, \cdots, \hat{b}_n$ *has a system of neighbourhoods.*

Proof. The proof is by induction on n. Suppose $n = 1$. Let $U(\hat{b}_1) \subset g'(A)$ be any open neighbourhood of zero. By (7.1)(ii), there is an open neighbourhood $\hat{V}'$ of zero in $\hat{B}$ such that $g^{-1}(\hat{V}') \subset U(\hat{b}_1)$. Since $0 + (-0) = 0$ and since addition and additive inversion are continuous, it follows that there is an open neighbourhood $\hat{V}''$ of zero such that $\hat{V}'' - \hat{V}'' \subset \hat{V}'$. If $\hat{V}(\hat{b}_1) = \hat{V}''$ then condition (7.11)(i) is satisfied and since $n = 1$, there is nothing to verify for condition (7.11)(ii). Suppose that $\hat{b}_1, \cdots, \hat{b}_{n+1} \in \hat{B}$ and that $U(\hat{b}_1) \supset \cdots \supset U(\hat{b}_1, \cdots, \hat{b}_n)$ and $\hat{V}(\hat{b}_1) \supset \cdots \supset \hat{V}(\hat{b}_1, \cdots, \hat{b}_n)$ is a system of neighbourhoods for $\hat{b}_1, \cdots, \hat{b}_n$. We note that the neighbourhoods above will continue to form a system of neighbourhoods if

$\hat{V}(\hat{b}_1, \dots, \hat{b}_n)$ is replaced by any smaller open neighbourhood of zero. This will be done during the course of the proof. Since $0+0=0$ and addition is continuous, there is an open neighbourhood $U' \subset U(\hat{b}_1, \cdots, \hat{b}_n)$ of zero such that if $u, v \in U'$ then $u+v \in U(\hat{b}_1, \cdots, \hat{b}_n)$. Since $0 \cdot 0 = 0$ and multiplication is continuous, there is an open neighbourhood $U'' \subset U'$ of zero such that for all $u, v \in U''$, $uv \in U'$. By (7.1)(ii), there is an open neighbourhood $\hat{V}''$ of zero in $\hat{B}$ such that if $b, c \in B$ such that $\hat{b}_n - g(b)$, $\hat{b}_n - g(c) \in \hat{V}''$ then $b - c \in U''$. We replace now $\hat{V}(\hat{b}_1, \cdots, \hat{b}_n)$ above by $\hat{V}'' \cap \hat{V}(\hat{b}_1, \cdots, \hat{b}_n)$. Let $b \in B$ such that $\hat{b}_n - g(b) \in \hat{V}'' \cap \hat{V}(\hat{b}_1, \cdots, \hat{b}_n)$. Since $b \cdot 0 = 0 = 0 \cdot b$ and multiplication is continuous, there is an open neighbourhood $U''' \subset U'$ of zero such that bU''' and $U'''b \subset U'$. By (7.1)(iii), $U'' \cap U'''$ contains an open neighbourhood of zero such that each element of U'''' can be written as a sum $\sum_{j=1}^{k} c_j d_j$ such that $c_j, d_j \in U'' \cap U'''$ $(1 \le j \le k)$. Let $U(\hat{b}_1, \cdots, \hat{b}_{n+1}) = U''''$ and using (7.1)(ii), choose an open neighbourhood $\hat{V}(\hat{b}_1, \cdots, \hat{b}_{n+1}) \subset \hat{V}'' \cap \hat{V}(\hat{b}_1, \cdots, \hat{b}_n)$ such that if $b, c \in B$ such that $\hat{b}_{n+1} - g(b), \hat{b}_{n+1} - g(c) \in \hat{V}(\hat{b}_1, \cdots, \hat{b}_{n+1})$ then $b - c \in U(\hat{b}_1, \cdots, \hat{b}_{n+1})$. It is straightforward now to verify that $U(\hat{b}_1) \supset \cdots \supset U(\hat{b}_1, \cdots, \hat{b}_{n+1})$ and $\hat{V}(\hat{b}_1) \supset \cdots \supset \hat{V}(\hat{b}_1, \cdots, \hat{b}_{n-1}) \supset \hat{V}'' \cap \hat{V}(\hat{b}_1, \cdots, \hat{b}_n) \supset V(\hat{b}_1, \cdots, \hat{b}_{n+1})$ form a system of neighbourhoods for $\hat{b}_1, \cdots, \hat{b}_{n+1}$.

Let

$$Y(\hat{B})$$

denote the set of all symbols $\prod_{i=1}^{n} y_{\alpha(i)\beta(i)}(\hat{b}_i)$ such that $n > 0$, $\alpha(i) \ge 1$, $\beta(i) \ge 1$, $\alpha(i) \neq \beta(i)$ are natural numbers and $\hat{b}_i \in \hat{B}$ $(i = 1, \cdots, n)$.

DEFINITION 7.13. An element $\prod_{i=1}^{n} s_{\alpha(i)\beta(i)}(b_i) \in St(B)$ is called a *good approximation* to $\prod_{i=1}^{n} y_{\alpha(i)\beta(i)}(\hat{b}_i) \in Y(\hat{B})$ if there is a system of neighbourhoods $U(\hat{b}_1) \supset \cdots \supset U(\hat{b}_1, \cdots, \hat{b}_n)$ and $\hat{V}(\hat{b}_1) \supset \cdots \supset \hat{V}(\hat{b}_1, \cdots, \hat{b}_n)$ for $\hat{b}_1, \cdots, \hat{b}_n$ such that $\hat{b}_i - g(b_i) \in \hat{V}(\hat{b}_1, \cdots, \hat{b}_i)$ $(1 \leq i \leq n)$.

LEMMA 7.14. *If* $y \in Y(\hat{B})$ *and if* s_1 *and* s_2 *are good approximations to* y *then* $S(A)s_1 = S(A)s_2$.

Proof. Let $y = \prod_{i=1}^{n} y_{\alpha(i)\beta(i)}(\hat{b}_i)$. Let $s_1 = \prod_{i=1}^{n} s_{\alpha(i)\beta(i)}(b_i)$ be a good approximation to y with respect to the system of neighbourhoods $U_1 \supset \cdots \supset U_n$ and $\hat{V}_1 \supset \cdots \supset \hat{V}_n$ for $\hat{b}_1, \cdots, \hat{b}_n$ and let $s_2 = \prod_{i=1}^{n} s_{\alpha(i)\beta(i)}(c_i)$ be a good approximation to y with respect to the system of neighbourhoods $U_1' \supset \cdots \supset U_n'$ and $\hat{V}_1' \supset \cdots \supset \hat{V}_n'$ for $\hat{b}_1, \cdots, \hat{b}_n$. It is clear that $U_1 \supset \cdots \supset U_n$ and $\hat{V}_1 \cap \hat{V}_1' \supset \cdots \supset \hat{V}_n \cap \hat{V}_n'$ is a system of neighbourhoods for $\hat{b}_1, \cdots, \hat{b}_n$ and that $U_1' \supset \cdots \supset U_n'$ and $\hat{V}_1 \cap \hat{V}_1' \supset \cdots \supset \hat{V}_n \cap \hat{V}_n'$ is a system of neighbourhoods for $\hat{b}_1, \cdots, \hat{b}_n$. By (7.1)(ii), we can find an element $s_3 \in St(B)$ which is a good approximation to y for each of the previous two systems of neighbourhoods. Thus, we can assume that s_1 and s_2 are good approximations with respect to the same system of neighbourhoods. By condition (7.10)(i), we can write $c_i = b_i + a_i$ for some $a_i \in U_i$ $(1 \leq i \leq n)$. If $s = s_{\alpha(1)\beta(1)}(a_1)\left({}^{s_{\alpha(1)\beta(1)}(b_1)}s_{\alpha(2)\beta(2)}(a_2)\right) \ldots \left({}^{s_{\alpha(1)\beta(1)}(b_1)\cdots s_{\alpha(n-1)\beta(n-1)}(b_{n-1})}s_{\alpha(n)\beta(n)}(a_n)\right)$, then $s_2 = ss_1$ and by Lemma 7.9, $s \in S(A)$.

LEMMA 7.15. *Let*

$$\varphi : Y(\hat{B}) \to St(B)/S(A)$$

$$y \mapsto S(A)\ (\textit{good approximation}\ (y)).$$

Let $Y_0(\hat{B}) = \{y_{\alpha\beta}(\hat{b})y_{\alpha\beta}(-\hat{b}), y_{\alpha\beta}(\hat{b}+\hat{c})y_{\alpha\beta}(-\hat{b})y_{\alpha\beta}(-\hat{c}),$ $y_{\alpha\beta}(\hat{b})y_{\gamma\delta}(\hat{c})y_{\alpha\beta}(-\hat{b})y_{\gamma\delta}(-\hat{c}), y_{\alpha\beta}(\hat{b})y_{\beta\delta}(\hat{c})y_{\alpha\beta}(-\hat{b})y_{\beta\delta}(-\hat{c})y_{\alpha\delta}(-\hat{b}\hat{c}),$

$|\alpha \neq \delta, \beta \neq \gamma, \hat{b}, \hat{c} \in \hat{B}\}$. *If* $y_i \in Y(\hat{B})$ $(i=1,2)$ *and* $z \in Y_o(\hat{B})$ *then* $\varphi(y_1 z y_2) = \varphi(y_1 y_2)$.

Proof. We shall work out only the case $z = y_{\alpha\beta}(\hat{b})y_{\alpha\beta}(-\hat{b})$. The other cases are done similarly. Suppose that $y_1 = \prod_{i=1}^{m} y_{\alpha(i)\beta(i)}(\hat{b}_i)$ and $y_2 = \prod_{i=1}^{n} y_{\gamma(i)\delta(i)}(\hat{c}_i)$. Let $U_1 \supset \cdots \supset U_{m+n+2}$ and $\hat{V}_1 \supset \cdots \supset \hat{V}_{m+n+2}$ be a system of neighbourhoods for $\hat{b}_1, \cdots, \hat{b}_m, \hat{b}, -\hat{b}, \hat{c}_1, \cdots, \hat{c}_n$. Since additive inversion is continuous, there is an open neighbourhood $\hat{V} \subset \hat{V}_{m+1}$ of zero such that $-\hat{V} \subset \hat{V}_{m+2}$. It follows now from (7.1)(ii) that with respect to the system of neighbourhoods above, there is a good approximation s to $y_1 z y_2$ such that $s = \prod_{i=1}^{m} s_{\alpha(i)\beta(i)}(b_i) s_{\alpha\beta}(b) s_{\alpha\beta}(-b) \prod_{i=1}^{n} s_{\gamma(i)\delta(i)}(c_i)$. By the Steinberg relations, $s = \prod_{i=1}^{m} s_{\alpha(i)\beta(i)}(b_i) \prod_{i=1}^{n} s_{\gamma(i)\delta(i)}(c_i)$. Thus, s is a good approximation to $y_1 y_2$ with respect to the system of neighbourhoods $U_1 \supset \cdots \supset U_m \supset U_{m+3} \supset \cdots \supset U_{m+n+2}$ and $\hat{V}_1 \supset \cdots \supset \hat{V}_m \supset \hat{V}_{m+3} \supset \cdots \supset \hat{V}_{m+n+2}$ for $\hat{b}_1, \cdots, \hat{b}_m$, $\hat{c}_1, \cdots, \hat{c}_n$. Thus, $\varphi(y_1 z y_2) = \varphi(y_1 y_2)$.

THEOREM 7.16. *The map*

$$\varphi : Y(\hat{B}) \to St(B)/S(A)$$

$$y \mapsto S(A) \text{ (good approximation (y))}$$

induces a bijective map

$$St(\hat{B})/S(\hat{A}) \to St(B)/S(A)$$

which is inverse to the canonical map $\theta : St(B)/S(A) \to St(\hat{B})/S(\hat{A})$.

Proof. It suffices to show that θ is surjective and that φ induces a map which is a retract to θ.

First, we show that θ is surjective. Let $\hat{s} = \prod_{i=1}^{n} s_{\alpha(i)\beta(i)}(\hat{b}_i) \in St(\hat{B})$. Let $f(\hat{A}) \supset U(\hat{b}_1) \supset \cdots \supset U(\hat{b}_1, \cdots, \hat{b}_n)$ be a family of open neighbourhoods of zero as in Lemma 7.8. By (7.1)(ii), there are elements $b_1, \cdots, b_n \in B$ such that $g(b_i) = \hat{b}_i + \hat{a}_i$ and $\hat{a}_i \in U(\hat{b}_1, \cdots, \hat{b}_i)$ $(1 \le i \le n)$. If

$$s = \prod_{i=1}^{n} s_{\alpha(i)\beta(i)}(b_i) \text{ and } \hat{t} = s_{\alpha(1)\beta(1)}(\hat{a}_1)\left({}^{s_{\alpha(1)\beta(1)}(\hat{b}_1)} s_{\alpha(2)\beta(2)}(\hat{a}_2)\right) \cdots$$

$$\left({}^{s_{\alpha(1)\beta(1)}(\hat{b}_1) \cdots s_{\alpha(n-1)\beta(n-1)}(\hat{b}_{n-1})} s_{\alpha(n)\beta(n)}(\hat{a}_n)\right)$$ then $\hat{s} = \hat{t}s$ and by Lemma 7.9, $\hat{t} \in S(\hat{A})$. Thus, θ is surjective.

Next, we show that φ induces a retract to θ. It is clear that if φ induces a map $St(\hat{B})/S(\hat{A}) \to St(B)/S(A)$ then this map will be a retract to θ. We show now that φ induces a map. Let $Y^{\pm 1}(\hat{B})$ denote the set of all symbols $\prod_{i=1}^{n} (y_{\alpha(i)\beta(i)}(\hat{b}_i))^{\eta(i)}$ such that $\prod_{i=1}^{n} y_{\alpha(i)\beta(i)}(\hat{b}_i) \in Y(\hat{B})$ and $\eta(i) = \pm 1$ $(1 \le i \le n)$. Let $\phi: Y^{\pm 1}(\hat{B}) \to Y(\hat{B})$, $\prod_{i=1}^{n} y_{\alpha(i)\beta(i)}(\hat{b}_i)^{\eta(i)} \mapsto \prod_{i=1}^{n} y_{\alpha(i)\beta(i)}(\eta(i)\hat{b}_i)$ and let $\psi = \varphi\phi$. By Lemma 7.15, ψ induces a map $St(\hat{B}) \to St(B)/S(A)$. Furthermore, if $y = \prod_{i=1}^{n} y_{\alpha(i)\beta(i)}(\hat{a}_i) \in Y(\hat{B})$ such that each $a_i \in f(\hat{A})$ and if $z \in Y(\hat{B})$ then it is clear that there is a good approximation st for yz such that s (resp. t) is a good approximation to y (resp. z) and $s \in S(A)$. Thus, ψ induces a map $St(\hat{B})/S(\hat{A}) \to St(B)/S(A)$.

B. *Arithmetic and localization, completion squares*

In this section, we describe two useful examples of approximation squares, namely arithmetic squares and localization, completion squares. The latter were introduced into K-theory by M. Karoubi [16.2], and the former turn out, as we shall see, to be a special case of the latter. An appropriate reference for certain technical aspects of the section is Chapter II §A of J.-P. Serre, *Algèbre Locale, Multiplicites*, Lecture Notes in Mathematics, Springer-Verlag, New York (1965).

Let M be an abelian group. If I is a family of subgroups of M closed under finite intersections, then the cosets of members of I form a basis for a topology on M called the *I-adic topology*. Let A be a ring and let S be a multiplicative set in the center(A). If M is a group with a monoid action of S then the *S-adic topology* on M is the $\{sM \mid s \in S\}$-adic topology. We let $S^{-1}M = \{\frac{m}{s} \mid m \in M, s \in S\}$ denote the localization of M at S. Thus, $\frac{m}{s}+\frac{n}{t} = \frac{tm+sn}{st}$ and if M has a multiplication, then $\frac{m}{s}\frac{n}{t} = \frac{mn}{st}$. We note that if $T = \{s^2 \mid s \in S\}$ then the S-adic and T-adic topologies on M coincide and $S^{-1}M = T^{-1}M$. Whenever we are concerned with a form ring (A, Λ), we shall *assume* that the elements of S are squares s^2 such that $s \in \text{center}(A)$ and $s = \overline{s}$, where $a \mapsto \overline{a}$ denotes the involution on A. The advantage of the assumption is that Λ is then closed under the action of S. M is called *S-torsion free* if the action of each element of S on M is injective.

We give A and $\varprojlim_{s \in S} A/sA$ the S-adic topologies and we give respectively $S^{-1}A$ and $S^{-1} \varprojlim_{s \in S} A/sA$ the $\{sA \mid s \in S\}$-adic and $\{s \varprojlim_{s \in S} A/sA \mid s \in S\}$-adic topologies. Thus, the canonical maps $A \to S^{-1}A$ and $\varprojlim_{s \in S} A/sA \to S^{-1} \varprojlim_{s \in S} A/sA$ are open (cf. (7.5)).

LEMMA 7.17. *If A is S-torsion free then the square*

$$\begin{array}{ccc} A & \longrightarrow & S^{-1}A \\ \downarrow & & \downarrow \\ \varprojlim_{s \in S} A/sA & \longrightarrow & S^{-1} \varprojlim_{s \in S} A/sA \end{array}$$

is an approximation square of rings.

Proof. We shall prove only that the square is fibred. The remaining details will be left to the reader. We shall begin by showing that the canonical homomorphism $\varprojlim_{s \in S} A/sA \to S^{-1} \varprojlim_{s \in S} A/sA$ is injective. It suffices to show that $\varprojlim_{s \in S} A/sA$ is S-torsion free. Let $\prod_{t \in S} a_t$ be a representative in $\prod_{t \in S} A$ for an element $\hat{a} \in \varprojlim_{s \in S} A/sA$ such that $s\hat{a} = 0$ for some $s \in S$. For each $t \in S$, $sa_{st} = sta$ for some $a \in A$. Since A is S-torsion free, it follows that $a_{st} = ta$. Since $a_{st} \equiv a_t \bmod (tA)$, it follows that $a_t \equiv 0 \bmod (tA)$. Thus, $\hat{a} = 0$. Suppose now that $\hat{a} \in \varprojlim_{s \in S} A/sA$ and $\frac{a}{s} \in S^{-1}A$ such that the equation $\hat{a} = \frac{a}{s}$ holds in $S^{-1} \varprojlim_{s \in S} A/sA$. We must show that $\hat{a}$ and $\frac{a}{s}$ have a unique, common preimage in A. Since the map $\varprojlim_{s \in S} A/sA \to S^{-1} \varprojlim_{s \in S} A/sA$ is injective, it follows that the equation $s\hat{a} = a$ holds in $\varprojlim_{s \in S} A/sA$. If $\prod_{t \in S} a_t \in \prod_{t \in S} A$ is a representative for $\hat{a}$ then $sa_s = a + sc$ for some $c \in A$. Thus, $s(a_s - c) = a$, and it follows that $a_s - c \in A$ is a preimage for $\frac{a}{s} \in S^{-1}A$. Furthermore, since $\varprojlim_{s \in S} A/sA \to S^{-1} \varprojlim_{s \in S} A/sA$ is injective and since in $S^{-1} \varprojlim_{s \in S} A/sA$, one can write

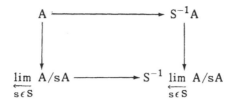

$a_s - c = \frac{a}{s} = \hat{a}$, it follows that $a_s - c$ is also a preimage for $\hat{a}$. Since A is S-torsion free, the map $A \to S^{-1}A$ is injective. Thus, $a_s - c$ is the unique, common preimage for $\hat{a}$ and $\frac{a}{s}$.

LEMMA 7.18. *Let* A *be a ring with involution* $a \mapsto \bar{a}$. *Let* $\Lambda \subset \Gamma$ *be form parameters on* A. *If* R *is a subring of the center*(A) *such that* $R\Lambda \subset \Lambda$ *and* $R\Gamma \subset \Gamma$ *then the rule*

$$A \times R \times \Gamma/\Lambda \to \Gamma/\Lambda$$
$$(a, r, [x]) \mapsto [arx\bar{a}]$$

defines an $A \otimes_{\mathbf{Z}} R$*-module structure on* Γ/Λ.

Proof. The proof is routine, except possibly for the fact that if $n \in \mathbf{Z}$ then $[a(nr)x\bar{a}] = [(an)(rx)(\overline{na})]$. But $[a(nr)x\bar{a}] - [anrx\overline{na}] = [(n-n^2)arx\bar{a}] = 0$, because $n-n^2 \in 2\mathbf{Z}$ and $2(arx\bar{a}) = arx\bar{a} - \lambda\overline{arx\bar{a}} \in \min \subset \Lambda$.

In the lemma above, if $\Lambda = \min$ and $\Gamma = \max$ then one can take $R = \{r \mid r \in \text{center}(A), r = \bar{r}\}$. For arbitrary Λ and Γ, the ring R generated by all $r\bar{r}$ and $r+\bar{r}$ such that $r \in \text{center}(A)$ will always satisfy the hypotheses of the lemma.

The significance of the lemma above is that Γ/Λ may be a noetherian $A \otimes_{\mathbf{Z}} R$-module, but not a noetherian A-module. This is the case, for example, if A is a field of characteristic 2 with trivial involution, $\Lambda = 0$, $\Gamma = A$, and A^2 has infinite index in A.

It will be assumed below that $\max/\Lambda$ is a noetherian $A \otimes_{\mathbf{Z}} R$-module where R is such that $S \subset R$ and $R\Lambda \subset \Lambda$. The assumption will be used to guarantee that the subspace topology on Λ which is inherited from the S-adic topology on A coincides with the S-adic topology on Λ. To establish the guarantee, one can argue as follows. We assume as usual that A is S-torsion free. It suffices to show that the subspace topology on max which is inherited from the S-adic topology on A coincides with the S-adic topology on max, because the noetherian assumption will then imply (cf. Serre reference above) that the subspace topology on Λ inherited from the S-adic topology on max coincides with the S-adic topology on Λ. It suffices to show that if $s \in S$ then $sA \cap \max = s \max$. It is clear that $s \max \subset sA \cap \max$. Conversely, if $a \in A$ and $sa \in \max$ then $sa = -\lambda\overline{sa} = -s\lambda\bar{a}$. Thus, $s(a+\lambda\bar{a}) = 0$. Since A is S-torsion free, it follows that $a = -\lambda\bar{a}$. Thus, $a \in \max$. Thus, $sa \in s \max$.

THEOREM 7.19. *Let* (A,Λ) *be a form ring. Let* R *be a subring of the center*(A) *such that* $S \subset R$ *and* $R\Lambda \subset \Lambda$. *If* A *is* S*-torsion free and if* $\max/\Lambda$ *is noetherian over* $A \otimes_{\mathbf{Z}} R$ *then the square*

$$\begin{array}{ccc} (A,\Lambda) & \longrightarrow & (S^{-1}A, S^{-1}\Lambda) \\ \downarrow & & \downarrow \\ (\varprojlim_{s\in S} A/sA, \varprojlim_{s\in S} \Lambda/s\Lambda) & \longrightarrow & (S^{-1}(\varprojlim_{s\in S} A/sA), S^{-1}(\varprojlim_{s\in S} \Lambda/s\Lambda)) \end{array}$$

is an approximation square of form rings.

Proof. By Lemma 7.17, the square

(1)
$$\begin{array}{ccc} A & \longrightarrow & S^{-1}A \\ \downarrow & & \downarrow \\ \varprojlim_{s\in S} A/sA & \longrightarrow & S^{-1}(\varprojlim_{s\in S} A/sA) \end{array}$$

is an approximation square of rings. The proof that was used to show that (1) is a fibre product square (of sets) works equally well to show that the square

(2)
$$\begin{array}{ccc} \Lambda & \longrightarrow & S^{-1}\Lambda \\ \downarrow & & \downarrow \\ \varprojlim_{s\in S} \Lambda/s\Lambda & \longrightarrow & S^{-1}(\varprojlim_{s\in S} \Lambda/s\Lambda) \end{array}$$

is a fibre product square. Give Λ and $\varprojlim_{s\in S} \Lambda/s\Lambda$ the S-adic topologies and give respectively $S^{-1}\Lambda$ and $S^{-1}(\varprojlim_{s\in S} \Lambda/s\Lambda)$ the $\{s\Lambda | s \in S\}$-adic and $\{s(\varprojlim_{s\in S} \Lambda/s\Lambda) | s \in S\}$-adic topologies. To complete the proof, it suffices to show that the canonical map (2) $\to$ (1) takes (2) homeomorphically onto its image. From the paragraph preceding the lemma, we know that the map $\Lambda \to A$ takes Λ homeomorphically onto its image. It follows automatically (cf. Serre reference above) that the map $\varprojlim_{s\in S} \Lambda/s\Lambda \to \varprojlim_{s\in S} A/sA$ takes $\varprojlim_{s\in S} \Lambda/s\Lambda$ homeomorphically onto its image. Since S^{-1} is an exact functor, it follows that the maps $S^{-1}\Lambda \to S^{-1}A$ and $S^{-1}(\varprojlim_{s\in S} \Lambda/s\Lambda) \to S^{-1}(\varprojlim_{s\in S} A/sA)$ are injective. To show that the map $S^{-1}\Lambda \to S^{-1}A$ takes $S^{-1}\Lambda$ homeomorphically onto its image, it suffices now to show that the subspace topology on $S^{-1}\Lambda$ inherited from $S^{-1}A$ coincides with the $\{s\Lambda | s \in S\}$-adic topology on $S^{-1}\Lambda$. It is clear that the latter is at least as fine as the former. To show the converse, we must show that given $s \in S$, there is a $t \in S$ such that $tA \cap S^{-1}\Lambda \subset s\Lambda$. Let $M = \{m | m \in \max, sm \in \Lambda$ for some $s \in S\}$. Let $M_s = \{m | m \in \max, sm \in \Lambda\}$. Clearly $\Lambda \subset M_s$. From the noetherian condition on $\max/\Lambda$, it follows that there is a $t_o \in S$ such that $M_{t_o} = M$. For any $s \in S$, it is clear that $st_oA \cap S^{-1}(\max) = st_o \max$ and since $\max/M$ is S-torsion free, it follows that $st_o \max \cap S^{-1}M = st_oM$. Thus, $st_oA \cap S^{-1}M = st_oM$. Since $t_oM \subset \Lambda$, it follows that $st_oA \cap S^{-1}\Lambda = st_oM \subset s\Lambda$. The proof that the map $S^{-1}(\varprojlim_{s\in S} \Lambda/s\Lambda) \to S^{-1}(\varprojlim_{s\in S} A/sA)$ takes $S^{-1}(\varprojlim_{s\in S} \Lambda/s\Lambda)$ homeomorphically onto its image becomes analogous to the one above, after one notes that $(\varprojlim_{s\in S} \max/s \max)/(\varprojlim_{s\in S} \Lambda/s\Lambda)$ is noetherian over $(\varprojlim_{s\in S} A/sA) \otimes_{\mathbf{Z}} (\varprojlim_{s\in S} R/sR)$.

DEFINITION 7.20. Let I be an index set. For each $i \in I$ let $f_i : H_i \to G_i$ be a homomorphism of groups, rings, etc. If F is a finite subset of I, let $T_F = \prod_{i \in F} G_i \times \prod_{i \notin F} H_i$. If $F \subseteq F_1$ then there is a canonical map $\psi : T_F \to T_{F_1}$ such that $\psi|_{G_i} = 1_{G_i}$ for $i \in F$, $\psi|_{H_i} = f_i$ for $i \in F_1 - F$, and $\psi|_{H_i} = 1_{H_i}$ for $i \notin F_1$. The family of finite subsets of I is partially ordered by inclusion. The *restricted direct product* $\prod_{i \in I} (G_i, H_i)$ is defined by

$$\prod_{i \in I} (G_i, H_i) = \varinjlim_{F} T_F .$$

Let R be a commutative ring. Let I be a set of maximal ideals of R. If M is an R-module then the I-*adic topology* on M is the $\{\mathfrak{q}M \mid \mathfrak{q} = \mathfrak{p}_1 \cdots \mathfrak{p}_n, \mathfrak{p}_i \in I\}$-adic topology on M. If $\mathfrak{p} \in I$, we let

$$M_{\mathfrak{p}} = \varprojlim_{i} M/\mathfrak{p}^i M .$$

Let R and I be as in the paragraph above. Let S be a multiplicative subset of R such that the S-adic and I-adic topologies on R coincide. Let A be a finite R-algebra. The finiteness condition on A guarantees that for any ideal $\mathfrak{q}$ of R, $\mathfrak{q}(\prod_{\mathfrak{p} \in I} A_{\mathfrak{p}}) = \prod_{\mathfrak{p} \in I} \mathfrak{q}A_{\mathfrak{p}}$ and that $S^{-1}(\prod_{\mathfrak{p} \in I} A_{\mathfrak{p}}) = \prod_{\mathfrak{p} \in I} (S^{-1}A_{\mathfrak{p}}, A_{\mathfrak{p}})$. Give A and $\prod_{\mathfrak{p} \in I} A_{\mathfrak{p}}$ the I-adic topologies and give respectively $S^{-1}A$ and $\prod_{\mathfrak{p} \in I} (S^{-1}A_{\mathfrak{p}}, A_{\mathfrak{p}})$ the $\{\mathfrak{q}A \mid \mathfrak{q} = \mathfrak{p}_1 \cdots \mathfrak{p}_n, \mathfrak{p}_i \in I\}$-adic and $\{\mathfrak{q}(\prod_{\mathfrak{p} \in I} A_{\mathfrak{p}}) \mid \mathfrak{q} = \mathfrak{p}_1 \cdots \mathfrak{p}_n, \mathfrak{p}_i \in I\}$-adic topologies. Thus, the maps $A \to S^{-1}A$ and $\prod_{\mathfrak{p}} A_{\mathfrak{p}} \to \prod_{\mathfrak{p}} (S^{-1}A_{\mathfrak{p}}, A_{\mathfrak{p}})$ are open.

LEMMA 7.21. *Let*

R = *commutative ring*

I = *set of maximal ideals of* R

S = *multiplicative set in* R

A = *finite,* S*-torsion free* R*-algebra*

and suppose that the S*-adic and* I*-adic topologies on* R *coincide. Then the canonical map of squares*

$$\begin{array}{ccc} A & \longrightarrow & S^{-1}A \\ \downarrow & & \downarrow \\ \varprojlim_{s\in S} A/sA & \longrightarrow & S^{-1}(\varprojlim_{s\in S} A/sA) \end{array} \quad \xrightarrow{\cong} \quad \begin{array}{ccc} A & \longrightarrow & S^{-1}A \\ \downarrow & & \downarrow \\ \prod_{\mathfrak{p}\in I} A_{\mathfrak{p}} & \longrightarrow & \prod_{\mathfrak{p}\in I} (S^{-1}A_{\mathfrak{p}}, A_{\mathfrak{p}}) \end{array}$$

is an isomorphism of approximation squares of rings.

Proof. Let $J = \{\mathfrak{q} \mid \mathfrak{q} = \mathfrak{p}_1 \cdots \mathfrak{p}_r, \mathfrak{p}_i \in I\}$. Since the S-adic and I-adic topologies on A coincide, we can canonically identify $\varprojlim_{s\in S} A/sA = \varprojlim_{\mathfrak{q}\in J} A/\mathfrak{q}A$. Using the Chinese Remainder Theorem, we can further identify $\varprojlim_{\mathfrak{q}\in J} A/\mathfrak{q}A = \prod_{\mathfrak{p}\in I} (\varprojlim_i A/\mathfrak{p}^i A) = \prod_{\mathfrak{p}\in I} A_{\mathfrak{p}}$. Thus, also $S^{-1}(\varprojlim_{s\in S} A/sA) = S^{-1}(\prod_{\mathfrak{p}\in I} A_{\mathfrak{p}}) = \prod_{\mathfrak{p}\in I} (S^{-1}A_{\mathfrak{p}}, A_{\mathfrak{p}})$.

LEMMA 7.22. *Let*

R = *commutative ring with trivial involution*

I = *set of maximal ideals of* R

S = *multiplicative set in* R

A = *finite,* S*-torsion free* R*-algebra with compatible involution*

Λ = *form parameter on* A *such that* $R\Lambda \subset \Lambda$

and suppose that the S*-adic and* I*-adic topologies on* R *coincide and that* $\max/\Lambda$ *is noetherian over* $A \otimes_{\mathbf{Z}} R$. *Then the canonical map of squares*

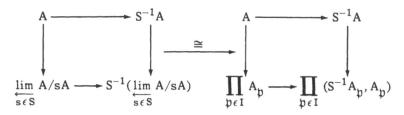

$$\begin{array}{ccc}
(A,\Lambda) & \longrightarrow & (S^{-1}A, S^{-1}\Lambda) \\
\downarrow & & \downarrow \\
(\varprojlim_{s\in S} A/sA, \varprojlim_{s\in S} \Lambda/s\Lambda) & \longrightarrow & (S^{-1}(\varprojlim_{s\in S} A/sA), S^{-1}(\varprojlim_{s\in S} \Lambda/s\Lambda))
\end{array}$$

$$\downarrow \cong$$

$$\begin{array}{ccc}
(A,\Lambda) & \longrightarrow & (S^{-1}A, S^{-1}\Lambda) \\
\downarrow & & \downarrow \\
(\prod_{\mathfrak{p}\in I} A_{\mathfrak{p}}, \prod_{\mathfrak{p}\in I} \Lambda_{\mathfrak{p}}) & \longrightarrow & (\prod_{\mathfrak{p}\in I} (S^{-1}A_{\mathfrak{p}}, A_{\mathfrak{p}}), \prod_{\mathfrak{p}\in I} (S^{-1}\Lambda_{\mathfrak{p}}, \Lambda_{\mathfrak{p}}))
\end{array}$$

is an isomorphism of approximation squares of form rings.

The proof of Lemma 7.22 is similar to that of Lemma 7.21. Details will be left to the reader.

C. *Conductor and related squares*

In this section, we discuss a discrete analogy to an approximation square. Namely, we consider a fibre product square of rings

$$\begin{array}{ccc}
A & \longrightarrow & B \\
\downarrow & & \downarrow \\
\hat{A} & \longrightarrow & \hat{B}
\end{array}$$

such that the vertical maps are surjective. It turns out that with an additional condition, one can make the same assertions as for approximation squares; namely, that the associated squares of general linear groups and general quadratic groups are S-exact, S-surjective, fibre product squares of groups. We prepare now for this result.

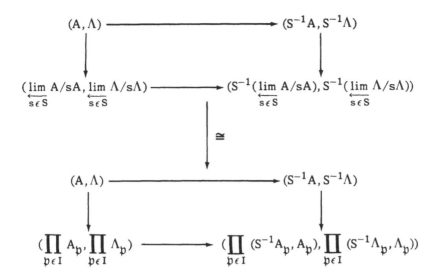

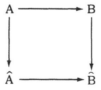

DEFINITION 7.23. Let

$$\begin{array}{ccc} A & \xrightarrow{f_2} & A_2 \\ {\scriptstyle f_1}\downarrow & & \downarrow{\scriptstyle f_2'} \\ A_1 & \xrightarrow[f_1']{} & \hat{A} \end{array}$$

be a commutative square of sets, groups, rings, topological groups, topological rings, etc. The square is called *exact* if given $a_i \in A_i$ $(i=1,2)$ such that $f_1'(a_1) = f_2'(a_2)$, there is an $a \in A$ such that $f_i(a) = a_i$ $(i=1,2)$. The square is called *surjective* if given $\hat{a} \in \hat{A}$, there are $a_i \in A_i$ $(i=1,2)$ such that $\hat{a} = f_1'(a_1)f_2'(a_2)$.

DEFINITION 7.24. A commutative square

$$(1) \qquad \begin{array}{ccc} A & \xrightarrow{f_2} & A_2 \\ {\scriptstyle f_1}\downarrow & & \downarrow{\scriptstyle f_2'} \\ A_1 & \xrightarrow[f_1']{} & \hat{A} \end{array}$$

of sets, groups, rings, topological groups, topological rings, etc., is called a *cofibre product square* if it solves the following

UNIVERSAL PROBLEM: Given a commutative diagram

$$\begin{array}{ccc} A & \xrightarrow{f_2} & A_2 \\ {\scriptstyle f_1}\downarrow & & \downarrow{\scriptstyle g_2} \\ A_1 & \xrightarrow[g_1]{} & \hat{B} \end{array}$$

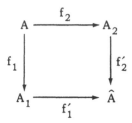

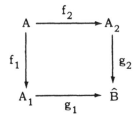

there is a *unique* map $f : \hat{A} \to \hat{B}$ such that $ff_i' = g_i$ $(i=1,2)$. If (1) is a cofibre product square then $\hat{A}$ is called the *cofibre product* of the following diagram

(2)
$$\begin{array}{ccc} A & \xrightarrow{f_2} & A_2 \\ {\scriptstyle f_1}\downarrow & & \\ A_1 & & \end{array}$$

Using the usual arguments, cf. 5.10, one can show that if a cofibre product exists then it is unique up to isomorphism. However, it is not true, as for fibre products, that each diagram (2) has a cofibre product.

LEMMA 7.25. *Let (7.24)(2) be a diagram of groups. If* f_1 *is surjective then the diagram has a cofibre product. Furthermore, if (7.24)(1) is a cofibre product diagram and* f_1 *is surjective then (7.24)(1) is surjective and it is exact* $\Longleftrightarrow f_2(\ker f_1)$ *is normal in* A_2.

Proof. Let C = normal closure of $f_2(\ker f_1)$ in A_2. Considering A, A_i $(i=1,2)$, and C as sets, we define $A_1 \times_{AC} A_2 = A_1 \times A_2 / \{(a_1, a_2) = (a_1 f_1(a), f_2(a)^{-1} c a_2) | a \in A, c \in C\}$. Define a binary operation on $A_1 \times_{AC} A_2$ by $(a_1, a_2)(b_1, b_2) = (a_1, a_2 f_2(b) b_2)$ where b is any element of A such that $f_1(b) = b_1$. One checks routinely that this operation is well defined and makes $A_1 \times_{AC} A_2$ into a group such that the maps $f_1' : A_1 \to A_1 \times_{AC} A_2$, $a_1 \mapsto (a_1, 1)$, and $f_2' : A_2 \to A_1 \times_{AC} A_2$, $a_2 \mapsto (1, a_2)$, are group homomorphisms. It is straightforward to check that the square

$$\begin{array}{ccc} A & \xrightarrow{f_2} & A_2 \\ {\scriptstyle f_1}\downarrow & & \downarrow{\scriptstyle f_2'} \\ A_1 & \xrightarrow[f_1']{} & A_1 \times_{AC} A_2 \end{array}$$

is a cofibre product square. The square is clearly surjective, and if

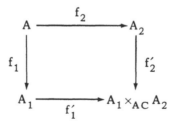

$C = f_2(\ker f_1)$ then it is clearly exact. On the other hand, if $C \neq f_2(\ker f_1)$, choose $c \in C - f_2(\ker f_1)$. Although $f_2'(c) = 1 = f_1'(1)$, there is no $a \in A$ such that $f_2(a) = c$ and $f_1(a) = 1$. Thus, the square is not exact.

THEOREM 7.26. a) *If*

$$\begin{array}{ccc} A & \xrightarrow{g'} & B \\ {\scriptstyle f'}\downarrow & & \downarrow{\scriptstyle g} \\ \hat{A} & \xrightarrow[f]{} & \hat{B} \end{array}$$

is a fibre product square of rings such that f' *and* g *are surjective and such that* $\mathrm{St}(g')(\ker(\mathrm{St}(A) \to \mathrm{St}(\hat{A}))$ *is normal in* $\mathrm{St}(B)$ *then the square*

$$\begin{array}{ccc} GL(A) & \longrightarrow & GL(B) \\ \downarrow & & \downarrow \\ GL(\hat{A}) & \longrightarrow & GL(\hat{B}) \end{array}$$

is an S*-exact,* S*-surjective, fibre product square of groups.*

b) *If*

$$\begin{array}{ccc} (A,\Lambda) & \xrightarrow{g'} & (B,\Gamma) \\ {\scriptstyle f'}\downarrow & & \downarrow{\scriptstyle g} \\ (\hat{A},\hat{\Lambda}) & \xrightarrow[f]{} & (\hat{B},\hat{\Gamma}) \end{array}$$

is a fibre product square of form rings such that f', $f'|_{\Lambda} : \Lambda \to \hat{\Lambda}$, g, *and* $g|_{\Gamma} : \Gamma \to \hat{\Gamma}$ *are surjective and such that* $\mathrm{StQ}(g')(\ker(\mathrm{StQ}(A,\Lambda) \to \mathrm{StQ}(\hat{A},\hat{\Lambda}))$ *is normal in* $\mathrm{StQ}(B,\Gamma)$ *then the square*

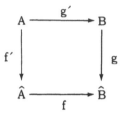

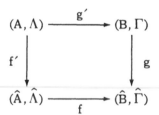

$$\begin{array}{ccc} GQ(A,\Lambda) & \longrightarrow & GQ(B,\Gamma) \\ \downarrow & & \downarrow \\ GQ(\hat{A},\hat{\Lambda}) & \longrightarrow & GQ(\hat{B},\hat{\Gamma}) \end{array}$$

is an S*-exact,* S*-surjective, fibre product square of groups.*

Proof. The proof for parts a) and b) is similar, so we shall give only the details for part a). It is clear that the square of general linear groups is a fibre product square. Let $s_{\alpha\beta}(d)$ denote a typical Steinberg generator. Consider the diagrams

(1)
$$\begin{array}{ccc} St(A) & \xrightarrow{St(g')} & St(B) \\ {\scriptstyle St(f')}\downarrow & & \downarrow{\scriptstyle St(g)} \\ St(\hat{A}) & \xrightarrow[St(f)]{} & St(\hat{B}) \end{array}$$

(2)
$$\begin{array}{ccc} St(A) & \xrightarrow{St(g')} & St(B) \\ {\scriptstyle St(f')}\downarrow & & \\ St(\hat{A}) & & \end{array}$$

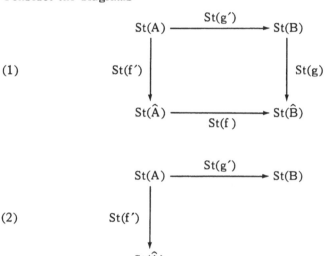

Since f' is surjective, it follows that $St(f')$ is surjective. Thus, by Lemma 7.25, the diagram (2) has a cofibre product. Let $T = St(g')(\ker St(f'))$ and construct the cofibre product $St(\hat{A}) \times_{St(A)T} St(B)$ as in the proof of Lemma 7.25. Define $h : St(\hat{A}) \to St(\hat{A}) \times_{St(A)T} St(B)$, $s_{\alpha\beta}(\hat{a}) \mapsto [s_{\alpha\beta}(\hat{a}), 1]$, and $k : St(B) \to St(\hat{A}) \times_{St(A)T} St(B)$, $s_{\alpha\beta}(b) \mapsto [1, s_{\alpha\beta}(b)]$, as in the proof of 7.25. By universality, there is a homomorphism $\phi : St(\hat{A}) \times_{St(A)T} St(B) \to St(\hat{B})$ such that $\phi St(g) = h$ and $\phi St(f) = k$. Thus, $\phi[s_{\alpha\beta}(\hat{a}), 1] = St(f)(s_{\alpha\beta}(\hat{a}))$ and $\phi[1, s_{\alpha\beta}(b)] = St(g)(s_{\alpha\beta}(b))$. Since g is surjective, it follows that St(g) is surjective, and so ϕ is also surjective. Let $\pi : St(\hat{B}) \to E(\hat{B})$ denote the canonical homomorphism to the elementary group $E(\hat{B})$. Recall from either Milnor [20] or Swan [29.1], the proof that π is a universal, perfect, central covering. One can adapt this proof easily to the map $\phi\pi$ and prove that $\phi\pi$ is a universal, perfect, central covering. Thus, ϕ must be an isomorphism. Thus, by Lemma 7.25, the square (1) is exact, surjective. But this implies by definition that the square of general linear groups is S-exact, S-surjective.

REMARK 7.27. a) *Below are two examples of fibre product squares which satisfy the hypotheses of Theorem 6.26 a).*

(i) Let A be a subring of a finite product $B = \prod_i A$ of rings A_i and let $\mathfrak{q}$ be a two-sided ideal of A and B. If each composite $A \to B \to A_i$ is surjective then the square

$$\begin{array}{ccc} A & \longrightarrow & B \\ \downarrow & & \downarrow \\ A/\mathfrak{q} & \longrightarrow & B/\mathfrak{q} \end{array}$$

satisfies the hypotheses of 6.26 a).

(ii) Let B be a commutative ring and let $\mathfrak{q}$ be an ideal contained in the Jacobson radical [10, III §2] of B. Let C be a subring of $B/\mathfrak{q}$ and let A be its inverse image in B. Then

$$\begin{array}{ccc} A & \longrightarrow & B \\ \downarrow & & \downarrow \\ C & \longrightarrow & B/\mathfrak{q} \end{array}$$

satisfies the hypotheses of 6.26 a).

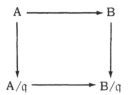

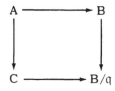

b) *Below are two examples of fibre squares which satisfy the hypotheses of Theorem 6.26 b).*

(i) The analogy of (i) above for form rings.

(ii) The analogy of (ii) above for form rings.

To prove (7.27) a) (i), one can adapt to the Steinberg group the elementary matrix group technics of H. Bass [10, IX (5.8)]. The analogous technics in the quadratic Steinberg group can be used to prove (7.27)b) (i).

D. *Fibre product categories*

In the previous sections, it was shown that for certain fibre product squares of rings, the corresponding square of general linear groups is S-surjective. And by Lemma 5.22, we know that S-surjective implies E-surjective. By a result [10, IX (5.1)] of Bass and Milnor, one knows that if the square of general linear groups is E-surjective then the corresponding square of categories of finitely generated, projective modules is a fibre product square. In this section, we start with a fibre product square of form rings such that the corresponding square of categories of finitely generated, projective modules is a fibre product square. Then we show that the corresponding squares of categories defined by nonsingular quadratic, hermitian, based quadratic and based hermitian forms are fibre product squares. The key to our results is the following lemma.

Call a functor $F : \mathcal{C}' \to \mathcal{C}$ *forgetful* if it is faithful (cf. [10, I]) and if it satisfies the condition that if c' is a morphism in $\mathcal{C}'$ such that $F(c')$ is an isomorphism then c' is an isomorphism.

LEMMA 7.28 (Preservation of extra structure.) *Let*

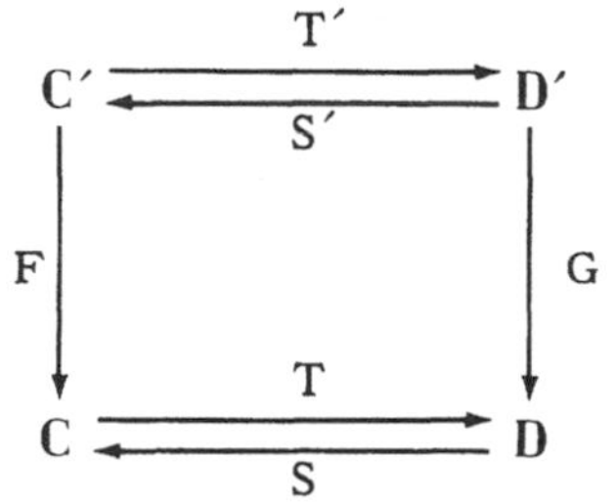

be a commutative diagram of categories and functors. Suppose that F *and* G *are faithful and that* T , S *forms an adjoint pair (cf.* [*10, I* §7]). *If* $\gamma = \gamma_{M,N} : \mathrm{Hom}_{\mathbf{C}}(M, SN) \to \mathrm{Hom}_{\mathbf{D}}(TM, N)$ *is the natural isomorphism which defines the adjoint pair, suppose that* $\gamma_{FM',GN'}(\mathrm{Hom}_{\mathbf{C}}(FM', FS'N')) = \mathrm{Hom}_{\mathbf{D}}(GT'M', GN')$. *Then* F *(resp.* G) *forgetful implies that* T′, S′ *forms an adjoint pair such that if the canonical morphism* $\alpha : 1 \to ST$ *(resp.* $\beta : TS \to 1$) *is an isomorphism then the canonical morphism* $\alpha' : 1 \to S'T'$ *(resp.* $\beta' : T'S' \to 1$) *is an isomorphism. In particular, if* F *and* G *are forgetful then the condition that* T *and* S *form adjoint, inverse equivalences implies the condition that* T′ *and* S′ *form adjoint, inverse equivalences.*

Proof. The composite $\mathrm{Hom}_{\mathbf{C}'}(M', SN') \xrightarrow{F_{M',SN'}} \mathrm{Hom}_{\mathbf{C}}(FM', FSN') \xrightarrow{\gamma_{FM',GN'}} \mathrm{Hom}_{\mathbf{D}}(GT'M', GN')$ has as its image $X = \text{image } G_{T'M',N'} : \mathrm{Hom}_{\mathbf{D}'}(T'M', N') \to \mathrm{Hom}_{\mathbf{D}}(GT'M', GN')$. The composite $\mathrm{Hom}_{\mathbf{C}'}(M', SN') \xrightarrow{\gamma_{FM',GN'} F_{M',SN'}} X \xrightarrow{G^{-1}_{T'M',N'}} \mathrm{Hom}_{\mathbf{D}'}(T'M', N')$ defines a natural isomorphism $\gamma' = \gamma_{M',N'} : \mathrm{Hom}_{\mathbf{C}'}(M', S'N') \to \mathrm{Hom}_{\mathbf{D}'}(T'M', N')$ which makes T′, S′ an adjoint pair. By definition, $\alpha_M = \gamma^{-1}_{M,TM}(1_{TM}) : M \to STM$ and $\alpha'_{M'} = \gamma'^{-1}_{M',T'M'}(1_{T'M'}) : M' \to S'T'(M')$. Hence, $F(\alpha'_{M'}) = \gamma^{-1}_{FM',GTM'}(1_{GTM'}) = \alpha_{FM'}$. Hence, if F is forgetful then $\alpha_{FM'}$ an isomorphism implies $\alpha'_{M'}$ an isomorphism. Similarly, if G is forgetful then $\beta_{GN'}$ an isomorphism implies $\beta'_{N'}$ an isomorphism.

For the sake of completeness, we quote the following result of Bass and Milnor.

THEOREM 7.29 (Bass-Milnor). *Let*

$$\begin{array}{ccc} A & \longrightarrow & A_2 \\ \downarrow & & \downarrow f_2 \\ A_1 & \xrightarrow[f_1]{} & \hat{A} \end{array}$$

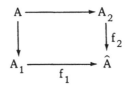

be a fibre product square of rings. If the corresponding square of general linear groups is E*-surjective then the corresponding square of categories of finitely generated, projective modules is a fibre product square.*

Theorem 7.29 is proved in [10, IX (5.1)] under the assumption that one of the homomorphisms $A_i \to \hat{A}$ $(i=1,2)$ is surjective. However, the proof is constructed so that one needs only a weaker assumption implied by the one above, namely the assumption that the corresponding square of general linear groups is E-surjective. Furthermore, although the notion of E-surjectivity in [10] is stronger than ours (cf. 5.20), it turns out that our weaker notion suffices for the applications required in the proof of [10, IX (5.1)].

THEOREM 7.30. *Let*

$$\begin{array}{ccc} (A,\Lambda) & \longrightarrow & (A_2,\Lambda_2) \\ \downarrow & & \downarrow \\ (A_1,\Lambda_1) & \longrightarrow & (\hat{A},\hat{\Lambda}) \end{array}$$

be a fibre product square of form rings such that the corresponding square of categories of finitely generated, projective modules is a fibre product square.

a) *If* $F = Q$ *or* H *then the square*

$$\begin{array}{ccc} F(A,\Lambda) & \longrightarrow & F(A_2,\Lambda_2) \\ \downarrow & & \downarrow \\ F(A_1,\Lambda_2) & \longrightarrow & F(\hat{A},\hat{\Lambda}) \end{array}$$

is a fibre product square.

b) *Let* $X \subset P(A)$, $X_i \subset P(A_i)$ $(i=1,2)$, *and* $\hat{X} \subset P(\hat{A})$ *be full subcategories closed under direct sums and dualization and containing the free module of rank* 1. *Let* X *denote the subgroup of* $K_0(A)$ *generated by* X. *Define* X_i $(i=1,2)$ *and* $\hat{X}$ *similarly. Let* $F = Q$ *or* H. *If the square*

$$\begin{array}{ccc} X & \longrightarrow & X_2 \\ \downarrow & & \downarrow \\ X_1 & \longrightarrow & \hat{X} \end{array}$$

is a fibre product square then so is the square

$$\begin{array}{ccc} F(A,\Lambda)_X & \longrightarrow & F(A_2,\Lambda_2)_{X_2} \\ \downarrow & & \downarrow \\ F(A_1,\Lambda_1)_{X_1} & \longrightarrow & F(\hat{A},\hat{\Lambda})_{\hat{X}} \end{array} .$$

c) *Let* $Y \subset K_1(A), Y_i \subset K_1(A_i)$ $(i=1,2)$, *and* $\hat{Y} \subset K_1(\hat{A})$ *be subgroups closed under the involution and containing the classes of the invertible elements* -1 *and* $-\lambda$ (Λ *is defined with respect to* λ). *Let* $Y_i(n) = \ker(GL_n(A_i) \to K_1(A_i)/Y_i)$ $(i=1,2)$ *and* $\hat{Y}(n) = \ker(GL_n(\hat{A}) \to K_1(\hat{A})/\hat{Y})$. *Let* $F = Q$ *or* H. *If the sequence*

$$Y \to Y_1 \oplus Y_2 \to \hat{Y}$$

is exact and if each $\sigma \in \hat{Y}(n)$ *is a product* $\sigma = \hat{\sigma}_2\hat{\sigma}_1$ *such that* $\hat{\sigma}_i \in \mathrm{image}(Y_i(n) \to \hat{Y}(n))$ $(i=1,2)$ *then the square*

$$\begin{array}{ccc} F(A,\Lambda)_{\text{based-}Y} & \longrightarrow & F(A_2,\Lambda_2)_{\text{based-}Y_2} \\ \downarrow & & \downarrow \\ F(A_1,\Lambda_1)_{\text{based-}Y_1} & \longrightarrow & F(\hat{A},\hat{\Lambda})_{\text{based-}\hat{Y}} \end{array}$$

is a fibre product square.

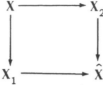

Proof. a) - b). Since the proofs of the different cases in a) and b) are similar to one another, we shall prove only the case $Q(A,\Lambda)$. The assertion will be proved by using Lemma 7.28 to compare quadratic modules with projective modules. Let $X = P(A_1)\times_{P(\hat{A})}P(A_2)$ and $Y = Q(A_1,\Lambda_1)\times_{Q(\hat{A},\hat{\Lambda})}Q(A_2,\Lambda_2)$. Identify (A,Λ) with its fibre product construction 5.10 and identify X and Y with their respective fibre product constructions 6.5. Define $T: P(A) \to X, N \mapsto (N\otimes_A A_1, 1_{N\otimes_A \hat{A}}, N\otimes_A A_2)$, and $S: X \to P(A), M = (M_1,\sigma,M_2) \mapsto \{(m_1,m_2) | m_i \in M_i, \sigma(m_1\otimes_{A_1} 1_{\hat{A}}) = m_2 \otimes_{A_2} 1_{\hat{A}}\}$. By our hypotheses, T, S form adjoint inverse equivalences (cf. proof of [10, IX (5.1)]). Define $T': Q(A,\Lambda) \to Y, (N,C) \mapsto ((N_1\otimes_A A_1, C), 1_{N_1\otimes_A \hat{A}}, (N_2\otimes_A A_2, C))$, and $S': Y \to Q(A,\Lambda), ((M_1,B_1),\sigma,(M_2,B_2)) \mapsto (S(M_1,\sigma,M_2), S(B_1,B_2))$ where $S(B_1,B_2)((m_1,m_2),(m_1',m_2')) = (B_1(m_1,m_1'), B_2(m_2,m_2'))$. We must show $(S(M_1,\sigma,M_2), S(B_1,B_2))$ is nonsingular. Let $d: S(M_1,\sigma,M_2) \to S(M_1,\sigma,M_2)^*, (m_1,m_2) \mapsto \langle (m_1,m_2), \;\rangle_{S(B_1,B_2)}$. Suppose $d(m_1,m_2) = 0$. Then $\langle m_1, m_1'\rangle_{B_1} = \langle m_2, m_2'\rangle_{B_2} = 0$ for all $(m_1',m_2') \in S(M_1,\sigma,M_2)$. If P_i = projection of $S(M_1,\sigma,M_2)$ on M_i $(i=1,2)$ then P_i generates M_i over A_i. Hence, $\langle m_i, m_i'\rangle_{B_i} = 0$ for all $m_i' \in M_i$. Now the fact that (M_i,B_i) is non-singular implies that $m_i = 0$. Therefore d is injective. If $g \in S(M_1,\sigma,M_2)^*$ then g determines elements $g_i \in M_i^*$. The non-singularity of (M_i,B_i) implies that we can find $m_i \in M_i$ such that $g_i = \langle m_i, \;\rangle_{B_i}$. If (m_1,m_2) were in $S(M_1,\sigma,M_2)$ then $d(m_1,m_2) = g$ and we could conclude that d is surjective. But $g_1\otimes_{A_1} 1_{\hat{A}} = (g_2\otimes_{A_2} 1_{\hat{A}})\sigma$ implies $\sigma(m_1\otimes_{A_1} 1_{\hat{A}}) = m_2\otimes_{A_2} 1_{\hat{A}}$, because $\langle\;,\;\rangle_{B' = B_1\otimes_{A_1}\hat{A}}$ is non-singular. Hence $(m_1,m_2) \in S(M_1,\sigma,M_2)$.

We want to show now that S' and T' form adjoint, inverse equivalences. To do this we shall show that the conditions in Lemma 7.28 are satisfied. The natural isomorphism which defines the adjoint pair T, S is the composite of the standard identities $\mathrm{Hom}_A(N, SM) = \mathrm{Hom}_{A_1}(N\otimes_A A_1, M_1) \times_{\mathrm{Hom}_A(N\otimes_A \hat{A}, M_2\otimes_{A_2}\hat{A})} \mathrm{Hom}_{A_2}(N\otimes_A A_2, M_2) =$

$\{(h_1, h_2) \mid h_i \in \mathrm{Hom}_{A_i}(N \otimes_A A_i, M_i),\ \sigma(h_1 \otimes_{A_1} 1_{\hat{A}}) = h_2 \otimes_{A_2} 1_{\hat{A}}\} =$ $\mathrm{Hom}_X(TN, M)$. We have noted above already that T, S form adjoint, inverse equivalences. If $F : Q(A, \Lambda) \to P(A)$, $(N, C) \mapsto N$, and $G : Y \to X$, $((M_1, B_1), \sigma, (M_2, B_2)) \mapsto (M_1, \sigma, M_2)$ then all the conditions of 7.28 are satisfied.

c) We could prove part c) as we did parts a) and b), by using Lemma 7.28 to compare based quadratic and hermitian modules with finitely generated, projective modules. However, to lessen the impact of the basings, we shall insert an introductory step into the proof. Namely, we shall show that the square of categories of based projective modules is a fibre product square. The rest of the proof, namely to compare via Lemma 7.28 based quadratic and based hermitian modules with based projective modules, is accomplished exactly like its counterpart in parts a) and b) and will be left to the reader.

Define the category with product

$$P(A)_{\text{based-}Y}$$

of *based*-Y *projective modules* (or simply *based*-Y *modules*) as follows: Its objects are pairs $(M, \{e_1, \cdots, e_m\})$ consisting of a free A-module M and a preferred basis $\{e_1, \cdots, e_m\}$ for M of finite rank m. A morphism between objects is an A-linear isomorphism f such that the matrix obtained for f by using the preferred basis vanishes in $K_1(A)/Y$. The product is defined by $(M, \{e_1, \cdots, e_m\}) \perp (N, \{f_1, \cdots, f_n\}) =$ $(M \oplus N, \{e_1, \cdots, e_m, f_1, \cdots, f_n\})$.

We want to prove now that the square

$$\begin{array}{ccc} P(A)_{\text{based-}Y} & \longrightarrow & P(A_2)_{\text{based-}Y_2} \\ \downarrow & & \downarrow \\ P(A_1)_{\text{based-}Y_1} & \longrightarrow & P(\hat{A})_{\text{based-}\hat{Y}} \end{array}$$

is a fibre product square. The proof will be analogous to the one in part

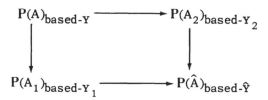

a) above. Identify A with its standard fibre product construction 5.10. Define T and S as in the proof of part a). Define $T' : P(A)_{\text{based-}Y} \to P(A_1)_{\text{based-}Y_1} \times_{P(\hat{A})_{\text{based-}\hat{Y}}} P(A_2)_{\text{based-}Y_2}$, $N \mapsto (N \otimes_A A_1, 1_{N \otimes_A \hat{A}}, N \otimes_A A_2)$, such that if $e_1, \cdots, e_n$ is a preferred basis for N then $e_1 \otimes 1_{A_1}, \cdots, e_n \otimes 1_{A_1}$ is a preferred basis for $N \otimes_A A_1$ and $e_1 \otimes 1_{A_2}, \cdots, e_n \otimes 1_{A_2}$ is a preferred basis for $N \otimes_A A_2$. Define $S' : P(A_1)_{\text{based-}Y_1} \times_{P(\hat{A})_{\text{based-}\hat{Y}}} P(A_2)_{\text{based-}Y_2} \to P(A)_{\text{based-}Y}$, $M = (M_1, \sigma, M_2) \mapsto \{(m_1, m_2) | m_i \in M_i, \sigma(m_1 \otimes_{A_1} 1_{\hat{A}}) = m_2 \otimes_{A_2} 1_{\hat{A}}\}$. Now the problem is to show that the module on the right is free (over $A_1 \times_{\hat{A}} A_2 = A$) and to choose a preferred basis for it. Let $e_1, \cdots, e_n$ be a preferred basis for M_1 and $f_1, \cdots, f_n$ a preferred basis for M_2. Let $\rho : M_1 \otimes_{A_1} \hat{A} \to M_2 \otimes_{A_2} \hat{A}$ denote the homomorphism whose matrix is the identity matrix. Then $S'(M_1, \rho, M_2)$ is a free module with basis $(e_1, f_1), \cdots, (e_n, f_n)$. By hypothesis, there are $\sigma_i \in Y_i(n)$ $(i = 1, 2)$ such that $\sigma = \hat{\sigma}_2 \hat{\sigma}_1$. The map $(\sigma_1, \sigma_2^{-1}) : M_1 \times M_2 \to M_1 \times M_2$ induces an isomorphism $S'(M_1, \sigma, M_2) \to S'(M_1, \rho, M_2)$ and we choose $(\sigma_1^{-1} e_1, \sigma_2 f_1), \cdots, (\sigma_1^{-1} e_n, \sigma_2 f_n)$ as our preferred basis for $S'(M_1, \sigma, M_2)$. If F and G denote respectively the forgetful functors $P(A)_{\text{based-}Y} \to P(A)$ and $P(A_1)_{\text{based-}Y_1} \times_{P(\hat{A})_{\text{based-}\hat{Y}}} P(A_2)_{\text{based-}Y_2} \to P(A_1) \times_{P(\hat{A})} P(A_2)$ then the conditions of 7.28 are satisfied. Hence, the square above is a fibre product square.

E. *Restricted direct products*

In this section, we show how the functors K_i and KQ_i $(i = 0, 1)$ behave with respect to restricted direct products of orders and semisimple algebras. The definition of restricted direct product which we require is given in 7.20.

If C is a ring, we let C^{op} denote its *opposite* ring; thus, C^{op} has the same elements and addition as C, but its multiplication is defined by $C \times C \to C$, $(c, c') \mapsto c'c$, where $c'c$ is the product in C. A ring C with involution is called *hyperbolic*, if it is isomorphic to a ring $D \times D^{op}$ with involution $(d_1, d_2) \mapsto (d_2, d_1)$.

The following notation is adopted for the rest of the section.

R = Dedekind ring with finite residue fields

I = a set of maximal ideals of R

S = multiplicative set in R

K = field of fractions of R

B = finite, separable, semisimple K-algebra

A = R-order on B (cf. [29])

$A_{\mathfrak{p}} = \varprojlim_i A/\mathfrak{p}^i A \quad (\mathfrak{p} \in I)$.

If B has an involution which leaves A invariant and the elements of R fixed then

Λ = form parameter on A

$\Lambda_{\mathfrak{p}}$ = closure of Λ in the $\mathfrak{p}$-adic topology on $A_{\mathfrak{p}}$

$$S^{-1}\Lambda = \left\{ \frac{x}{s^2} \,|\, x \in \Lambda, s \in S \right\}$$

$$S^{-1}\Lambda_{\mathfrak{p}} = \left\{ \frac{x}{s^2} \,|\, x \in \Lambda_{\mathfrak{p}}, s \in S \right\} .$$

THEOREM 7.31. *The canonical map below is an isomorphism*

$$K_1\Big(\coprod_{\mathfrak{p} \in I} (S^{-1}A_{\mathfrak{p}}, A_{\mathfrak{p}})\Big) \to \coprod_{\mathfrak{p} \in I} (K_1(S^{-1}A_{\mathfrak{p}}), K_1(A_{\mathfrak{p}})) .$$

THEOREM 7.32. *The canonical map below is an isomorphism*

$$KQ_1\Big(\coprod_{\mathfrak{p} \in I} ((S^{-1}A_{\mathfrak{p}}, S^{-1}\Lambda_{\mathfrak{p}}), (A_{\mathfrak{p}}, \Lambda_{\mathfrak{p}}))\Big) \to \coprod_{\mathfrak{p} \in I} ((KQ_1(S^{-1}A_{\mathfrak{p}}, S^{-1}\Lambda_{\mathfrak{p}}), KQ_1(A_{\mathfrak{p}}, \Lambda_{\mathfrak{p}})) .$$

LEMMA 7.33. *If* $S = 1$ *then in the theorems above* $\coprod_{\mathfrak{p}} = \prod_{\mathfrak{p}}$.

The proof of the lemma is trivial.

Proof of Theorem 7.31. If C is a semilocal ring then by a result of H. Bass [10, V (9.1)], the canonical map $GL_1(C) \to K_1(C)$ is surjective. Consider the commutative diagram

$$\begin{array}{ccc}
GL_1(\prod(S^{-1}A_{\mathfrak{p}}, A_{\mathfrak{p}})) & \longrightarrow & \prod(GL_1(S^{-1}A_{\mathfrak{p}}), GL_1(A_{\mathfrak{p}})) \\
\downarrow & & \downarrow \\
K_1(\prod(S^{-1}A_{\mathfrak{p}}, A_{\mathfrak{p}})) & \longrightarrow & \prod(K_1(S^{-1}A_{\mathfrak{p}}), K_1(A_{\mathfrak{p}})) .
\end{array}$$

The top map is clearly an isomorphism and since $A_{\mathfrak{p}}$ and $S^{-1}A_{\mathfrak{p}}$ are semilocal, it follows from the opening remark that the right map is surjective. Thus, the bottom map is surjective.

Next, we establish that the bottom map is injective. From the definition of the restricted direct product, it follows that it is sufficient to show that for some finite subset $F \subset I$, the map $K_1(\prod_{\mathfrak{p} \in I\text{-}F} A_{\mathfrak{p}}) \to \prod_{\mathfrak{p} \in I\text{-}F} K_1(A_{\mathfrak{p}})$ is injective. From [29, 5.1 and 5.29] and [10, III 8.2], one can deduce that $A_{\mathfrak{p}}$ is a maximal order on $B_{\mathfrak{p}} = K \otimes_R A_{\mathfrak{p}}$ for almost all $\mathfrak{p}$. Furthermore, by a well-known result of Deuring, $B_{\mathfrak{p}}$ is a product of matrix rings over fields for almost all $\mathfrak{p}$. Deuring's result is the only point in the proof where the assumption that the residue fields of R are finite is used. Let F be a finite subset of I such that if $\mathfrak{p} \in I\text{-}F$ then $B_{\mathfrak{p}} = \prod_j M_{n_{\mathfrak{p}},j}(L_{\mathfrak{p},j})$ is a product of matrix rings $M_{n_{\mathfrak{p}},j}(L_{\mathfrak{p},j})$ of rank $n_{\mathfrak{p}}$ over complete discrete valued fields $L_{\mathfrak{p},j}$, and $A_{\mathfrak{p}}$ is a maximal order on $B_{\mathfrak{p}}$. Let $S_{\mathfrak{p},j}$ denote the valuation ring of $L_{\mathfrak{p},j}$. Since each $S_{\mathfrak{p},j}$ is a discrete valuation ring, it follows from [29, 9.12] that all maximal orders on $B_{\mathfrak{p}}$ are isomorphic (by an inner automorphism of $B_{\mathfrak{p}}$). Thus, one can assume that $A_{\mathfrak{p}} = \prod_j M_{n_{\mathfrak{p}},j}(S_{\mathfrak{p},j})$. If $S_{\mathfrak{p}} = \prod_j S_{\mathfrak{p},j}$ then the canonical map $K_1(A_{\mathfrak{p}}) \to K_1(S_{\mathfrak{p}})$ is an isomorphism. Furthermore, since the set $\{n_{\mathfrak{p},j} | \mathfrak{p} \in I\text{-}F\}$ is bounded (by the number of elements required to generate A as an R-module), the canonical map $K_1(\prod A_{\mathfrak{p}}) \to K_1(\prod S_{\mathfrak{p}})$ is also an isomorphism. Thus, it suffices to show that the map $K_1(\prod S_{\mathfrak{p}}) \to \prod K_1(S_{\mathfrak{p}})$ is injective. By [10, V 3.4 and 4.2], any element of $K_1(\prod S_{\mathfrak{p}})$ can be represented by an element $x \in GL_1(\prod S_{\mathfrak{p}})$. Write $x = \prod x_{\mathfrak{p}}$ such that $x_{\mathfrak{p}} \in GL_1(S_{\mathfrak{p}})$. The image of x in $\prod K_1(S_{\mathfrak{p}})$ is trivial $\Longleftrightarrow$ the image of each $x_{\mathfrak{p}}$ in $K_1(S_{\mathfrak{p}})$ is trivial $\Longleftrightarrow$ [10, V 9.6] each $x_{\mathfrak{p}} = 1$.

Proof of Theorem 7.32. If (C,Ω) is a form ring such that C is semilocal then by [1] (cf. [11]), the map $GQ_2(C,\Omega) \to KQ_1(C,\Omega)$ is surjective. Using this fact, one can deduce as in the proof of the previous theorem that the map in the current theorem is surjective.

Next, we establish that the map in the theorem is injective. From the definition of the restricted direct product, it follows that it is sufficient to show that for some finite subset $F \subset I$, the map $KQ_1(\prod_{\mathfrak{p} \notin F} (A_{\mathfrak{p}}, \Lambda_{\mathfrak{p}})) \to \prod_{\mathfrak{p} \notin F} KQ_1(A_{\mathfrak{p}}, \Lambda_{\mathfrak{p}})$ is injective. We shall pick F conveniently later in the proof. Let $WH_0(A,\Lambda)_{\text{d-based-}K_1(A)}$ denote the functor $WH_0(A,\Lambda)_{\text{discr-based-}K_1(A)}$ defined in §1C. If G is an abelian group with a $\mathbf{Z}/2\mathbf{Z}$-action $g \mapsto \bar{g}$, let $N(G) = \{g\bar{g} \mid g \in G\}$. Consider the commutative diagram

$$\begin{array}{ccc}
WH_0(\prod_{\mathfrak{p} \notin F} (A_{\mathfrak{p}},\Lambda_{\mathfrak{p}}))_{\text{d-based-}K_1(\prod A_{\mathfrak{p}})} & \xrightarrow{\ f\ } & \prod_{\mathfrak{p} \notin F} WH_0(A_{\mathfrak{p}},\Lambda_{\mathfrak{p}})_{\text{d-based-}K_1(A_{\mathfrak{p}})} \\
\Big\downarrow \text{discr} & & \Big\downarrow \text{discr} \\
K_1(\prod_{\mathfrak{p} \notin F} A_{\mathfrak{p}})/N(K_1(\prod_{\mathfrak{p} \notin F} A_{\mathfrak{p}})) & \xrightarrow[g]{\ \simeq\ } & \prod_{\mathfrak{p} \notin F} K_1(A_{\mathfrak{p}})/N(K_1(A_{\mathfrak{p}})) \\
\Big\downarrow H & & \Big\downarrow H \\
KQ_1(\prod_{\mathfrak{p} \notin F} (A_{\mathfrak{p}},\Lambda_{\mathfrak{p}})) & \xrightarrow[h]{\qquad} & \prod_{\mathfrak{p} \notin F} KQ_1(A_{\mathfrak{p}},\Lambda_{\mathfrak{p}}) .
\end{array}$$

The columns are exact by Theorem 8.19 a) and g is an isomorphism by Theorem 7.31. Suppose that we know that

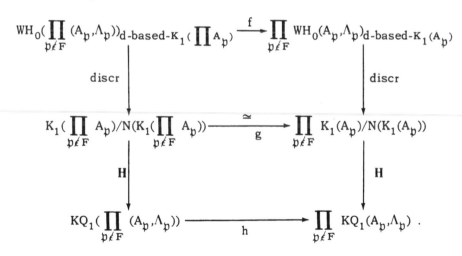

(i) f is surjective,

(ii) $\ker h \subset \text{image } \mathbf{H} : K_1(\prod_{\mathfrak{p} \notin F} (A_{\mathfrak{p}})) \to KQ_1(\prod_{\mathfrak{p} \notin F} (A_{\mathfrak{p}}, \Lambda_{\mathfrak{p}}))$.

Then, by chasing the diagram above, one deduces that h is injective.

We prove (i). It suffices to show that for all $\mathfrak{p} \notin F$, every element of $WH_0(A_{\mathfrak{p}}, \Lambda_{\mathfrak{p}})_{d\text{-based-}K_1(A_{\mathfrak{p}})}$ can be represented by a $\Lambda_{\mathfrak{p}}$-hermitian module of rank 2. Let K denote the field of fractions of R and let $B = K \otimes_R A$. Pick F so that for all $\mathfrak{p} \notin F$, $B_{\mathfrak{p}}$ is a product of matrix rings over fields and $A_{\mathfrak{p}}$ is a maximal order on $B_{\mathfrak{p}}$. As in the proof of the previous theorem, we can identify $A_{\mathfrak{p}}$ with a product $A_{\mathfrak{p}} = \prod_j A_{\mathfrak{p},j}$ of involution invariant rings $A_{\mathfrak{p},j}$ such that $A_{\mathfrak{p},j}$ is either hyperbolic or a matrix ring $M_{n,j}(S_{\mathfrak{p},j})$ over the ring of integers $S_{\mathfrak{p},j}$ in a complete local field. By Lemma 11.6, the form parameter $\Lambda_{\mathfrak{p}}$ has a corresponding decomposition $\Lambda_{\mathfrak{p}} = \prod_j \Lambda_{\mathfrak{p},j}$. Thus, $WH_0(A_{\mathfrak{p}}, \Lambda_{\mathfrak{p}})_{d\text{-based-}K_1(A_{\mathfrak{p}})} = \prod_j WH_0(A_{\mathfrak{p},j}, \Lambda_{\mathfrak{p},j})_{d\text{-based-}K_1(A_{\mathfrak{p},j})}$. If $A_{\mathfrak{p},j}$ is hyperbolic then it is easy to show that $WH_0(A_{\mathfrak{p},j}, \Lambda_{\mathfrak{p},j}) = 0$. If $A_{\mathfrak{p},j} = M_{n,j}(S_{\mathfrak{p},j})$ then by the Morita Theorem 9.2, there is, for a suitable involution on $S_{\mathfrak{p},j}$ and for a suitable form parameter $\Gamma_{\mathfrak{p},j}$ on $S_{\mathfrak{p},j}$, an isomorphism $WH_0(A_{\mathfrak{p},j}, \Lambda_{\mathfrak{p},j})_{d\text{-based-}K_1(A_{\mathfrak{p},j})} \xrightarrow{\cong} WH_0(S_{\mathfrak{p},j}, \Gamma_{\mathfrak{p},j})_{d\text{-based-}K_1(S_{\mathfrak{p},j})}$. Let $L_{\mathfrak{p},j}$ denote the quotient field of $S_{\mathfrak{p},j}$. It is a routine exercise to show that $WH_0(S_{\mathfrak{p},j}, \Gamma_{\mathfrak{p},j})_{d\text{-based-}K_1(S_{\mathfrak{p}})} \to WH_0(L_{\mathfrak{p},j}, \max)_{d\text{-based-}K_1(L_{\mathfrak{p},j})}$ is injective. But every element of the latter group is determined by its Hasse-Witt invariant and by its discriminant with values in $H^2(\text{units}(L_{\mathfrak{p},j}))$ and the Hasse-Witt invariant vanishes on the former group. Thus, every element of the former group can be represented by a module of rank 2.

We prove (ii). F will play no role in the proof. Consider the commutative diagram

$$\begin{array}{ccc}
K_1(\prod_{\mathfrak{p}\in I} A_\mathfrak{p}) & & \\
\Big\downarrow{\scriptstyle H} & & \\
KQ_1(\prod_{\mathfrak{p}\in I}(A_\mathfrak{p},\Lambda_\mathfrak{p})) & \xrightarrow{\ h\ } & \prod_{\mathfrak{p}\in I} KQ_1(A_\mathfrak{p},\Lambda_\mathfrak{p}) \\
\Big\downarrow & & \Big\downarrow \\
WQ_1(\prod_{\mathfrak{p}\in I}(A_\mathfrak{p},\Lambda_\mathfrak{p})) & \xrightarrow[\ k\ ]{} & \prod_{\mathfrak{p}\in I} WQ_1(A_\mathfrak{p},\Lambda_\mathfrak{p})\,.
\end{array}$$

The left-hand column is exact by definition. Thus, it suffices to show that the map k is injective. Let $\mathfrak{q}_\mathfrak{p}$ denote the Jacobson radical $(A_\mathfrak{p})$. If $O_{\Pi\Lambda_\mathfrak{p}}$ denotes the number of involution invariant, simple components $M_n(K)$ of $\prod_{\mathfrak{p}\in I} A_\mathfrak{p}/\mathfrak{q}_\mathfrak{p}$ such that the categories $Q(M_n(K)$, image $(\Pi\Lambda_\mathfrak{p} \to M_n(K))$ and $Q(K,0)$ are Morita equivalent in the sense of §9, then there is a commutative diagram

$$\begin{array}{ccc}
WQ_1(\prod_{\mathfrak{p}\in I}(A_\mathfrak{p},\Lambda_\mathfrak{p})) & \xrightarrow{\ k\ } & \prod_{\mathfrak{p}\in I} WQ_1(A_\mathfrak{p},\Lambda_\mathfrak{p}) \\
\searrow{\scriptstyle \cong} & & \swarrow{\scriptstyle \cong} \\
& (Z/2Z)^{O_{\Pi\Lambda_\mathfrak{p}}} &
\end{array}$$

such that the bottom two maps are isomorphisms. Thus, k is injective.

THEOREM 7.34 a) *In the commutative square*

$$\begin{array}{ccc}
K_0(\prod_{\mathfrak{p}\in I}(S^{-1}A_\mathfrak{p},A_\mathfrak{p})) & \longrightarrow & \prod_{\mathfrak{p}\in I}(K_0(S^{-1}A_\mathfrak{p}),K_0(A_\mathfrak{p})) \\
\Big\downarrow & & \Big\downarrow \\
K_0(\prod_{\mathfrak{p}\in I}(S^{-1}A_\mathfrak{p})) & \xrightarrow[\ g\ ]{} & \prod_{\mathfrak{p}\in I}(K_0(S^{-1}A_\mathfrak{p}))
\end{array}$$

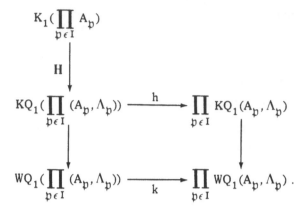

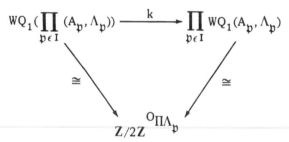

the vertical maps are isomorphisms and the horizontal maps are injective.

b) *In the analogous commutative square for* KQ_0, *the horizontal maps are injective.*

c) *If* A *is generated by* n *elements as an* R*-module then* $K_0(S^{-1}A_{\mathfrak{p}})$ *is a free group of rank* $r_{\mathfrak{p}} \le n$. *If one identifies* $K_0(S^{-1}A_{\mathfrak{p}}) = \mathbf{Z}_1 \times \cdots \times \mathbf{Z}_{r_{\mathfrak{p}}}$ *and denotes the coordinate decomposition of* $x_{\mathfrak{p}} \in K_0(S^{-1}A_{\mathfrak{p}})$ *by* $x_{\mathfrak{p}} = (x_{\mathfrak{p},1}, \cdots, x_{\mathfrak{p},r_{\mathfrak{p}}})$ *then an element* $x = \Pi x_{\mathfrak{p}}$ *lies in the image* $g \Longleftrightarrow$ *the set* $\left\{\sum_{i=1}^{r_{\mathfrak{p}}} |x_{\mathfrak{p},i}| \,\middle|\, \mathfrak{p} \in I\right\}$ *is bounded.*

The details of the proof of Theorem 7.34 are routine and will be left to the reader.

REMARK 7.35. The purpose of the remark is to comment on extensions of Theorems 7.31 and 7.32 to other rings. Let J be a set. Let $\{A_j \mid j \in J\}$ and $\{B_j \mid j \in J\}$ be families of rings such that there is a fixed homomorphism $A_j \to B_j$. Suppose that there is an integer n such that for all j the rings A_j and B_j satisfy the stability condition SR_n of H. Bass [10, V §3]. Let f denote the canonical map $f: K_1(\prod_{j \in J} (B_j, A_j)) \to \prod_{j \in J} (K_1(B_j), K_1(A_j))$. From the surjective stability theorem [10, V §4] for K_1, it follows that f is surjective. Suppose that there is an integer k such that for every $j \in J$, each element of $E_{n+2}(A_j)$ can be written as a product of fewer than k elementary matrices. Then it follows from the injective stability theorem [10, V §4] for K_1 that the map f is injective. Similar remarks hold for KQ_1.

F. *Orders*

In this section, we record a $K_2 - K_1$ exact sequence for orders over Hasse domains, which is well suited for making fine arithmetic computations.

THEOREM 7.36. a) *Let*

R = *Dedekind ring with finite residue class fields*

K = *field of fractions of* R

B = *finite, semisimple* K-*algebra*

A = R-*order* [29] *on* B

I = *set of all maximal ideals on* R

$A_{\mathfrak{p}}$ = *localization or completion of* A *with respect to* $\mathfrak{p} \in I$

$B_{\mathfrak{p}} = K \otimes_R A_{\mathfrak{p}}$

$\mathfrak{q}$ = *two-sided ideal of* A *such that* $K\mathfrak{q} = B$

$\mathfrak{q}_{\mathfrak{p}} = \mathfrak{q}A_{\mathfrak{p}}$.

Then there is an exact sequence

$$K_2(A,\mathfrak{q}) \longrightarrow K_2(B) \longrightarrow \coprod_{\mathfrak{p}\in I} \mathrm{coker}(K_2(A_{\mathfrak{p}},\mathfrak{q}_{\mathfrak{p}}) \longrightarrow K_2(B_{\mathfrak{p}})) \xrightarrow{\partial} K_1(A) \longrightarrow$$
$$K_1(B) \oplus \prod_{\mathfrak{p}\in I} K_1(A_{\mathfrak{p}}) \longrightarrow \coprod_{\mathfrak{p}\in I} (K_1(B_{\mathfrak{p}}), K_1(A_{\mathfrak{p}}))$$

where ∂ *is as in Theorem 6.32.*

b) *Adopt the notation in part a). Suppose further that* B *has an involution which leaves each element of* K *fixed and which leaves* A *and* $\mathfrak{q}$ *invariant. Let*

Λ = *form parameter on* A *such that* $R\Lambda \subset \Lambda$

$\Lambda_{\mathfrak{p}}$ = *localization or completion of* Λ *at* $\mathfrak{p}$.

Then there is an exact sequence

$$KQ_2(A,\Lambda,\mathfrak{q}) \longrightarrow KQ_2(B,K\Lambda) \longrightarrow \coprod_{\mathfrak{p}\in I} \mathrm{coker}(KQ_2(A_{\mathfrak{p}},\Lambda_{\mathfrak{p}},\mathfrak{q}_{\mathfrak{p}}) \longrightarrow KQ_2(B_{\mathfrak{p}},K\Lambda_{\mathfrak{p}})) \xrightarrow{\partial}$$
$$KQ_1(A,\Lambda,\mathfrak{q}) \longrightarrow KQ_1(B,K\Lambda) \oplus \prod_{\mathfrak{p}\in I} KQ_1(A_{\mathfrak{p}},\Lambda_{\mathfrak{p}},\mathfrak{q}_{\mathfrak{p}}) \longrightarrow$$
$$\coprod_{\mathfrak{p}\in I} (KQ_1(B_{\mathfrak{p}},K\Lambda_{\mathfrak{p}}), KQ_1(A_{\mathfrak{p}},\Lambda_{\mathfrak{p}},\mathfrak{q}_{\mathfrak{p}}))$$

where ∂ *is as in Theorem 6.32.*

REMARK 7.37. *Above, it is often useful to compare* A *with some over order* $\mathcal{O}$ *instead of with* B. *In fact, if one substitutes for* B *(resp.* $B_{\mathfrak{p}}, K\Lambda, \Lambda_{\mathfrak{p}}$*) in the exact sequences above an involution invariant* R-order $\mathcal{O}$ *(resp.* $\mathcal{O}_{\mathfrak{p}} = R_{\mathfrak{p}}\mathcal{O}, \Lambda(\mathcal{O})$ = *form parameter on* $\mathcal{O}$ *generated by* $\Lambda, \Lambda(\mathcal{O})_{\mathfrak{p}} = R_{\mathfrak{p}}\Lambda(\mathcal{O})$*) such that* $A \subset \mathcal{O}$, *then the sequences remain exact. Furthermore, if* $\mathfrak{q}'$ *is any involution invariant ideal of* $\mathcal{O}$ *such that* $\mathfrak{q} \subset \mathfrak{q}'$ *and if* $\mathfrak{q}'_{\mathfrak{p}} = R_{\mathfrak{p}}\mathfrak{q}'$ *then in the exact sequences above, one may substitute respectively for the groups* $K_i(B)$, $KQ_i(B, K\Lambda)$, $K_i(B_{\mathfrak{p}})$, *and* $KQ_i(B_{\mathfrak{p}}, K\Lambda_{\mathfrak{p}})$ *the groups* $K_i(\mathcal{O}, \mathfrak{q}')$, $KQ_i(\mathcal{O}, \Lambda(\mathcal{O}), \mathfrak{q}')$, $K_i(\mathcal{O}_{\mathfrak{p}}, \mathfrak{q}'_{\mathfrak{p}})$, *and* $KQ_i(\mathcal{O}_{\mathfrak{p}}, \Lambda(\mathcal{O})_{\mathfrak{p}}, \mathfrak{q}'_{\mathfrak{p}})$.

Proof of Theorem 7.36. Since the proofs of parts a) and b) are similar to one another, we shall give only the proof of part a).

We consider first the case that $A_{\mathfrak{p}}$ is a completion. Let $\hat{A} = \prod_{\mathfrak{p}\in I} A_{\mathfrak{p}}$ and $\hat{B} = \prod_{\mathfrak{p}\in I} (B_{\mathfrak{p}}, A_{\mathfrak{p}})$. By 7.17, the arithmetic square

$$\text{(i)}\qquad \begin{array}{ccc} A & \longrightarrow & B \\ \downarrow & & \downarrow \\ \hat{A} & \longrightarrow & \hat{B} \end{array}$$

is an approximation square of rings. Thus, by 6.32, 7.6 a), and 7.29, there is an exact Mayer-Vietoris sequence

$$K_2(A) \longrightarrow K_2(B) \oplus K_2(\hat{A}) \longrightarrow K_2(\hat{B}) \xrightarrow{\partial} K_1(A) \longrightarrow K_1(B) \oplus K_1(\hat{A}) \longrightarrow K_1(\hat{B}).$$

It follows on formal grounds that there is an exact sequence

$$K_2(A) \to K_2(B) \to \operatorname{coker}(K_2(\hat{A}) \to K_2(\hat{B})) \xrightarrow{\partial} K_1(A) \to K_1(B) \oplus K_1(\hat{A}) \to K_1(\hat{B}).$$

It is an easy exercise to show that K_2 commutes with direct limits of rings. Thus, from the definition of the restricted directed product as a certain direct limit, it follows that $\operatorname{coker}(K_2(\hat{A}) \to K_2(\hat{B})) = \coprod_{\mathfrak{p}\in I} \operatorname{coker}(K_2(A_{\mathfrak{p}}) \to K_2(B_{\mathfrak{p}}))$. Furthermore, by 7.32 and 7.33, $K_1(\prod_{\mathfrak{p}\in I} A_{\mathfrak{p}}) = \prod_{\mathfrak{p}\in I} K_1(A_{\mathfrak{p}})$ and by 7.32, $K_1(\prod_{\mathfrak{p}\in I}(B_{\mathfrak{p}}, A_{\mathfrak{p}})) = \prod_{\mathfrak{p}\in I}(K_1(B_{\mathfrak{p}}), K_1(A_{\mathfrak{p}}))$. Thus, we have established an exact sequence

$$\text{(1)}\qquad \begin{aligned} &K_2(A) \to K_2(B) \to \prod_{\mathfrak{p}\in I} \operatorname{coker}(K_2(A_{\mathfrak{p}}) \to K_2(B_{\mathfrak{p}})) \overset{\partial}{\to} K_1(A) \to \\ &\qquad K_1(B) \oplus \prod_{\mathfrak{p}\in I} K_1(A_{\mathfrak{p}}) \to \prod_{\mathfrak{p}\in I}(K_1(B_{\mathfrak{p}}), K_1(A_{\mathfrak{p}})) . \end{aligned}$$

Let $\hat{\mathfrak{q}} = \mathfrak{q}\hat{A} = \prod_{\mathfrak{p}\in I} \mathfrak{q}_{\mathfrak{p}}$. By 7.7 a) or 7.17, the square

$$\text{(ii)}\qquad \begin{array}{ccc} A \ltimes \mathfrak{q} & \longrightarrow & B \ltimes B \\ \downarrow & & \downarrow \\ \hat{A} \ltimes \hat{\mathfrak{q}} & \longrightarrow & \hat{B} \ltimes \hat{B} \end{array}$$

is an approximation square of rings. Thus, as above, we get a K-theory exact sequence

$$\text{(2)}\qquad K_2(A \ltimes \mathfrak{q}) \to \cdots \qquad \cdots \to \prod_{\mathfrak{p}\in I}(K_1(B_{\mathfrak{p}} \ltimes B_{\mathfrak{p}}), K_1(A_{\mathfrak{p}} \ltimes \mathfrak{q}_{\mathfrak{p}})) .$$

By §4C, there is a split homomorphism (ii) → (i) (e.g. $A \ltimes \mathfrak{q} \to A$, $(a, q) \mapsto a$, with splitting $A \to A \ltimes \mathfrak{q}$, $a \mapsto (a, 0)$). By the naturality properties of Mayer-Vietoris sequences, the split map (ii) → (i) induces a split map (2) → (1) whose kernel is the exact sequence of the theorem.

The case that $A_{\mathfrak{p}}$ is a localization can be handled ad hoc, as follows: First, one shows directly that the squares (i) and (ii) above are approximation squares of ring. Next, one shows that the analog of Theorem 7.31 is valid. Then all of the general machinery used above applies.

§8. COMPARISON EXACT SEQUENCES

In §A, we establish an exact sequence 8.3 which measures the change in certain K-theory groups of forms caused by a change in K_1-torsion. The sequence shows that if the K-theory groups of forms are tensored first with $Z\left[\frac{1}{2}\right]$ then no change occurs. The sequence generalizes the Rothenberg exact sequence (cf. Shaneson [24, 4.13]).

In §B, we establish an exact sequence 8.17 similar to the one above, except that it measures change caused by a change in K_0-torsion. The sequence shows that no change occurs when the K-theory groups of forms are tensored first with $Z\left[\frac{1}{2}\right]$.

In §C, we collect in convenient form results appearing in §A and §B on kernels and cokernels of hyperbolic and metabolic maps.

In Ranicki [22, 5.7 and 4.3], there are exact sequences which correspond to the exact sequences in Corollaries 8.4 and 8.18 below. Ranicki's work takes place in the setting of λ-forms and λ-formations. In [8.1], we translate the language of λ-forms and λ-formations into the language used in this book. In §14, we give for quick reference purposes a table which includes a comparison of the K-theory groups of λ-forms and λ-formations with those developed in this book.

A. *Change of* K_1*-torsion*

Let $\mathcal{C}$ be a category with product in which the isomorphism classes form a set. Let $\mathcal{C}'$ be a cofinal subcategory closed under products. Let $M_0(\mathcal{C})$ be the abelian monoid whose elements are the isomorphism classes $[M]$ of objects M of $\mathcal{C}$ and whose addition is defined by $[M]+[N] = [M \perp N]$. (Later in the section we shall make the elements of $M_0(\mathcal{C})$ the objects of a category which we shall denote also by $M_0(\mathcal{C})$). Let

$$K_0(\mathcal{C}/\mathcal{C}')$$

denote the equivalence classes on $M_0(C)$ defined by the equivalence relation $[M] \sim [N] \Longleftrightarrow \exists M', N' \in C'$ such that $M \perp M' \simeq N \perp N'$.

LEMMA 8.1. $K_0(C/C')$ *is a group and the canonical map* $K_0(C/C') \to \operatorname{coker}(K_0(C') \to K_0(C))$ *is an isomorphism.*

EXAMPLE. $K_0(Q(A,\Lambda)/H(P(A))) \simeq WQ_0(A,\Lambda)$.

COROLLARY 8.2. *Every element of* $\operatorname{coker}(K_0(C') \to K_0(C))$ *can be represented by the class of an element of* C.

We leave the proofs of 8.1 and 8.2 as easy exercises.

If G is an abelian group with involution, i.e. $\mathbf{Z}/2\mathbf{Z}$-action, $g \mapsto \bar{g}$, we let

$$\hat{H}^0(G) = \{g \in G \mid g = \bar{g}\}/\{g + \bar{g} \mid g \in G\}$$

$$\hat{H}_0(G) = \{g \in G \mid g = -\bar{g}\}/\{g - \bar{g} \mid g \in G\}$$

be respectively the reduced homology and cohomology groups of the involution.

If X is an involution invariant subgroup of $K_1(A)$ which contains the class of $\pm\lambda$, we define

$$WQ_1(A,\Lambda)_{\text{based-}X} = \ker(KQ_1(A,\Lambda) \to K_1(A)/X) \text{ modulo image } (\Lambda\text{-}H : X \to KQ_1(A,\Lambda))$$

$$WH_1(A,\Lambda)_{\text{based-}X} = \ker(KH_1(A,\Lambda) \to K_1(A)/X) \text{ modulo image } (\Lambda\text{-}M : K_1(S(A,\Lambda))_{\text{based-}X} \to KH_1(A,\Lambda)).$$

If $F = H$ or Q, we ask the reader to recall the groups

$$KF_0(A,\Lambda)_{\text{even-based-}X}$$

$$WF_0(A,\Lambda)_{\text{even-based-}X}$$

defined in §1B and §1C.

THEOREM 8.3. *Let* $X \subset Y$ *be involution invariant subgroups of* $K_1(A)$ *such that the class of* $\pm\lambda$ *is contained in* X. *Then there is an exact sequence*

$$\begin{aligned}
&\hat{H}^0(K_1(S(A,\Lambda)_{\text{based-}Y})/K_1(S(A,\Lambda)_{\text{based-}X})) \xrightarrow{M} WH_1^{-\lambda}(A,\Lambda)_{\text{based-}X} \to \\
&\quad WH_1^{-\lambda}(A,\Lambda)_{\text{based-}Y} \to \hat{H}_0(Y/X) \to WH_0^{-\lambda}(A,\Lambda)_{\text{even-based-}X} \to \\
&\quad WH_0^{-\lambda}(A,\Lambda)_{\text{even-based-}Y} \to \hat{H}^0(Y/X) \xrightarrow{H} WQ_1^{\lambda}(A,\Lambda)_{\text{based-}X} \to \\
&\quad WQ_1^{\lambda}(A,\Lambda)_{\text{based-}Y} \to \hat{H}_0(Y/X) \to WQ_0^{\lambda}(A,\Lambda)_{\text{even-based-}X} \to \\
&\qquad\qquad WQ_0^{\lambda}(A,\Lambda)_{\text{even-based-}Y} \to \hat{H}^0(Y/X)\,.
\end{aligned}$$

COROLLARY 8.4. *There is an exact sequence*

$$\begin{array}{ccc}
\multicolumn{3}{c}{\to \hat{H}^0(Y/X) \to WQ_1^{\lambda}(A,\min)_{\text{based-}X} \to WQ_1^{\lambda}(A,\min)_{\text{based-}Y} \to \hat{H}_0(Y/X) \to} \\
\uparrow & & \downarrow \\
WQ_0^{-\lambda}(A,\min)_{\text{even-based-}Y} & & WQ_0^{\lambda}(A,\min)_{\text{even-based-}X} \\
\uparrow & & \downarrow \\
WQ_0^{-\lambda}(A,\min)_{\text{even-based-}X} & & WQ_0^{\lambda}(A,\min)_{\text{even-based-}Y} \\
\uparrow & & \downarrow \\
\multicolumn{3}{c}{\leftarrow \hat{H}_0(Y/X) \leftarrow WQ_1^{-\lambda}(A,\min)_{\text{based-}Y} \leftarrow WQ_1^{-\lambda}(A,\min)_{\text{based-}X} \leftarrow \hat{H}^0(Y/X) \leftarrow}
\end{array}$$

8.4 is an immediate consequence of 8.3, except for perhaps exactness at $\hat{H}^0(Y/X)$. Here, exactness follows from 8.3 and the additional facts that $WQ_0^{-\lambda}(A,\max)_{\text{even-based-}Y} = WH_0^{-\lambda}(A,\min)_{\text{even-based-}Y}$ (1.3) and the canonical map $WQ_0^{-\lambda}(A, \min)_{\text{even-based-Y}} \to WQ_0^{-\lambda}(A, \max)_{\text{even-based-Y}}$ is surjective.

The maps in 8.3 are defined as follows:

$$\hat{H}^0(Y/X) \xrightarrow{H} WQ_1(A,\Lambda)_{\text{based-X}} (\text{resp. } \hat{H}^0(K_1(S(A,\Lambda)_{\text{based-Y}})$$
$$/K_1(S(A,\Lambda)_{\text{based-X}})) \xrightarrow{M} WH_1(A,\Lambda)_{\text{based-X}}), [\alpha] \mapsto [\alpha \oplus \alpha^{*-1}].$$

$WQ_0(A,\Lambda)_{\text{even-based-Y}} \to \hat{H}^0(Y/X)$ (resp. $WH_0(A,\Lambda)_{\text{even-based-Y}} \to \hat{H}^0(Y/X)$), $[M,B] \mapsto [\langle e_i, e_j \rangle_B]$ (resp. $[M,B] \mapsto [B(e_i,e_j)]$) where $e_1, \cdots, e_m$ is a preferred ordered basis for M.

$WQ_0(A,\Lambda)_{\text{even-based-X}} \to WQ_0(A,\Lambda)_{\text{even-based-Y}}$ (resp. $WH_0(A,\Lambda)_{\text{even-based-X}} \to WH_0(A,\Lambda)_{\text{even-based-Y}}$) is the canonical map.

$\hat{H}_0(Y/X) \to WQ_0(A,\Lambda)_{\text{even-based-X}}$ (resp. $\hat{H}_0(Y/X) \to WH_0(A,\Lambda)_{\text{even-based-X}}$). Let $x \in \hat{H}_0(Y/X)$ and let $\alpha \in \text{Aut}(A^{2n})$ which represents x. If $h_1, \cdots, h_{2n}$ is the standard preferred basis for $\overset{n}{\perp} H(A)_{\text{based}}$ then we send $x \mapsto [\overset{n}{\perp} H(A)]$ where $\overset{n}{\perp} H(A)$ has the preferred basis $\alpha(h_1), \cdots, \alpha(h_{2n})$.

$WQ_1(A,\Lambda)_{\text{based-Y}} \to \hat{H}_0(Y/X)$ (resp. $WH_1(A,\Lambda)_{\text{based-Y}} \to \hat{H}_0(Y/X)$) is the canonical map.

$WQ_1(A,\Lambda)_{\text{based-X}} \to WQ_1(A,\Lambda)_{\text{based-Y}}$ (resp. $WH_1(A,\Lambda)_{\text{based-X}} \to WH_1(A,\Lambda)_{\text{based-Y}}$) is the canonical map.

Exactness at $\hat{H}^0(Y/X)$. Suppose $[\alpha] \mapsto 0$. After multiplying α by a suitable matrix whose class belongs to X, we can assume that the matrix $\begin{pmatrix} \alpha & 0 \\ 0 & \bar{\alpha}^{-1} \end{pmatrix} \in EQ(A,\Lambda)$. But, by 3.10, every element of $EQ(A,\Lambda)$ has a product decomposition $\begin{pmatrix} I & U \\ 0 & I \end{pmatrix} \omega \begin{pmatrix} I & B \\ 0 & I \end{pmatrix}\begin{pmatrix} I & 0 \\ L & I \end{pmatrix}\begin{pmatrix} \varepsilon & 0 \\ 0 & \bar{\varepsilon}^{-1} \end{pmatrix}$ where $\omega = \begin{pmatrix} 0 & \pi \\ -\pi^{-1} & 0 \end{pmatrix}$ and the classes of π and ε belong to X. Equating $\begin{pmatrix} a & 0 \\ 0 & \bar{a}^{-1} \end{pmatrix}$ with the product decomposition above, we obtain $a = \pi L \varepsilon$. Since $L = -\lambda \bar{L}$, it follows that $[a] = [\bar{L}]$. Hence, the class of the Λ-hermitian module $(A^n, B_{\bar{L}})$ where $B_{\bar{L}}(x,y) = \bar{x}\bar{L}y$ is a preimage for $[a]$.

If B is a nonsingular matrix such that $B = -\lambda\bar{B}$ and such that the diagonal coefficients of B lie in Λ then the equation

$$\left(\begin{array}{c|c} -\bar{\pi} \quad \bar{B}^{-1} & \\ \hline & -\pi^{-1} \quad B \end{array}\right) = \left(\begin{array}{c|c} I & \pi B^{-1}\pi \\ \hline & I \end{array}\right)\left(\begin{array}{c|c} 0 & \pi \\ \hline -\pi^{-1} & 0 \end{array}\right)\left(\begin{array}{c|c} I & B \\ \hline 0 & I \end{array}\right)\left(\begin{array}{c|c} I & 0 \\ \hline -B^{-1} & I \end{array}\right)$$

implies that the composite at $\hat{H}^0(Y/X)$ is trivial.

Exactness at $WQ_0(A,\Lambda)_{\text{even-based-Y}}$ (resp. $WH_0(A,\Lambda)_{\text{even-based-Y}}$). Let (M, B) be a quadratic (resp. hermitian) module with preferred basis $e_1, \cdots, e_m$. If the class of the m×m-matrix $(\langle e_i, e_j \rangle_B)$ (resp. $(B(e_i, e_j))$) in $\hat{H}^0(Y/X)$ vanishes then for some invertible matrix a whose class in $K_1(A)/Y$ vanishes, the class of $\bar{a}(\langle e_i, e_j\rangle_B)a$ (resp. $\bar{a}(B(e_i,e_j))a$) in $K_1(A)/X$ vanishes. After stabilizing (M, B) by a suitable number of standard based hyperbolic planes, we can assume that a determines an automorphism of M. But then (M, B) with the preferred basis $a^{-1}(e_1), \cdots, a^{-1}(e_m)$ is an element of $Q(A,\Lambda)_{\text{based-X}}$ (resp. $H(A,\Lambda)_{\text{based-X}}$) whose class in $WQ_0(A,\Lambda)_{\text{even-based-X}}$ (resp. $WH_0(A,\Lambda)_{\text{even-based-X}}$) is a preimage for the class of (M, B) in $WQ_0(A,\Lambda)_{\text{even-based-Y}}$ (resp. $WH_0(A,\Lambda)_{\text{even-based-Y}}$).

Exactness at $WQ_0(A,\Lambda)_{\text{even-based-X}}$ (resp. $WH_0(A,\Lambda)_{\text{even-based-X}}$). Suppose $[M,B] \mapsto 0$. After stabilizing (M, B) by a suitable number of standard based hyperbolic planes (resp. Λ-metabolic planes), we can assume there is an isomorphism in $Q(A,\Lambda)_{\text{based-Y}}$ (resp. $H(A,\Lambda)_{\text{based-Y}}$) $\overset{n}{\perp} H(A)_{\text{based}} \cong (M,B)$ (resp. $\overset{n}{\perp} \Lambda\text{-}M(A)_{\text{based}} \cong (M,B)$). If α denotes the matrix associated with the isomorphism above then the class of α in $\hat{H}_0(Y/X)$ is a preimage for $[M,B] \in WQ_0(A,\Lambda)_{\text{based-X}}$ (resp. $WH_0(A,\Lambda)_{\text{based-X}}$, because $[\Lambda\text{-}M(A)_{\text{based}}] = [H(A)_{\text{based}}]$ (2.12)).

Exactness at $\hat{H}_0(Y/X)$. Let $\alpha \in \text{Aut}(A^{2m})$ whose class in $K_1(A)/Y$ is trivial. Let M be the quadratic module (resp. hermitian module) $\overset{m}{\perp} H(A)$ with preferred basis $\alpha(h_1),\cdots,\alpha(h_{2m})$ where $h_1,\cdots,h_{2m}$ is the standard basis for $\overset{m}{\perp} H(A)_{\text{based}}$. If $[M] = 0$ then there is an isomorphism in $Q(A,\Lambda)_{\text{based-X}}$ (resp. $H(A,\Lambda)_{\text{based-X}}$) $\varepsilon : M \perp (\overset{n}{\perp} H(A)_{\text{based}})$ $\cong \overset{m}{\perp} H(A)_{\text{based}} \perp (\overset{n}{\perp} H(A)_{\text{based}})$ (resp. $M \perp (\overset{n}{\perp} \Lambda\text{-}M(A)_{\text{based}})$ $\cong \overset{m}{\perp} \Lambda\text{-}M(A)_{\text{based}} \perp (\overset{n}{\perp} \Lambda\text{-}M(A)_{\text{based}})$). $\varepsilon(\alpha \perp 1_{\overset{n}{\perp} H(A)}) \in \text{Aut}(\overset{m+n}{\perp} H(A))$ (resp. $\varepsilon(\alpha \perp 1_{\overset{n}{\perp} \Lambda\text{-}M(A)}) \in \text{Aut}(\overset{m+n}{\perp} \Lambda\text{-}M(A))$) and the class of $\varepsilon(\alpha \perp 1_{\overset{n}{\perp} H(A)})$ in $WQ_1(A,\Lambda)_{\text{based-Y}}$ (resp. $\varepsilon(\alpha \perp 1_{\overset{n}{\perp} \Lambda\text{-}M(A)})$ in $WH_1(A,\Lambda)_{\text{based-Y}}$) is a preimage for the class of α in $\hat{H}_0(Y/X)$.

Exactness at $WQ_1(A,\Lambda)_{\text{based-Y}}$ (resp. $WH_1(A,\Lambda)_{\text{based-Y}}$). If $[\alpha] \mapsto 0$ then there is an invertible matrix γ such that the class of $\sigma = \alpha \begin{pmatrix} \gamma & \\ & \bar{\gamma}^{-1} \end{pmatrix}$ in $K_1(A)/X$ vanishes. The class of σ in $WQ_1(A,\Lambda)_{\text{based-X}}$ (resp. $WH_1(A,\Lambda)_{\text{based-X}}$) is a preimage for $[\alpha]$.

Exactness at $WQ_1(A,\Lambda)_{\text{based-X}}$ (resp. $WH_1(A,\Lambda)_{\text{based-X}}$). If $[\alpha] \mapsto 0$ then $[\alpha] = \left[\begin{pmatrix} \gamma & \\ & \bar{\gamma}^{-1} \end{pmatrix}\right]$ for some $\gamma \in \text{Aut}(A^n)_{\text{based-Y}}$ (resp. $\gamma \in \text{Aut}(A^n,\rho)$

where $(A^n, \rho) \in S(A, \Lambda)_{\text{based-}Y})$. The class of γ in $\hat{H}^0(Y/X)$ (resp. $\hat{H}^0(K_1(S(A,\Lambda)_{\text{based-}Y})/K_1(S(A,\Lambda)_{\text{based-}X}))$ is a preimage for $[\alpha]$.

B. *Change of* K_0*-torsion*

To prepare for the proof of the main result 8.17, we shall establish several interpretations of the relative Grothendieck group.

Let

$$\begin{array}{ccc} & & \mathbf{C} \\ & & \big\downarrow F \\ \mathbf{C} & \xrightarrow[F]{} & \mathbf{C}' \end{array}$$

be a diagram of categories with product. Assume that the isomorphism classes of objects of $\mathbf{C}$ and $\mathbf{C}'$ form a set. Recall the construction in §6A of the fibre product

$$\mathbf{C} \times_{\mathbf{C}'} \mathbf{C}$$

of the diagram above. Assume now that all the morphisms in $\mathbf{C}$ and $\mathbf{C}'$ are isomorphisms. Assume that $\mathbf{C}$ and $\mathbf{C}'$ are subcategories of a common category $\mathbf{D}$ and that there is a natural transformation t of the identity functor on $\mathbf{D}$ restricted to $\mathbf{C}$ to the functor F, i.e. given objects M and N in $\mathbf{C}$ we have morphisms $t_M : M \to F(M)$ and $t_N : N \to F(N)$ in $\mathbf{D}$ such that the diagram

$$\begin{array}{ccc} M & \xrightarrow{\alpha} & N \\ {\scriptstyle t_M}\big\downarrow & & \big\downarrow{\scriptstyle t_N} \\ F(M) & \xrightarrow{F(\alpha)} & F(N) \end{array}$$

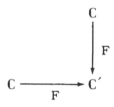

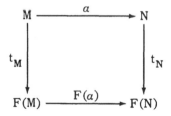

commutes for any morphism $\alpha : M \to N$ in $\mathbf{C}$. Assume also that if ψ is a morphism in $\mathbf{D}$ then the expression image(ψ) has been defined such that if α is an isomorphism in $\mathbf{D}$ (and α and ψ compose) then image $(\psi\alpha)$ = image (ψ) (e.g. the underlying elements of $\mathbf{D}$ are sets and set theoretic maps). We define

$$(\mathbf{C} \times_{\mathbf{C}'} \mathbf{C})^+$$

to be the category whose objects are the same as those of $\mathbf{C} \times_{\mathbf{C}'} \mathbf{C}$. Define a morphism $\phi : (M_1, \sigma, M_2)^+ \to (N_1, \rho, N_2)^+$ in $(\mathbf{C} \times_{\mathbf{C}'} \mathbf{C})^+$ to be a pair of morphisms $\phi_i : F(M_i) \to F(N_i)$ in $\mathbf{C}'$ such that the diagram

$$\begin{array}{ccc} F(M_1) & \xrightarrow{\ \sigma\ } & F(M_2) \\ {\scriptstyle \phi_1}\downarrow & & \downarrow{\scriptstyle \phi_2} \\ F(N_1) & \xrightarrow[\ \rho\ ]{} & F(N_2) \end{array}$$

commutes, and

$$\text{image } \phi_i t_{M_i} = \text{image } t_{N_i} .$$

The morphisms of $(\mathbf{C} \times_{\mathbf{C}'} \mathbf{C})^+$ include those of $\mathbf{C} \times_{\mathbf{C}'} \mathbf{C}$, and we have a canonical functor

$$+ : \mathbf{C} \times_{\mathbf{C}'} \mathbf{C} \to (\mathbf{C} \times_{\mathbf{C}'} \mathbf{C})^+ .$$

EXAMPLE. Let $A \to B$ be a homomorphism of rings. Then every right B-module can be viewed as a right A-module. Let $\mathbf{C} = \mathbf{P}(A)$, $\mathbf{C}' = \mathbf{P}(B)$, and $\mathbf{D}$ = mod-A = category of finitely generated, right A-modules. Let $F : \mathbf{P}(A) \to \mathbf{P}(B)$, $P \mapsto P \otimes_A B$, and let $t_P : P \to P \otimes_A B$, $p \mapsto p \otimes 1$.

EXAMPLE. Let $\mathbf{C} = \mathbf{P}(A)$, $\mathbf{C}' = \mathbf{Q}(A, \Lambda)$, $\mathbf{D} = \mathbf{P}(A)$, F = hyperbolic functor $\mathbf{H}$, and $t_P : P \to \mathbf{H}(P)$, $p \mapsto p \oplus 0$.

LEMMA 8.5. *Suppose that if* $\psi : F(M) \to F(N)$ *is a morphism in* $\mathbf{C}'$ *such that* image $t_M \psi$ = image t_N *then there exists a morphism* $\alpha : M \to N$ *in*

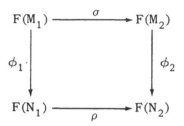

$\mathbf{C}$ *and an object* P *in* $\mathbf{C}$ *such that* $F(\alpha^{-1})\psi \perp 1_{F(P)}$ *lies in the commutator subgroup of* $\mathrm{Aut}_{\mathbf{C}'}(F(M) \perp F(P))$. *Then the canonical map*

$$K_0(\mathbf{C} \times_{\mathbf{C}'} \mathbf{C})/\mathfrak{M}(\mathbf{C} \times_{\mathbf{C}'} \mathbf{C}) \xrightarrow{K_0(+)} K_0((\mathbf{C} \times_{\mathbf{C}'} \mathbf{C})^+)/\mathfrak{M}(\mathbf{C} \times_{\mathbf{C}'} \mathbf{C})^+$$

is an isomorphism where $\mathfrak{M}(\mathbf{C} \times_{\mathbf{C}'} \mathbf{C})$ *is the subgroup of* $K_0(\mathbf{C} \times_{\mathbf{C}'} \mathbf{C})$ *generated by all* $[M, \sigma, N] + [N, \rho, P] - [M, \rho\sigma, P]$ *and* $\mathfrak{M}(\mathbf{C} \times_{\mathbf{C}'} \mathbf{C})^+$ *denotes its image in* $K_0((\mathbf{C} \times_{\mathbf{C}'} \mathbf{C})^+)$.

Proof. To construct an inverse to $K_0(+)$, it suffices to show that if $\phi: (M_1, \sigma, M_2)^+ \to (N_1, \rho, N_2)^+$ is a morphism in $(\mathbf{C} \times_{\mathbf{C}'} \mathbf{C})^+$ then the class $[M_1, \sigma, M_2]$ of (M_1, σ, M_2) in $K_0(\mathbf{C} \times_{\mathbf{C}'} \mathbf{C})/\mathfrak{M}(\mathbf{C} \times_{\mathbf{C}'} \mathbf{C})$ is the class $[N_1, \rho, N_2]$ of (N_1, ρ, N_2). Our hypotheses imply that we can find morphisms $\alpha_i: M_i \to N_i$ in $\mathbf{C}$ and an object P in $\mathbf{C}$ such that $\phi_1 F(\alpha_1)^{-1} \perp 1_{F(P)} \in$ commutator subgroup of $\mathrm{Aut}_{\mathbf{C}'}(F(N_1) \perp F(P))$ and $F(\alpha_2)\phi_2^{-1} \perp 1_{F(P)} \in$ commutator subgroup of $\mathrm{Aut}_{\mathbf{C}'}(F(N_2) \perp F(P))$. The commutative diagram

$$\begin{array}{ccc} F(M_1) & \xrightarrow{\quad \sigma \quad} & F(M_2) \\ {\scriptstyle F(\alpha_1)}\big\downarrow & & \big\downarrow{\scriptstyle F(\alpha_2)} \\ F(N_1) & \xrightarrow[F(\alpha_2)\phi_2^{-1}\rho\phi_1 F(\alpha_1)^{-1}]{} & F(N_2) \end{array}$$

defines an isomorphism $(M_1, \sigma, M_2) \cong (N_1, F(\alpha_2)\phi_2^{-1}\rho\phi_1 F(\alpha_1)^{-1}, N_2)$ in $\mathbf{C} \times_{\mathbf{C}'} \mathbf{C}$. Hence, $[M_1, \sigma, M_2] = [N_1, (F(\alpha_2)\phi_2^{-1})\rho(\phi_1 F(\alpha_1)^{-1}), N_2] = [N_1, F(\alpha_2)\phi_2^{-1}, N_1] + [N_1, \rho, N_2] + [N_2, \phi_1 F(\alpha_1)^{-1}, N_2]$. But, $[N_1, F(\alpha_2)\phi_2^{-1}, N_1] = [N_1 \perp P, F(\alpha_2)\phi_2^{-1} \perp 1_{F(P)}, N_1 \perp P]$ and $[N_2, \phi_1 F(\alpha_1)^{-1}, N_2] = [N_2 \perp P, \phi_1 F(\alpha_1)^{-1} \perp 1_{F(P)}, N_2 \perp P]$ vanish, because $F(\alpha_2)\phi_2^{-1} \perp 1_{F(P)}$ and $\phi_1 F(\alpha_1)^{-1} \perp 1_{F(P)}$ are products of commutators.

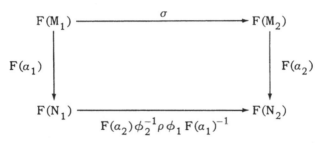

Assume now that the objects of $\mathbf{C}$, $\mathbf{C}'$, and $\mathbf{D}$ are sets with extra structure, and that morphisms between objects are set theoretic maps which preserve the extra structure. Maintain the previous assumptions on C, C', D, and t. Call a subset $S \subset V \in \mathrm{Obj}(C')$ *distinguished* if there is an $M \in \mathrm{Obj}(C)$ and an isomorphism $\psi : F(M) \to V$ (in C') such that image $\psi t_M = S$. Next, we give a useful alternative description of $(C \times_{C'} C)^+$. Define

$$C \times^{C'} C$$

to be the category whose objects are triples (V, M_1, M_2) where $V \in \mathrm{Obj}(C')$ and M_1 and M_2 are distinguished subsets of V. A morphism $(V, M_1, M_2) \to (W, N_1, N_2)$ is a morphism $\phi : V \to W$ in C' such that $\phi(M_i) = N_i$. $C \times^{C'} C$ has a natural product defined by $(V, M_1, M_2) \perp (W, N_1, N_2) = (V \perp W, M_1 \perp M_2, N_1 \perp N_2)$.

LEMMA 8.6. *The product preserving functors below form inverse equivalences*

$$(C \times_{C'} C)^+ \to C \times^{C'} C, (M_1, \sigma, M_2)^+ \mapsto (F(M_1), t_{M_1}(M_1), \sigma^{-1} t_{M_2}(M_2))$$

$$C \times^{C'} C \to (C \times_{C'} C)^+, (V, N_1, N_2) \mapsto (N_1', \sigma, N_2')$$

where N_i = image $\psi_i t_{N_i'}$ *and* $\sigma = \psi_2^{-1} \psi_1$. *(It is possible that there is more than one choice for* N_i' *and* ψ_i, *in which case we simply pick one of the choices.)*

Proof. Clear.

COROLLARY 8.7. *Let* $\mathfrak{M}(C \times^{C'} C)$ = *subgroup of* $K_0(C \times^{C'} C)$ *generated by all* $[V, M_1, M_2] + [V, M_2, M_3] - [V, M_1, M_3]$. *If the hypotheses of 8.5 are satisfied then the canonical map below is an isomorphism*

$$K_0(C \times_{C'} C)/\mathfrak{M}(C \times_{C'} C) \xrightarrow{\simeq} K_0(C \times^{C'} C)/\mathfrak{M}(C \times^{C'} C) .$$

Proof. Follows from 8.4 and 8.5.

If C is a category with product, we define the *monoid category on* C

$$M_0(C)$$

to be the category whose objects are the isomorphism classes [M] of the objects M of $\mathbf{C}$. A morphism $\alpha: [M] \to [N]$ is defined by giving for each pair of objects $M' \in [M]$ and $N' \in [N]$ a morphism $\alpha_{M',N'}: M' \to N'$ in $\mathbf{C}$ such that if $M', M'' \in [M]$ and $N', N'' \in [N]$ then there are isomorphisms $M' \to M''$ and $N' \to N''$ such that the diagram

$$\begin{array}{ccc} M' & \xrightarrow{\alpha_{M',N'}} & N' \\ \downarrow & & \downarrow \\ M'' & \xrightarrow[\alpha_{M'',N''}]{} & N'' \end{array}$$

commutes. If the objects of $M_0(\mathbf{C})$ form a set then it was shown at the beginning of the section how one makes $\mathrm{Obj}(M_0(\mathbf{C}))$ into an abelian monoid.

If $P \in \mathbf{P}(A)$ we let

$$S(P, \Lambda)$$

denote the group of all homomorphisms $\alpha: P \to P^*$ such that $\alpha = -\lambda\alpha^*$ and $\alpha(p)(p) \in \Lambda$ for all $p \in P$. We denote a typical element of $S(P, \Lambda)$ by (P, α). We let $(P, [\alpha])$ denote the residue class of (P, α) in $S(P, \Lambda)/S(P, \min)$. We define

$$s(A, \Lambda)$$

to be the category whose objects are all pairs $(P, [\alpha])$. A morphism $\sigma: (P, [\alpha]) \to (Q, [\beta])$ is an isomorphism $\sigma: P \to Q$ such that $(P, [\alpha]) = (P, [\sigma^*\beta\sigma])$. The Λ-metabolic functor defined in §1C induces a product preserving functor

$$\Lambda - M: s(A, \Lambda) \to M_0(\mathbf{H}(A, \Lambda))$$

$$(P, [\alpha]) \mapsto \left[P \oplus P^*, \begin{pmatrix} \alpha & -\lambda I_{P^*} \\ I_P & 0 \end{pmatrix}\right]$$

$$(\sigma: (P, [\alpha]) \to (Q, [\beta])) \mapsto \sigma \oplus \sigma^{*-1}$$

which we call the Λ-metabolic functor on $s(A, \Lambda)$.

For the rest of this section we fix the following notation. We let

$$X \subset Y$$

be full, cofinal subcategories of $P(A)$ closed under direct sums and dualization such that A is an object of X. We assume X is cofinal in Y. We let

$$(Y/X)^0 \quad (\text{resp. } (Y/X)_0)$$

denote the full subcategory of Y of all P such that $P \oplus X_1 \simeq P^* \oplus X_2$ for some $X_1, X_2 \in X$ (resp. $P \oplus P^* \in X$). Note that if $Y = X$ then $(Y/X)_0 = X$. Since X is cofinal in $P(A)$, it follows that $K_0(X) \hookrightarrow K_0(A)$. Similarly, $K_0(Y) \hookrightarrow K_0(A)$. We shall confuse sometimes X and Y with their images $K_0(X)$ and $K_0(Y)$ in $K_0(A)$, so that it makes sense to write $Q(A,\Lambda)_X$, $WQ_0(A,\Lambda)_X$, $\hat{H}^0(Y/X) = \hat{H}^0(K_0(Y)/K_0(X))$, etc.

With respect to the functors $\Lambda - H : Y \to Q(A,\Lambda)$ and $\Lambda - M : s(A,\Lambda)_Y \to M_0(H(A,\Lambda))$, we define

$$WQ_1(A,\Lambda)_Y = K_0(Y \times_{Q(A,\Lambda)} Y)/\mathfrak{M}(Y \times_{Q(A,\Lambda)} Y)$$

$$WH_1(A,\Lambda)_Y = K_0(s(A,\Lambda)_Y \times_{M_0(H(A,\Lambda))} s(A,\Lambda)_Y)/\mathfrak{M}(s(A,\Lambda)_Y \times_{M_0(H(A,\Lambda))} s(A,\Lambda)_Y).$$

Since the functors above are cofinal (2.10), it follows from (6.20) and (6.22) that we have exact sequences

8.8 $$K_1(A) \xrightarrow{H} KQ_1(A,\Lambda) \to WQ_1(A,\Lambda)_Y \to K_0(Y) \xrightarrow{H} KQ_0(A,\Lambda)$$

8.9 $$K_1(s(A,\Lambda)) \xrightarrow{M} KH_1(A,\Lambda) \to WH_1(A,\Lambda)_Y \to K_0(s(A,\Lambda)_Y) \xrightarrow{M} KH_0(A,\Lambda).$$

COROLLARY 8.10. *Let* $F(A)$ = *full subcategory of free modules of* $P(A)$. *Then there are canonical isomorphisms*

$$WQ_1(A,\Lambda) \simeq WQ_1(A,\Lambda)_{F(A)}$$

$$WH_1(A,\Lambda) \simeq WH_1(A,\Lambda)_{F(A)}.$$

REMARK. If in the definition of $WH_1(A,\Lambda)_{F(A)}$, one replaced $s(A,\Lambda)_{F(A)}$ by its counterpart $S(A,\Lambda)_{F(A)}$ defined in §1C then the corollary would not be true.

Proof. By definition, $WQ_1(A,\Lambda) = \operatorname{coker} \mathbf{H} : K_1(A) \to KQ_1(A,\Lambda)$ and $WH_1(A,\Lambda) = \operatorname{coker} \mathbf{M} : K_1(s(A,\Lambda)) \to KH_1(A,\Lambda)$. Thus, in 8.8 (resp. 8.9), it suffices to show that the map $WQ_1(A,\Lambda)_{F(A)} \to K_0(F(A))$ (resp. $WH_1(A,\Lambda)_{F(A)} \to K_0(s(A,\Lambda)_{F(A)})$) is trivial. In 8.8, $[A^n, \sigma, A^n] \mapsto [A^n] - [A^n] = 0$. In 8.9, $[(A^n,[\alpha]), \sigma, (A^n,[\beta])] \mapsto [A^n,[\alpha]] - [A^n,[\beta]]$. Thus, it suffices to show $(A^n,[\alpha]) \perp (A^n,[0_{A^n}]) \simeq (A^n,[\beta]) \perp (A^n,[0_{A^n}])$. Let $\sigma = \begin{pmatrix} a & b \\ c & d \end{pmatrix} : A^n \oplus (A^n)^* \to A^n \oplus (A^n)^*$. If we pick suitably representatives (P,α) for $(P,[\alpha])$ and (Q,β) for $(Q,[\beta])$ then

$$\begin{pmatrix} \alpha & -\lambda I \\ I & 0 \end{pmatrix} = \begin{pmatrix} a^* & c^* \\ b^* & d^* \end{pmatrix} \begin{pmatrix} \beta & -\lambda I \\ I & 0 \end{pmatrix} \begin{pmatrix} a & b \\ c & d \end{pmatrix}.$$

Hence

$$\left(\begin{array}{cc|cc} \alpha & 0 & -\lambda I & \\ 0 & 0 & & -\lambda I \\ \hline I & & 0 & \\ & I & & 0 \end{array}\right) = \left(\begin{array}{cc|cc} a^* & & c^* & \\ & I & & 0 \\ \hline b^* & & d^* & \\ & 0 & & I \end{array}\right) \left(\begin{array}{cc|cc} \beta & & -\lambda I & \\ & 0 & & -\lambda I \\ \hline I & & 0 & \\ & I & & 0 \end{array}\right) \left(\begin{array}{cc|cc} a & & b & \\ & I & & 0 \\ \hline c & & d & \\ & 0 & & I \end{array}\right).$$

If we multiply both sides of the equation above by

$$\left(\begin{array}{cc|cc} I & & & \\ & I & c+d & -I-c \\ \hline & & I & \\ & & & I \end{array}\right) \left(\begin{array}{cc|cc} I & & & \\ & I & & \\ \hline I & I & I & \\ I & 0 & & I \end{array}\right)$$

on the right and by

$$\left(\begin{array}{cc|cc} I & & I & I \\ & I & I & 0 \\ \hline & & I & \\ & & & I \end{array}\right) \left(\begin{array}{cc|cc} I & & & \\ & I & & \\ \hline & c^*+d^* & I & \\ & -I-c^* & & I \end{array}\right)$$

on the left then we obtain

$$\left(\begin{array}{c|c} \begin{pmatrix}\alpha & 0\\ 0 & 0\end{pmatrix} + R-\lambda R^* & S \\ \hline T & U\end{array}\right) = \left(\begin{array}{cc|c} a^*+b^* & c^*+d^* & \\ b^* & d^* & Y^* \\ \hline X^* & & Z^*\end{array}\right)\left(\begin{array}{cc|cc} \beta & & -\lambda I & \\ & 0 & & -\lambda I \\ \hline I & & 0 & \\ & I & & 0\end{array}\right)$$

$$\left(\begin{array}{cc|c} a+b & b & X \\ c+d & d & \\ \hline Y & & Z\end{array}\right)$$ for suitable R, S, T, U, X, Y, Z. Hence, if

$W = \begin{pmatrix} a+b & b\\ c+d & d\end{pmatrix}$ then $\begin{pmatrix}\alpha & 0\\ 0 & 0\end{pmatrix} + R - \lambda R^* = W^*\begin{pmatrix}\beta & 0\\ 0 & 0\end{pmatrix} W + Y^*W - \lambda W^*Y$.

Hence, $(A^{2n}, [\alpha \oplus 0_{A^n}]) = (A^{2n}, [W^*\begin{pmatrix}\beta & 0\\ 0 & 0\end{pmatrix} W]) \simeq$ (because

$W = \begin{pmatrix} a & b\\ c & d\end{pmatrix}\begin{pmatrix} I & 0\\ I & I\end{pmatrix}$ is invertible) $(A^{2n}, [\beta \oplus 0_{A^n}])$.

We define

$$\varepsilon : S(A, \bar{\Lambda}) \to P(A) \times_{Q(A,\Lambda)} P(A)$$

$$(P, \gamma) \mapsto (P, \varepsilon(\gamma) = \begin{pmatrix} I_P & 0\\ \gamma & I_{P^*}\end{pmatrix}, P)$$

$$\sigma: (P, \gamma) \to (Q, \delta) \mapsto (\sigma_1 = \sigma \oplus \sigma^{*-1}, \sigma_2 = \sigma \oplus \sigma^{*-1}) .$$

ε is product preserving. Choose for each isomorphism class $[P]$ of $P \in (Y/X)^0$ an element $-P$ such that $P \oplus -P$ and $P^* \oplus -P \in X$. To see how this is done, see the discussion of the maps following 8.18.

We define

$$h : (Y/X)^0 \to X \times_{Q(A,\Lambda)_X} X$$

$$P \mapsto (P, \sigma_P, P^*) \perp (-P, 1, -P)$$

$$(\sigma: P \to Q) \mapsto (\sigma \oplus \sigma^{*-1}, \sigma \oplus \sigma^{*-1}) \perp (1, 1)$$

where $\sigma_P = \begin{pmatrix} 0 & \lambda I_{P^*}\\ I_P & 0\end{pmatrix} : P \oplus P^* \to P^* \oplus P$.

h is product preserving providing the $-P$'s can be chosen such that $-P \oplus Q = -(P \oplus Q)$. Otherwise, we have to pass to stable isomorphism classes which we shall do later on.

LEMMA 8.11. *The diagram*

$$\begin{array}{ccccc}
 & [P,B] & \longmapsto & [(P,\gamma_B) \perp (-P, 0_P)] & \\
[P,B] & M_0(H(A,\bar{\Lambda})_Y) & \longrightarrow & M_0(S(A,\bar{\Lambda})_X) & \\
\big\downarrow & \big\downarrow & & \big\downarrow & \varepsilon(\;)^+ \\
[P] & M_0((Y/X)^0) & \xrightarrow{h(\;)^+} & M_0(X \times_{Q(A,\Lambda)} X)^+ &
\end{array}$$

is fibred where $\gamma_B(p) = B(p,\;)$.

Proof. First, we show the diagram is commutative up to isomorphism. It suffices to show $(P, \sigma_P, P^*)^+ \simeq (P, \varepsilon(\gamma), P)^+$ for any $(P,\gamma) \in S(A,\bar{\Lambda})_Y$ such that $\gamma: P \to P^*$ is an isomorphism. But the commutative diagram

$$\begin{array}{ccc}
H(P) & \xrightarrow{\varepsilon(\gamma)} & H(P) \\
\begin{pmatrix} I_P & \gamma^{-1} \\ 0 & I_{P^*} \end{pmatrix}\Big\downarrow & & \Big\downarrow \begin{pmatrix} I_{P^*} & -\gamma^* \\ 0 & I_P \end{pmatrix}\begin{pmatrix} \gamma^* & 0 \\ 0 & \gamma^{-1} \end{pmatrix} \\
H(P) & \xrightarrow[\sigma_P]{} & H(P)
\end{array}$$

defines such an isomorphism.

Next, suppose $\phi = (\phi_1, \phi_2): (P \oplus -P, \sigma_P \perp 1_{H(P)}, P^* \oplus -P)^+ \xrightarrow{\simeq} (Q, \varepsilon(\gamma), Q)^+$ where $P \in Y^0$ and $(Q,\gamma) \in S(A,\Lambda)_X$. Let $\phi_1 = \begin{pmatrix} a & b \\ 0 & a^{*-1} \end{pmatrix}$ and $\phi_2 = \begin{pmatrix} c & d \\ 0 & c^{*-1} \end{pmatrix}$. Then

$$\begin{pmatrix} c^{-1} & -d \\ 0 & c^* \end{pmatrix}\begin{pmatrix} 1 & 0 \\ \gamma & 1 \end{pmatrix}\begin{pmatrix} a & b \\ 0 & a^{*-1} \end{pmatrix} = \left(\begin{array}{cc|cc} & I_{-P} & \lambda I_{P^*} & \\ \hline I_P & & & I_{-P^*} \end{array}\right).$$

Hence,

$$\begin{pmatrix} * & * \\ c^*\gamma a & * \end{pmatrix} = \left(\begin{array}{cc|cc} & I_{-P} & \lambda I_{P*} & \\ \hline I_P & & & I_{-P*} \end{array}\right).$$

Hence, $c^*\gamma a = \begin{pmatrix} I_P & 0 \\ 0 & 0 \end{pmatrix}$. Hence, $a^*\gamma a = a^*c^{*-1}c^*\gamma a = a^*c^{*-1}\begin{pmatrix} I_P & 0 \\ 0 & 0 \end{pmatrix}$ $= \begin{pmatrix} x & 0 \\ y & 0 \end{pmatrix}$ for suitable x and y. But $\begin{pmatrix} x & 0 \\ y & 0 \end{pmatrix} = -\bar{\lambda}\begin{pmatrix} x & 0 \\ y & 0 \end{pmatrix}^*$ implies $y = 0$. Moreover, it is clear that $x: P \to P^*$ must be invertible. Hence, if $B_x(p, p') = x(p)(p')$ then $[P, B_x] \mapsto [P]$ and $[P, B_x] \mapsto [Q, \gamma]$. Furthermore, if $[P, B]$ and $[Q, C] \in M_0(\mathbf{H}(A, \Lambda)_Y)$ go to the same element in $M_0(S(A, \Lambda)_X)$ then it is easy to see that $[P, B] = [Q, C]$. Hence, the diagram is fibred.

LEMMA 8.12. *The functor* $\varepsilon: S(A, \bar{\Lambda})_{free} \to P(A) \times_{Q(A,\Lambda)} P(A)$ *is cofinal.*

Proof: Since $S(A, \bar{\Lambda})_{free}$ is a cofinal subcategory of $S(A, \bar{\Lambda})$, it suffices to show that $\varepsilon: S(A, \Lambda) \to P(A) \times_{Q(A,\Lambda)} P(A)$ is cofinal. An object $(P_1, \sigma, P_2) \in P(A) \times_{Q(A,\Lambda)} P(A)$ is isomorphic to an element in image $\varepsilon: S(A, \Lambda) \to P(A) \times_{Q(A,\Lambda)} P(A)$ if and only if $P_1 \simeq P_2$, and $\sigma^{-1}(P_2)$ and P_1 have a common, totally isotropic, direct complement in $\mathbf{H}(P_1)$. If $\mathbf{H}(P) = (P \oplus P^*, B_P)$, let $-\mathbf{H}(P) = (P \oplus P^*, -B_P)$. Let $(P, \rho, Q) \in (P(A) \times_{Q(A,\Lambda)} P(A))$. In $\mathbf{H}(P) \perp -\mathbf{H}(P)$, the totally isotropic subspaces $P \perp P^*$ and $\rho^{-1}(Q) \perp \rho^{-1}(Q^*)$ have a common, totally isotropic, direct complement, namely the diagonal subspace $\Delta = \{((p, f), (p, f)) | p \in P, f \in P^*\}$ of $\mathbf{H}(P) \perp -\mathbf{H}(P)$. Hence, if $\varphi: -\mathbf{H}(P^*) \to \mathbf{H}(P^*)$, $(f, p) \mapsto (-f, p)$, and if $\psi: -\mathbf{H}(Q^*) \to \mathbf{H}(Q^*)$, $(g, q) \mapsto (-g, q)$, then the commutative diagram

$$\begin{array}{ccc} \mathbf{H}(P) \perp -\mathbf{H}(P^*) & \xrightarrow{\rho \perp \rho} & \mathbf{H}(Q) \perp -\mathbf{H}(Q^*) \\ \Big\downarrow{\scriptstyle 1 \perp \varphi} & & \Big\downarrow{\scriptstyle 1 \perp \psi} \\ \mathbf{H}(P) \perp \mathbf{H}(P^*) & \xrightarrow[\rho \perp \psi\rho\varphi^{-1}]{} & \mathbf{H}(Q) \perp \mathbf{H}(Q^*) \end{array}$$

shows that $P \perp P^* = (1 \perp \varphi)(P \perp P^*)$ and $(\rho \perp \psi\rho\varphi^{-1})^{-1}(Q \perp Q^*) = (1 \perp \varphi)(\rho \perp \rho)^{-1}(Q \perp Q^*)$ have a common, totally isotropic, direct complement, namely $(1 \perp \varphi)\Delta$. Hence, ε is cofinal.

LEMMA 8.13. *The canonical map below is an isomorphism*

$$K_0(X \times_{Q(A,\Lambda)} X/\varepsilon(S(A,\bar{\Lambda})_X)) \xrightarrow{\simeq} WQ_1(A,\Lambda)_X .$$

Let

$$\varepsilon(X \times {}^{Q(A,\Lambda)}X)$$

be the full subcategory of $X \times {}^{Q(A,\Lambda)}X$ of all (V, P, Q) such that P and Q have a common, totally isotropic, direct complement.

COROLLARY 8.14. *There is a canonical isomorphism*

$$K_0(X \times {}^{Q(A,\Lambda)}X/\varepsilon(X \times {}^{Q(A,\Lambda)}X)) \simeq WQ_1(A,\Lambda)_X .$$

8.14 follows from 8.13 and 8.7.

Proof. Let $(P, \varepsilon(\gamma), P) \in$ image $\varepsilon : S(A,\bar{\Lambda})_X \to X \times_{Q(A,\Lambda)} X$. For some $Q \in X$, $\varepsilon(\gamma \oplus 0_Q) \in$ commutator subgroup of $\mathrm{Aut}(\mathbf{H}(P \oplus Q))$. Thus, the class of $(P, \varepsilon(\gamma), P)$ in $WQ_1(A,\Lambda)_X$ vanishes. Thus, we have a canonical map $K_0(X \times_{Q(A,\Lambda)} X/\varepsilon(S(A,\bar{\Lambda})_X)) \to WQ_1(A,\Lambda)_X$. To construct an inverse to this map, it suffices to show that if (M,σ,N), $(N,\rho,P) \in X \times_{Q(A,\Lambda)} X$ then the class $[(M,\sigma,N) \perp (N,\rho,P)]$ of $(M,\sigma,N) \perp (N,\rho,P)$ in $K_0(X \times_{Q(A,\Lambda)} X/\varepsilon(S(A,\bar{\Lambda})_X))$ equals the class $[M,\rho\sigma,P]$ of $(M,\rho\sigma,P)$. Consider first the case that (M,σ,N) is isomorphic to an object in image ε. We can assume $M = N$. Let $\varphi : -\mathbf{H}(M^*) \to \mathbf{H}(M^*)$, $(f,m) \mapsto (-f,m)$, and $\psi : -\mathbf{H}(P^*) \to \mathbf{H}(P)$, $(g,p) \mapsto (-g,p)$. The proof that shows ε is cofinal shows also that $(M,\rho\sigma,P) \perp (M^*,\psi\rho\varphi^{-1},P^*)$ and $(M,\rho,P) \perp (M^*,\psi\rho\varphi^{-1},P^*)$ are isomorphic to objects in image ε. Hence, $[M,\rho\sigma,P] = [M,\rho,P] = [M,\sigma,M] + [M,\rho,P]$. Consider now the general case. Let φ be defined as above and define $\psi : -\mathbf{H}(N^*) \to \mathbf{H}(N)$, $(f,n) \mapsto (-f,n)$. Then $[M,\sigma,N] + [N,\rho,P] = [(M,\sigma,N) \perp (M^*,\psi\sigma\varphi^{-1},N^*)] + [(N,\rho,P) \perp (N,1,N)] -$

$[M^*, \psi\sigma\varphi^{-1}, N^*] =$ (by first case) $[M \oplus M^*, (\rho \perp 1)(\sigma \perp \psi\sigma\varphi^{-1}), P \oplus N^*] - [M^*, \psi\sigma\varphi^{-1}, N^*] = [M, \rho\sigma, P]$.

LEMMA 8.15. *Let* C *and* C' *be categories with product and let* $F : C \to C'$ *be a product preserving functor. Let* C_1 *be a full subcategory of* C *closed under products such that if* $M \in C_1$ *and* $N \in C$ *then* $F(M) \simeq F(N)$ *implies* $N \in C_1$. *Then the canonical map*

$$K_0(C_1 \times_{C'} C_1)/\mathfrak{M}(C_1 \times_{C'} C_1) \to K_0(C \times_{C'} C)/\mathfrak{M}(C \times_{C'} C)$$

is an isomorphism.

COROLLARY 8.16. *The canonical maps below are isomorphisms*

$$WQ_1(A,\Lambda)_{(Y/X)_0} \xrightarrow{\simeq} WQ_1(A,\Lambda)_Y$$

$$WH_1(A,\Lambda)_{(Y/X)_0} \xrightarrow{\simeq} WH_1(A,\Lambda)_Y .$$

Proof. Let $C = Y$ (resp. $s(A,\Lambda)_Y$), $C' = Q(A,\Lambda)$ (resp. $M_0(H(A,\Lambda))$), $F = \Lambda - H$ (resp. $\Lambda - M$), and $C_1 = (Y/X)_0$ (resp. $s(A,\Lambda)_{(Y/X)_0}$). Then 8.16 follows from 8.15.

Proof of 8.15. We shall construct an inverse to the canonical map. Let $(P, \sigma, Q) \in C \times_{C'} C$. If $Q' \in C$ such that $Q \perp Q' \in C_1$ then $(P \perp Q', \sigma \perp 1_{F(Q')}, Q \perp Q') \in C_1 \times_{C'} C_1$. We claim that the class $[P \perp Q', \sigma \perp 1, Q \perp Q']$ of $(P \perp Q', \sigma \perp 1, Q \perp Q')$ in $K_0(C_1 \times_{C'} C_1)/\mathfrak{M}(C_1 \times_{C'} C_1)$ is independent of the choice of Q'. Suppose Q'' satisfies also the conditions on Q'. Then $[P \perp Q', \sigma \perp 1, Q \perp Q'] = [P \perp Q' \perp Q'' \perp Q, \sigma \perp 1 \perp 1 \perp 1, Q \perp Q' \perp Q'' \perp Q] = [P \perp Q'', \sigma \perp 1, Q \perp Q'']$. Next, suppose that (M, σ, N), $(N, \rho, P) \in C \times_{C'} C$. Then $[M \perp P', \sigma \perp 1, N \perp P'] + [N \perp P', \rho \perp 1, P \perp P'] = [M \perp P', (\rho \perp 1)(\sigma \perp 1), P \perp P']$. Hence, the canonical map has an inverse.

THEOREM 8.17. *Let* $X \subset Y$ *be full, cofinal subcategories of* $P(A)$ *closed under direct sums and dualization such that* A *is an object of* X.

Assume X *is cofinal in* Y. *Then there is a sequence*

$$WH_1^{-\lambda}(A,\Lambda)_X \longrightarrow WH_1^{-\lambda}(A,\Lambda)_Y \longrightarrow K_0(s(A,\Lambda)_{(Y/X)_0}/s(A,\Lambda)_X) \xrightarrow{M}$$

$$WH_0^{-\lambda}(A,\Lambda)_X \longrightarrow WH_0^{-\lambda}(A,\Lambda)_Y \longrightarrow \hat{H}^0(Y/X) \xrightarrow{h} WQ_1^{\lambda}(A,\Lambda)_X/\{[P,\sigma_P,P^*] \mid P \in X\} \longrightarrow$$

$$WQ_1^{\lambda}(A,\Lambda)_Y/\{[P,\sigma_P,P^*] \mid P \in Y\} \longrightarrow \hat{H}_0(Y/X) \xrightarrow{H} WQ_0^{\lambda}(A,\Lambda)_X \longrightarrow WQ_0^{\lambda}(A,\Lambda)_Y \longrightarrow$$

$$\hat{H}^0(Y/X)$$

which is exact except perhaps at $WQ_1(A,\Lambda)_Y/\{[P,\sigma_P,P^*] \mid P \in Y\}$ *and* $WH_1(A,\Lambda)_Y$. If $K_0(X) \overset{H}{\hookrightarrow} KQ_0(A,\Lambda)$ (resp. $K_0(s(A,\Lambda)_X) \overset{M}{\hookrightarrow} KH_0(A,\Lambda)$) *then the sequence is exact at* $WQ_1(A,\Lambda)_Y/\{[P,\sigma_P,P^*] \mid P \in Y\}$ *(resp.* $WH_1(A,\Lambda)_Y$).

COROLLARY 8.18. *If* $K_0(X) \overset{H}{\hookrightarrow} KQ_0(A,\Lambda)$ *then there is an exact sequence*

$$\begin{array}{ll}
\hat{H}^0(Y/X) \xrightarrow{h} WQ_1^{\lambda}(A,\min)_X/\{[P,\sigma_P,P^*] \mid P\in X\} \to WQ_1^{\lambda}(A,\min)_Y/\{[P,\sigma_P,P^*] \mid P\in Y\} \to & \hat{H}_0(Y/X) \\
\uparrow & \downarrow H \\
WQ_0^{-\lambda}(A,\min)_Y & WQ_0^{\lambda}(A,\min)_X \\
\uparrow & \downarrow \\
WQ_0^{-\lambda}(A,\min)_X & WQ_0^{\lambda}(A,\min)_Y \\
H\uparrow & \downarrow \\
\hat{H}_0(Y/X) & \hat{H}^0(Y/X) \\
\uparrow & \downarrow h \\
WQ_1^{-\lambda}(A,\min)_Y/\{[P,\sigma_P,P^*] \mid P\in Y\} \leftarrow WQ_1^{-\lambda}(A,\min)_X/\{[P,\sigma_P,P^*] \mid P\in X\} \leftarrow &
\end{array}$$

8.18 is an immediate consequence of 8.17, except perhaps for exactness at $\hat{H}^0(Y/X)$. Here, exactness follows from 8.17 and the additional facts that $WQ_0(A,\max)_Y = WH_0(A,\min)_Y$ and the canonical map $WQ_0(A,\min) \to WQ_0(A,\max)$ is surjective.

The maps in 8.17 are defined as follows:

$WQ_0(A,\Lambda)_Y \to \hat{H}^0(Y/X)$, $WQ_0(A,\Lambda)_X \to WQ_0(A,\Lambda)_Y$, $WQ_1(A,\Lambda)_X/\{[P,\sigma_P,P^*] \mid P \in X\} \to WQ_1(A,\Lambda)_Y/\{[P,\sigma_P,P^*] \mid P \in Y\}$ (resp. $WH_0(A,\Lambda)_Y \to \hat{H}^0(Y/X)$, $WH_0(A,\Lambda)_X \to WH_0(A,\Lambda)_Y$, $WH_1(A,\Lambda)_X \to WH_1(A,\Lambda)_Y$) are the canonical maps.

$\hat{H}_0(Y/X) \xrightarrow{H} WQ_0(A,\Lambda)_X$ (resp. $K_0(s(A,\Lambda)_{(Y/X)_0}) \xrightarrow{M} WH_0(A,\Lambda)_X$) is is induced by the hyperbolic (resp. metabolic) functor.

$h : \hat{H}^0(Y/X) \to WQ_1(A,\Lambda)_X/\{[P,\sigma_P,P^*] \mid P \in X\}$, $[P] \mapsto$ class of $(P\oplus -P, \sigma_P \perp 1_{H(-P)}, P^*\oplus -P)$ where $-P$ is any object of Y such that $P\oplus -P \in X$, and $\sigma_P = \begin{pmatrix} 0 & \lambda I_{P^*} \\ I_P & 0 \end{pmatrix} : H(P) \to H(P^*)$. We explain the definition in detail. Since X is cofinal in Y, every element of $K_0(Y)/K_0(X)$ is represented by the class $[P]$ of an object P of Y. If $[P] = [P^*]$ then $P\oplus X_1 \simeq P^*\oplus X_2$ for suitable X_1 and $X_2 \in X$. Choose $-P' \in Y$ such that $P\oplus -P' \in X$. If $-P = X_2 \oplus -P'$ then $P\oplus -P$ and $P^*\oplus -P \in X$. Thus, if $[P] \in K_0(Y)/K_0(X)$ such that $[P] = [P^*]$ then there is a $-P \in Y$ such that $P\oplus -P$ and $P^*\oplus -P \in X$. Suppose $-P_1 \in Y$ such that $P\oplus -P_1$ and $P^*\oplus -P_1 \in X$. Then we claim that the class of $(P\oplus -P, \sigma_P \perp 1, P^*\oplus -P)$ and $(P\oplus -P_1, \sigma_P \perp 1, P^*\oplus -P_1)$ in $WQ_1(A,\Lambda)_X$ are equal. $[P\oplus -P_1, \sigma_P \perp 1, P^*\oplus -P] = [P\oplus -P\oplus -P_1\oplus P, \sigma_P \perp 1 \perp 1 \perp 1, P^*\oplus -P\oplus -P_1 \oplus P^*] = [P\oplus -P_1, \sigma_P \perp 1, P^*\oplus -P_1]$. Furthermore, $[P]+[P^*] \mapsto [P\oplus -P, \sigma_P \perp 1, P^*\oplus -P] + [P^*\oplus -P, \sigma_{P^*} \perp 1, P\oplus -P] = [P\oplus -P, (\sigma_{P^*} \perp 1)(\sigma_P \perp 1) = \lambda \perp 1, P\oplus -P] = 0$.

Exactness at $WQ_0(A,\Lambda)_Y$ (resp. $WH_0(A,\Lambda)_Y$). $[M,B] \mapsto 0$ implies there is a $P \in Y$ such that $M\oplus P\oplus P^* \in X$. Thus, $[M,B] = [M,B] \perp [H(P)]$ lies in the image of $WQ_0(A,\Lambda)_X$.

Exactness at $WQ_0(A,\Lambda)_X$ (resp. $WH_0(A,\Lambda)_X$). $[M,B] \mapsto 0$ implies there is a $P \in Y$ (resp. $(P,[a]) \in s(A,\Lambda)_{(Y/X)_0}$) such that $(M,B) \perp H(A^n) \simeq H(P)$ (resp. $(M,B) \perp \Lambda - M(A^n) \simeq \Lambda - M(P,[a])$). Thus, $[M,B] = [H(P)]$ (resp. $[\Lambda - M(P,[a])]$) lies in the image of $\hat{H}_0(Y/X)$ (resp. $K_0(s(A,\Lambda)_{(Y/X)_0})$).

Exactness at $\hat{H}_0(Y/X)$ (resp. $K_0(s(A,\Lambda)_{(Y/X)_0})$). 8.8 implies there is an exact sequence $WQ_1(A,\Lambda)_{(Y/X)_0} \longrightarrow K_0((Y/X)_0) \xrightarrow{H} KQ_0(A,\Lambda)$. By 8.16, we can replace $WQ_1(A,\Lambda)_{(Y/X)_0}$ by $WQ_1(A,\Lambda)_Y$. Since X is cofinal in $P(A)$, it follows that $KQ_0(A,\Lambda)_X \hookrightarrow KQ_0(A,\Lambda)$. Since the image of H above lies in $KQ_0(A,\Lambda)_X$, it follows that there is an exact sequence $WQ_1(A,\Lambda)_Y \to K_0((Y/X)_0) \to KQ_0(A,\Lambda)_X$. From this exact sequence, we deduce an exact sequence $WQ_1(A,\Lambda)_Y \to K_0((Y/X)_0/X) \to WQ_0(A,\Lambda)_X$. If ψ denotes the canonical map $K_0((Y/X)_0/X) \to \hat{H}_0(Y/X)$ then ψ is surjective and H kills the $\ker\psi$. Hence, there is an exact sequence $WQ_1(A,\Lambda)_Y \to \hat{H}_0(Y/X) \to WQ_0(A,\Lambda)_X$. Exactness at $\hat{H}_0(Y/X)$ follows. Exactness at $K_0(s(A,\Lambda)_{(Y/X)_0}/s(A,\Lambda)_X)$ is proved similarly.

Exactness at $WQ_1(A,\Lambda)_Y/\{[P,\sigma_P,P^*] | P \in Y\}$ (resp. $WH_1(A,\Lambda)_Y$). $[M_1,\sigma,M_2] \mapsto 0$ implies there are $X_1, X_2 \in X$ and $Q \in Y$ such that $M_1 \oplus Q \oplus X_1 \simeq M_2 \oplus Q^* \oplus X_2$ (resp. $X_1, X_2 \in s(A,\Lambda)_X$ such that $M_1 \oplus X_1 \simeq M_2 \oplus X_2$). Because of our special hypotheses, the equations $H(M_1 \oplus Q \oplus X_1) \simeq H(M_2 \oplus Q^* \oplus X_2)$, $\sigma\colon H(M_1) \simeq H(M_2)$, and $H(Q) \simeq H(Q^*)$ imply that stably $X_1 \simeq X_2$ (resp. $M(M_1 \oplus X_1) \simeq M(M_2 \oplus X_2)$ and $\sigma\colon M(M_1) \simeq M(M_2)$ imply that stably $X_1 \simeq X_2$). Hence after stabilizing X, we can write $M_1 \oplus Q \oplus X_1 \simeq M_2 \oplus Q^* \oplus X_1$ (resp. $M_1 \oplus X_1 \simeq M_2 \oplus X_1$). Choose $N \in Y$ (resp. $N \in s(A,\Lambda)_{(Y/X)_0}$) such that $M_1 \oplus Q \oplus X_1 \oplus N \in X$ (resp. $M_1 \oplus X_1 \oplus N_1 \in s(A,\Lambda)_X$). Then $[M_1,\sigma,M_2] = [M \oplus Q \oplus X_1 \oplus N, \sigma \perp \sigma_Q \perp 1_{H(X_1 \oplus N)}, M_2 \oplus Q^* \oplus X_1 \oplus N]$ lies in the image of $WQ_1(A,\Lambda)_X$ (resp. $[M_1,\sigma,M_2] = [M \oplus X_1 \oplus N, \sigma \perp 1_{M(X_1 \oplus N)}, M_2 \oplus X_1 \oplus N]$ lies in the image of $WH_1(A,\Lambda)_X$).

For the rest of the proof of 8.17, we shall identify $WQ_1(A,\Lambda)_Z$, $Z = X$ or Y, as in 8.14. Thus, the elements of $WQ_1(A,\Lambda)_Z$ are equivalence classes of triples (V,P,P_1) where $V \in Q(A,\Lambda)_Z$ and P and P_1 are totally isotropic, direct summands of V such that $P = P^\perp$ and $P_1 = P_1^\perp$; two triples (V,P,P_1), (W,Q,Q_1) are equivalent if there are elementary triples (S,E,E'), (T,F,F') (*elementary* means E and E′ have a common, totally isotropic, direct complement) such that $(V,P,P_1) \perp (S,E,E')$

$\simeq (W,Q,Q_1) \perp (T,F,F')$. A morphism of triples $(V,P,P_1) \to (W,Q,Q_1)$ is an isomorphism $\varphi: V \to W$ in $Q(A,\Lambda)_Z$ such that $\varphi(P) = Q$ and $\varphi(P_1) = Q_1$. The class of the triple $(H(P), P, P^*)$ corresponds to the element $[P, \sigma_P, P^*]$.

Exactness at $WQ_1(A,\Lambda)_X/\{[P,\sigma_P,P^*] | P \in X\}$. $[V,P,Q] \mapsto 0$ implies there are elementary triples (S,E,E') and (T,F,F') such that $(V,P,Q) \perp (S,E,E') \simeq (H(P),P,P^*) \perp (T,F,F')$ for some $P \in Y$. Choose $-P \in Y$ such that $P \oplus -P \in X$. Thus, if $(S_1,E_1,E_1') = (S,E,E') \perp (H(P),-P,-P)$ then $(V,P,Q) \perp (S_1,E_1,E_1') \simeq (H(P \oplus -P), P \oplus -P, P^* \oplus -P) \perp (T,F,F')$. Next, we show that we can assume that $E_1, E_1', F, F' \in X$. Since E_1 and E_1' are isomorphic (because they have a common direct complement) and since F and F' are isomorphic, it is enough to show that E_1 and $F \in X$. After stabilizing (S,E_1,E_1') and (T,F,F') by a suitable elementary triple, we can assume that $E_1 \in X$. The isomorphism above implies that $P \oplus E_1 \simeq P \oplus -P \oplus F$. Since $P \oplus E_1 \in X$, it follows that if we stabilize (S_1,E_1,E_1') and (T,F,F') by $(H(P \oplus -P), P \oplus -P, P \oplus -P)$ then we can assume that E_1 and $F \in X$. Next, we show that we can assume that $P \oplus -P$ and $P^* \oplus -P \in X$. Then it will follow that $[V,P,Q] = [H(P \oplus -P), P \oplus -P, P^* \oplus -P]$ lies in the image of h. The equation $Q \oplus E_1' \simeq P^* \oplus -P \oplus F'$ and the fact that $Q \oplus E_1' \in X$ shows that our assumption can be satisfied if we stabilize $(H(P \oplus -P), P \oplus -P, P^* \oplus -P)$ by $(H(F'), F', F')$.

Exactness at $\hat{H}^0(Y/X)$. Consider the fibre square 8.11. The functor $\varepsilon(\ )^+$ is product preserving, and cofinal by 8.12. Thus, if $h(\ )^+$ were product preserving and cofinal then we could deduce exactness at $\hat{H}^0(Y/X)$ from the formal exact sequence [10, VII 4.3] associated to a fibre square. We shall force $h(\ )^+$ to be product preserving and cofinal by considering the following quotient diagram of 8.11

$$\begin{array}{ccc} [P,B] & K_0(H(A,\bar{\Lambda})_Y) & \longrightarrow & 0 \\ \downarrow & \downarrow & & \downarrow \\ [P] & K_0((Y/X)^0) & \xrightarrow{h} & K_0(X \times^{Q(A,\Lambda)} X/\varepsilon(X \times^{Q(A,\Lambda)} X)) . \end{array}$$

We have used 8.6 to identify $(X \times_{Q(A,\Lambda)} X)^+$ with $X \times^{Q(A,\Lambda)} X$. Since the diagram is commutative, the sequence $K_0(H(A,\bar{\Lambda})_Y) \longrightarrow K_0((Y/X)^0) \xrightarrow{h} K_0(X \times^{Q(A,\Lambda)} X/\varepsilon(X \times^{Q(A,\Lambda)} X))$ is a 0-sequence. We show that it is exact.

Suppose $P \in (Y/X)^0$ such that $h[P] = [H(P \oplus -P), P \oplus -P, P^* \oplus -P] = 0$. Then there are elementary triples (V, E, E') and (W, F, F') such that $(H(P \oplus -P), P \oplus -P, P^* \oplus -P) \perp (V, E, E') \simeq (W, F, F')$.

We can replace in the equation isomorphism by equality. For the rest of the proof we follow almost word for word the proof in [22] of a related result. Let T be a totally isotropic, direct complement to both F and F'. Let Δ be the diagonal subspace $\{(x,x) \mid x \in W\}$ of $W \perp -W$ (if $W =$ the quadratic module (M, B) then $-W = (M, -B)$). If π is the projection $W = F \oplus F^* \to F$ then π sends $P \oplus -P \oplus E'$ onto the direct summand $P \oplus E$ of F. Hence, in $W \perp -W$, $F \perp F^*$ is a totally isotropic, direct complement to Δ such that the projection $W \perp -W = \Delta \oplus F \perp F^* \to \Delta$, $(x,y) \mapsto (\pi(x) + (1-\pi)y, \pi x + (1-\pi)y)(x \in W, y \in -W)$ sends $F' \perp T$ onto the submodule $\Delta_0 = \{(x,x) \in \Delta \mid x \in \pi(P \oplus -P \oplus E') \oplus (1-\pi)T\}$. Since T is a direct complement to F in W, it follows that $(1-\pi)T = F^*$ and that Δ_0 is a direct summand of Δ isomorphic to $P \oplus E \oplus F^*$, with direct complement isomorphic to $-P$. It follows that $(W \perp -W, \Delta, P \oplus -P \oplus E' \perp T) \simeq (H(\Delta_0), \Delta_0, \Delta_0') \perp (H(-P), -P, -P^*)$ where $(H(\Delta_0), \Delta_0, \Delta_0')$ is elementary (see [22, 2.3]). Since Δ and $P \oplus -P \oplus E' \perp T$ are totally isotropic, direct complements, it follows that $(W \perp -W, \Delta, P \oplus -P \oplus E' \perp T) \simeq (H(\Delta), \Delta, \Delta^*)$. Hence, if in the identification 8.6 of the categories $X \times^{Q(A,\Lambda)} X = (X \times_{Q(A,\Lambda)} X)^+$, $(H(\Delta_0), \Delta_0, \Delta_0')$ corresponds to $(H(\Delta_0), \varepsilon(\gamma), H(\Delta_0))^+$ then it follows that $\gamma : \Delta_0 \to \Delta_0^*$ is an isomorphism. Define the $\bar{\Lambda}$-hermitian form B on Δ_0 by $B(x_0, y_0) = \gamma(x_0)(y_0)$. Then, since $\Delta_0 = P \oplus E \oplus F^*$, and E and

$F^* \in X$, it follows that the class of (Δ_0, B) is a preimage for $[P]$. Thus, the sequence $K_0(H(A,\bar{\Lambda})_Y) \longrightarrow K_0((Y/X)^0) \xrightarrow{h} K_0(X \times^{Q(A,\Lambda)} X/\varepsilon(X \times^{Q(A,\Lambda)} X))$ is exact. Using the isomorphism $H(A,\bar{\Lambda})_Y \to H(A,\Lambda)_Y$, $(M,B) \mapsto (M,\bar{B})$, we can replace in the exact sequence above $K_0(H(A,\bar{\Lambda})_Y)$ by $K_0(H(A,\Lambda)_Y)$. From the exact sequence, we can deduce another exact sequence $K_0(H(A,\Lambda)_Y) \longrightarrow K_0((Y/X)^0)/K_0(X) \xrightarrow{h} K_0(X \times^{Q(A,\Lambda)} X/\varepsilon(X \times^{Q(A,\Lambda)} X))/\{[H(P),P,P^*] \mid P \in X\}$. Exactness at $\hat{H}^0(Y/X)$ follows now from the fact that the canonical map $K_0((Y/X)^0)/K_0(X) \to \hat{H}^0(Y/X)$ is surjective and h kills its kernel.

C. *Kernels and cokernels of hyperbolic and metabolic maps*

The purpose of this section is to provide a handy reference for the kernels of the hyperbolic and metabolic functors and the cokernels of the forgetful functors. All of the material will be extracted from the previous section.

Let F denote the forgetful functor.

THEOREM 8.19. *Let* Y *be an involution invariant subgroup of* $K_1(A)$ *and let* π *be the matrix* $\pi = \begin{pmatrix} 0 & -1 \\ \lambda & 0 \end{pmatrix}$. *The following sequences are exact.*

a) $$WH_0^{-\lambda}(A,\Lambda)_{\text{even-based-}Y} \longrightarrow Y \xrightarrow{H} KQ_1^{\lambda}(A,\Lambda).$$
$$[L] \longmapsto [\pi^{\frac{1}{2}\operatorname{rank} L}\bar{L}]$$

b) $$KQ_1(A,\Lambda)_{\text{based-}Y} \xrightarrow{F} \check{H}_0(Y) \longrightarrow KQ_0(A,\Lambda)_{\text{even-based-}0}$$

$$KH_1(A,\Lambda)_{\text{based-}Y} \xrightarrow{F} \check{H}_0(Y) \longrightarrow KH_0(A,\Lambda)_{\text{even-based-}0}$$

where $\check{H}_0(Y) = \{y \in Y \mid y = -\bar{y}\}$.

THEOREM 8.20. *Let* Y *be an involution invariant subgroup of* $K_0(A)$ *which contains the class of* A. *The following sequences are exact.*

a) $WQ_1(A,\Lambda)_Y \longrightarrow Y \xrightarrow{H} KQ_0(A,\Lambda)$

$$WH_1(A,\Lambda)_Y \longrightarrow K_0(s(A,\Lambda)_Y) \xrightarrow{M} KH_0(A,\Lambda)\ .$$

b) $KQ_0^{-\lambda}(A,\Gamma)_Y \xrightarrow{F} Y \xrightarrow{h} WQ_1^{\lambda}(A,\min)_Y$

$$[P] \longmapsto [P,\sigma_P,P^*]$$

$$KH_0^{-\lambda}(A,\Lambda)_Y \xrightarrow{F} Y \xrightarrow{h} WQ_1^{\lambda}(A,\Lambda)_Y$$

$$[P] \longmapsto [P,\sigma_P,P^*]\ .$$

Proof of 8.19. The proof of a) (resp. b)) is similar to the proof of exactness at $\hat{H}^0(Y/X)$ (resp. $\hat{H}_0(Y/X)$) in 8.3.

Proof of 8.20. The proof of a) (resp. the second sequence in b)) is similar to the proof of exactness at $\hat{H}_0(Y/X)$ and $K_0(s(A,\Lambda)_{(Y/X)^0}/s(A,\Lambda)_X)$ (resp. $H^0(Y/X)$) in 8.17. The first sequence in b) follows from the second, after one notes that $KH_0^{-\lambda}(A,\min)_Y = KQ_0^{-\lambda}(A,\max)_Y$ (1.3) and the canonical map $KQ_0^{-\lambda}(A,\Gamma)_Y \to KQ_0^{-\lambda}(A,\max)_Y$ is surjective.

§9. SCALING AND MORITA THEORY

In this chapter, we adapt to our situation the technic of scaling and the Morita theory of Fröhlich and McEvett [15, §8].

Let ${}^{\lambda}(A,\Lambda)$ be a form ring. Let

$$ {}^{\lambda}(A,\Lambda) - \textbf{quad} $$

$$ {}^{\lambda}(A,\Lambda) - \textbf{herm} $$

be respectively the categories with product of Λ-quadratic and Λ-hermitian modules. If u is a unit in the center(A) then the functors

$$ {}^{\lambda}(A,\Lambda) - \textbf{quad} \to {}^{u\bar{u}^{-1}\lambda}(A,u\Lambda) - \textbf{quad} $$

$$ {}^{-\lambda}(A,\Lambda) - \textbf{herm} \to {}^{-u\bar{u}^{-1}\lambda}(A,u\Lambda) - \textbf{herm} $$

$$ (M,\varphi) \mapsto (M,u\varphi) $$

$$ \text{morphism } f \mapsto f $$

called *scaling by* u are product preserving equivalences which preserve nonsingular, hyperbolic, and metabolic modules. They induce isomorphisms $KQ_i^{\lambda}(A,\Lambda) \cong KQ_i^{u\bar{u}^{-1}}(A,u\Lambda)$, $WQ_i^{\lambda}(A,\Lambda) \cong WQ_i^{u\bar{u}^{-1}\lambda}(A,u\Lambda)$, etc., $i = 0,1,2$. The homomorphism $GQ^{\lambda}(A,\Lambda) \to GQ^{u\bar{u}^{-1}\lambda}(A,u\Lambda)$, $\begin{pmatrix}\alpha & \beta\\ \gamma & \delta\end{pmatrix} \mapsto \begin{pmatrix}\alpha & u\beta\\ u^{-1}\gamma & \delta\end{pmatrix} = \begin{pmatrix}u & 0\\ 0 & 1\end{pmatrix}\begin{pmatrix}\alpha & \beta\\ \gamma & \delta\end{pmatrix}\begin{pmatrix}u^{-1} & 0\\ 0 & 1\end{pmatrix}$ induces the isomorphism $KQ_1^{\lambda}(A,\Lambda) \to KQ_1^{u\bar{u}^{-1}\lambda}(A,u\Lambda)$. This disposes of scaling.

Let k be a commutative ring with involution. Let $\lambda, \eta \in k$ such that $\lambda\bar{\lambda} = \eta\bar{\eta} = 1$. Let A and B be k-algebras with compatible involution and let ${}^{\lambda}(A,\Lambda)$ and ${}^{\eta\lambda}(B,\Gamma)$ be form rings. If M is an A-B bimodule, we let $\overline{M}$ be the B-A bimodule whose underlying abelian group is $\{\overline{m} \mid m \in M\}$ and whose bimodule structure is defined by $\overline{b}\,\overline{m}\,\overline{a} = \overline{a\,m\,b}$. If φ is a sesquilinear form on the right B-module M then one says that φ *admits* A if $\varphi(m\,a, m_1) = \varphi(\overline{a}\,m, m_1)$. If φ admits A then the pairing $\overline{M} \times M \to B$, $(\overline{m}, n) \mapsto \varphi(m,n)$, induces a homomorphism $\varphi : \overline{M} \otimes_A M \to B$.

THEOREM 9.1. *Let* P *be an* A-B *bimodule such that* $cp = pc$ *for all* $c \in k$ *and* $p \in P$. *Let* φ *and* ψ *be* η*-hermitian forms on respectively the right* B*-module* P *and the right* A*-module* $\overline{P}$ *such that* φ *admits* A *and* ψ *admits* B. *Assume that*

$$\varphi : \overline{P} \otimes_A P \xrightarrow{\cong} B$$

$$\psi : P \otimes_B \overline{P} \xrightarrow{\cong} A$$

$$\varphi(\overline{p}_1 \otimes p_2)\overline{p}_3 = \overline{p}_1 \psi(p_2 \otimes \overline{p}_3)$$

$$p_1 \varphi(\overline{p}_2 \otimes p_3) = \psi(p_1 \otimes \overline{p}_2) p_3$$

$$\varphi(\overline{p} \otimes \Lambda p) \subset \Gamma$$

$$\psi(p \otimes \Gamma \overline{p}) \subset \eta\Lambda \, .$$

Then the functors

$${}^{\lambda}(A,\Lambda)\text{-quad} \to {}^{\eta\lambda}(B,\Gamma)\text{-quad}$$

$${}^{-\lambda}(A,\Lambda)\text{-herm} \to {}^{-\eta\lambda}(B,\Gamma)\text{-herm}$$

$$(M,\phi) \mapsto (M \otimes_A P, \phi \otimes \varphi),$$

$$\phi \otimes \varphi(m \otimes p, m_1 \otimes p_1) = \varphi(p, \phi(m, m_1) p_1)$$

are product preserving equivalences. Both equivalences preserve non-singularity. Furthermore, under the first equivalence, Λ*-hyperbolic modules correspond to* Γ*-hyperbolic modules, and* Λ*-quadratic modules which* *become hyperbolic over* ${}^{\lambda}(A, \max)$ *correspond to* Γ*-quadratic modules which become hyperbolic over* ${}^{\eta\lambda}(B, \max)$; *under the second equivalence, hyperbolic modules correspond to hyperbolic modules, and* Λ*-metabolic modules correspond to* Γ*-metabolic modules. The equivalences, thus, induce isomorphisms* $KQ_i^{\lambda}(A,\Lambda) \cong KQ_i^{\eta\lambda}(B,\Gamma)$, $WQ_i^{\lambda}(A,\Lambda) \cong WQ_i^{\eta\lambda}(B,\Gamma)$, etc. $i = 0,1,2$.

The proof is similar to that of [15, 8.1]. To verify the first equivalence, one shows that the composite of ${}^{\lambda}(A,\Lambda)$-quad $\to {}^{\eta\lambda}(B,\Gamma)$-quad, $(M,\phi) \mapsto (M \otimes_A P, \phi \otimes \varphi)$, with ${}^{\eta\lambda}(B,\Gamma)$-quad $\to {}^{\eta^2\lambda}(A, \eta\Lambda)$-quad, $(N, \phi') \mapsto (N \otimes_B \overline{P}, \phi' \otimes \psi)$ is isomorphic to scaling by η. One does the same thing for hermitian modules. To verify that nonsingularity is preserved, one can give a simpler proof than the one in [15, 2.2]; namely a quadratic (resp. hermitian) module is nonsingular if and only if it is an orthogonal summand of a hyperbolic (resp. metabolic) module (2.2 and 2.9); thus, the fact that the equivalences preserve orthogonal sums and hyperbolic and metabolic modules implies that they preserve nonsingularity. In going through the details to verify 9.1, it is useful to keep in mind that $\overline{\phi \otimes \varphi} = \overline{\phi} \otimes \overline{\varphi}$. A good reference for the Morita theory of modules is [10, II 3.2, 3.4, 3.5, and 4.1 - 4.4].

Next we construct a Morita equivalence from a module with an η-hermitian form. The results extend [15, 8.2].

THEOREM 9.2. *Let* k *be a commutative ring with involution and let* B *be a* k*-algebra with compatible involution. Let* P *be a faithfully projective, right* B*-module* ([10, II 1.2 *and* 4.4c]). *If* $A = \mathrm{End}_B(P)$ *then* A *is a* k*-algebra and* P *is an* A-B *bimodule such that* $cp = pc$ *for all* $c \in k$ *and* $p \in P$. *Let* φ *be a nonsingular,* η*-hermitian form on the right* B*-module* P. *Using the nonsingularity of* φ, *we define an involution on* A *via* $\varphi(\overline{a}p_1, p_2) = \varphi(p_1, ap_2)$. *Then*

(i) A *is a* k*-algebra with compatible involution.*

(ii) φ *admits* A *and* $\overline{P} \otimes_A P \to B$ *is an isomorphism.*

(iii) *If* $\psi : \overline{P} \otimes P \to A$ *is defined by*

$$\varphi(\overline{p}_1, p_2)\overline{p}_3 = \overline{p}_1\psi(p_2, \overline{p}_3)$$

then

$$p_1\varphi(\overline{p}_2, p_3) = \psi(p_1, \overline{p}_2)p_3$$

ψ *is a nonsingular* η*-hermitian form admitting* B, *and* $P \otimes_B \overline{P} \to A$, $p \otimes \overline{p}_1 \mapsto \psi(p, p_1)$, *is an isomorphism.*

(iv) *If* ${}^{\eta\lambda}(B, \Gamma)$ *is a form ring and if* Λ = *the additive subgroup of* A *generated by all* $a - \lambda\overline{a}$ *and all* $\eta^{-1}\psi(p \otimes \gamma\overline{p})$ *where* $p \in P$ *and* $\gamma \in \Gamma$ *then* ${}^{\lambda}(A, \Lambda)$ *is a form ring and*

$$\varphi(\overline{p} \otimes \Lambda p) \subset \Gamma$$

$$\psi(p \otimes \Gamma\overline{p}) \subset \eta\Lambda .$$

Thus, the conditions of 9.1 for a Morita equivalence are satisfied.

Proof. (i) is trivial.

(ii) Clearly φ admits A. If one considers $\mathrm{Hom}_B(P, B)$ as a B-A bimodule via the definition $(bfa)(p) = b(f(ap))$ then the map $\overline{P} \to \mathrm{Hom}_B(P, B)$, $\overline{p} \mapsto \varphi(p, \ )$, is an isomorphism of B-A bimodules $(b\overline{p}a \mapsto \varphi(\overline{a}p\overline{b}, \_) = b\varphi(\overline{a}p, \_) = b\varphi(p, a\_))$. If we use this map to identify $\overline{P}$ with $\mathrm{Hom}_B(P, B)$ then it follows from [10, II 4.4b] that $\varphi : \overline{P} \otimes_A P \to B$ is an isomorphism.

(iii) Verification of the equation and of the assertion that ψ is a non-singular, η-hermitian form admitting B is straightforward. The rest follows from [10, II 4.4a].

(iv) If $a \in A$ and if $\gamma \in \Gamma$ then $\overline{a}\psi(p \otimes \gamma\overline{p})a = \overline{a}\psi(\overline{p}, \gamma\overline{p})a = \psi(\overline{p}a, \gamma\overline{p}a) = \psi(\overline{a}p \otimes \gamma(\overline{ap})) \in \Lambda$. Furthermore, it is routine to check that $\psi(p \otimes \gamma\overline{p}) = -\lambda\eta^2\overline{\psi(p \otimes \gamma\overline{p})}$. Hence, ${}^{\lambda}(A, \Lambda)$ is a form ring. That $\psi(p \otimes \Gamma\overline{p}) \subset \eta\Lambda$ follows from the definition of Λ. If $\gamma \in \Gamma$ and if $p, q \in P$ then $\varphi(\overline{p} \otimes \eta^{-1}\psi(q \otimes \gamma\overline{q})p) = \eta^{-1}\varphi(\overline{p} \otimes q\,\varphi(\gamma\overline{q} \otimes p)) = \eta^{-1}\varphi(\overline{p} \otimes q)\varphi(\gamma\overline{q} \otimes p) = \eta^{-1}\varphi(\overline{p} \otimes q)\gamma\,\eta\,\varphi(\overline{\overline{p} \otimes q}) = \varphi(\overline{p} \otimes q)\gamma\overline{\varphi(\overline{p} \otimes q)} \in \Gamma$. Hence, $\varphi(\overline{p} \otimes \Lambda p) \subset \Gamma$.

§10. REDUCTION MODULO A COMPLETE IDEAL

In this chapter, ${}^{\lambda}(A, \Lambda)$ is a form ring and $\mathfrak{q}$ is an involution invariant ideal of A such that A is complete in the $\mathfrak{q}$-adic topology. $\Lambda/\mathfrak{q}^n$ denotes the image of Λ in $A/\mathfrak{q}^n$. X and Y are involution invariant subgroups of $K_0(A)$ and $K_1(A)$ as in §1B. Since the canonical map $K_0(A) \to K_0(A/\mathfrak{q}^n)$ is an isomorphism [10, III 2.12], we can identify X with its image in $K_0(A/\mathfrak{q}^n)$. We shall assume that Y contains the kernel of the canonical map $K_1(A) \to K_1(A/\mathfrak{q})$ and we let $Y/\mathfrak{q}^n$ denote the image of Y in $K_1(A/\mathfrak{q}^n)$. To simplify notation, we shall let

based-Y

denote anyone of the following expressions defined in §1B and C

based-$\dot{Y}$
even-based-Y
discr-based-Y .

In an expression of the kind $Q(A, \Lambda)_{X,\text{based-}Y}$ the subscript X,based-Y means X or based-Y.

THEOREM 10.1. *Each of the canonical functors below induces a bijection on isomorphism classes of objects*

$$Q(A, \Lambda)_{X,\text{based-}Y} \to Q(A/\mathfrak{q}, \Lambda/\mathfrak{q})_{X,\text{based-}Y/\mathfrak{q}} .$$

COROLLARY 10.2. *The canonical maps below are isomorphisms*

$$KQ_0(A, \Lambda)_{X,\text{based-}Y} \to KQ_0(A/\mathfrak{q}, \Lambda/\mathfrak{q})_{X,\text{based-}Y/\mathfrak{q}}$$

$$WQ_0(A, \Lambda)_{X,\text{based-}Y} \to WQ_0(A/\mathfrak{q}, \Lambda/\mathfrak{q})_{X,\text{based-}Y/\mathfrak{q}} .$$

THEOREM 10.3. *Let* t *be an integer* ≥ 0. *If for all* $m \geq 2t+1$, $\Lambda \cap \mathfrak{q}^m \subset \{q + \eta\bar{q} \mid q \in \mathfrak{q}^{m-t}, \eta = -\lambda\}$ mod $\mathfrak{q}^{m+1}$ *then each of the canonical functors below induces a bijection on isomorphism classes of objects*

$$H(A,\Lambda)_{X,\text{based-}Y} \to H(A/\mathfrak{q}^{2t+1},\Lambda/\mathfrak{q}^{2t+1})_{X,\text{based-}Y/\mathfrak{q}^{2t+1}} .$$

COROLLARY 10.4. *The canonical maps below are isomorphisms*

$$KH_0(A,\Lambda)_{X,\text{based-}Y} \to KH_0(A/\mathfrak{q}^{2t+1},\Lambda/\mathfrak{q}^{2t+1})_{X,\text{based-}Y/\mathfrak{q}^{2t+1}}$$

$$WH_0(A,\Lambda)_{X,\text{based-}Y} \to WH_0(A/\mathfrak{q}^{2t+1},\Lambda/\mathfrak{q}^{2t+1})_{X,\text{based-}Y/\mathfrak{q}^{2t+1}} .$$

EXAMPLE. A = 2-adic integers $\mathbf{Z}_2$, $\Lambda = \mathbf{Z}_2$, $t = 1$.

EXAMPLE. More generally, A = ring of integers in a 2-adic local field, Λ any form parameter, t = valuation of the ideal generated by $\operatorname{tr}(A) = \{a+\bar{a} \mid a \in A\}$.

We use in our proofs the following standard identification.

LEMMA 10.5. *Let $\mathfrak{a}$ be an involution invariant ideal in A. If V is a finitely generated, projective, right A-module then the canonical homomorphism below is an isomorphism*

$$\operatorname{Hom}_A(V,A)/\operatorname{Hom}_A(V,A)\mathfrak{a} \to \operatorname{Hom}_{A/\mathfrak{a}}(V/V\mathfrak{a}, A/\mathfrak{a}) .$$

Proof. The proof is easy. It suffices to consider the case V is free. But here the isomorphism is obvious.

Proof of 10.1. We shall prove the assertions in the special case in which no subscripts are needed. The general case is handled exactly the same way. First, we shall show that every object (M',B') of $Q(A/\mathfrak{q},\Lambda/\mathfrak{q})$ lifts to an object of $Q(A,\Lambda)$. By 2.10, we can choose $(N',C') \in Q(A/\mathfrak{q},\Lambda/\mathfrak{q})$ such that $(M' \oplus N')$ is free. By [10, III 2,12], there are finitely generated, projective A-modules M and N such that $M' = M/M\mathfrak{q}$, $N' = N/N\mathfrak{q}$ and $M \oplus N$ is free. It is clear that we can lift $B' \oplus C'$ to a sesquilinear form D on $M \oplus N$. If $B = D|_M$ then $(M/M\mathfrak{q}, B) = (M',B')$.

We shall show that $(M, B) \in Q(A, \Lambda)$. Since the diagram below commutes

$$\begin{array}{ccc}
m & \longmapsto & \langle m, \ \rangle_B \\
M & \longrightarrow & \mathrm{Hom}_A(M, A) \\
\downarrow & & \downarrow \\
M/Mq & \longrightarrow \mathrm{Hom}_A(M/Mq, A/q) \cong & \mathrm{Hom}_A(M, A)/\mathrm{Hom}_A(M, A)q \\
m' & \longmapsto & \langle m', \ \rangle_{B'}
\end{array}$$

and the bottom map is an isomorphism, it follows from Nakayama's lemma [10, III 2.2] that the top map is an isomorphism. Hence, $(M, B) \in Q(A, \Lambda)$.

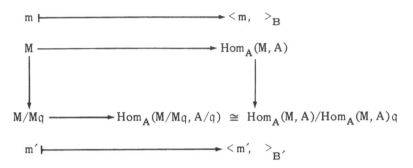

Suppose now that (M, B) and $(N, C) \in Q(A, \Lambda)$ such that $(M/Mq, B) \cong (N/Nq, C)$. Let α' denote this isomorphism. We want to show that $(M, B) \overset{?}{\cong} (N, C)$. Since M is projective, we can lift α' to a homomorphism $\alpha : M \to N$, and then using Nakayama's lemma we can conclude α must be an isomorphism. Thus, it suffices to consider the case that $M = N$ and $(M/Mq, B) = (M/Mq, C)$. Let $\Lambda_i = \Lambda/q^i$. Suppose that we have found isomorphisms $\alpha = \alpha_1, \cdots, \alpha_m : M \to M$ such that $\alpha_i \equiv \alpha_{i+1} \bmod Mq^i$ and such that α_i induces an isomorphism $(M/Mq^i, B) \to (M/Mq^i, C)$ of Λ_i-quadratic modules. We define next $\alpha_{m+1} : M \to M$ such that the sequence $\alpha_1, \cdots, \alpha_{m+1}$ has the same properties. It suffices to consider the case α_m = identity. Thus, $(M/Mq^m, B) = (M/Mq^m, C)$. Let $f : M \to M^*$, $m \mapsto B(m, \ )$, and let $g : M \to M^*$, $m \mapsto C(m, \ )$. Choose P such that $M \oplus P$ is a finitely generated, free module A^n. Then $(M \oplus P/(M \oplus P)q^m, (g-f) \oplus 0)$ is the trivial Λ_m-quadratic module. The crucial step is next. It is where the form parameter takes an active part. Since Λ covers Λ_m, it follows from 13.10 that we can cover $(g-f) \oplus 0 : M \oplus P/(M \oplus P)q^m \to (M \oplus P/(M \oplus P)q^m)^*$ by a homomorphism $k : M \oplus P \to (M \oplus P)^*$ such that $(M \oplus P, k)$ is the trivial Λ-quadratic module. Hence, if $\ell = k|_M$ then $(M, f) = (M, f + \ell)$. Moreover, $f + \ell \equiv g \bmod q^m$. Next, choose $\beta : M \to Mq^m$ such that $f + \ell \equiv g + \beta \bmod q^{m+1}$. It suffices now to find a $\gamma: M \to Mq^m$ such that $(M/Mq^{m+1}, (1+\gamma^*)g(1+\gamma)) = (M/Mq^{m+1}, g+\beta)$. Let $\gamma = (g+\lambda g^*)^{-1}\beta$. Then $(1+\gamma^*)g(1+\gamma) = g+g\gamma+\gamma^*g+\gamma^*g\gamma = g+(g+\lambda g^*)\gamma+(-\lambda g^*\gamma+\gamma^*g)+\gamma^*g\gamma \equiv (\bmod\ q^{m+1})$ $g+\beta-\lambda g^*\gamma+\gamma^*g$.

By induction, we can construct a countably infinite sequence of isomorphisms $\alpha = \alpha_1, \cdots \alpha_m, \cdots : M \to M$ such that α_m induces an isomorphism $(M/Mq^m, B) \to (M/Mq^m, C)$ and such that $\alpha_m \equiv \alpha_{m+1} \bmod Mq^m$. It follows that $\lim_{m\to\infty} \alpha_m$ defines an isomorphism $(M, B) \to (N, C)$.

Proof of 10.3. We shall prove the assertion in the special case in which there are no subscripts. The general case is handled exactly the same way. To begin, we note that every object of $H(A/q^{2t+1}, \Lambda/q^{2t+1})$ lifts to an object of $H(A, \Lambda)$. One establishes this similar to the way one establishes the corresponding fact for 10.1.

Suppose now that (M, B) and $(N, C) \in H(A, \Lambda)$. If $\alpha: M \to N$ is an isomorphism of A-modules such that α induces an isomorphism $(M/Mq^m, B) \to (N/Nq^m, C)$ for some $m \geq 2t+1$ then we shall construct an isomorphism $\alpha': M \to N$ of A-modules such that α' induces an isomorphism $(M/Mq^{m+1}, B) \to (N/Nq^{m+1}, C)$ and such that $\alpha \equiv \alpha' \bmod Mq^{m-t}$. As in the proof of 10.1, it suffices to consider the case $M = N$ and $\alpha =$ identity. Thus, $(M/Mq^m, B) = (M/Mq^m, C)$. Let $f: M \to M^*$, $m \mapsto B(m, \ )$, and let $g: M \to M^*$, $m \mapsto C(m, \ )$. It suffices to find a $\gamma: M \to M^*q^{m-t}$ such that $(M/Mq^{m+1}, (1+\gamma^*)g(1+\gamma)) = (M/Mq^{m+1}, f)$. Choose $\beta: M \to M^*q^m$ such that $g+\beta \equiv f \bmod q^{m+1}$. Since both g and f are Λ-hermitian, it follows that β is Λ-hermitian mod q^{m+1}. The crucial step is next. From the hypotheses on Λ, it follows that we can find a $\sigma: M \to Mq^{m-t}$ such that $\sigma+\lambda\sigma^* \equiv \beta \bmod q^{m+1}$. Choose $\gamma = g^{-1}\sigma$. Then $(1+\gamma^*)g(1+\gamma) = g+g\gamma+\gamma^*g+\gamma^*g\gamma = g+(g\gamma+\eta(g\gamma)^*)-\eta\gamma^*g^*+\gamma^*g+\gamma^*g\gamma = g+(\sigma+\eta\sigma^*)+\gamma^*g\gamma \equiv (\bmod\ q^{m+1})\, g+\beta \equiv (\bmod\ q^{m+1}) f$. Note that $\gamma^*g\gamma \equiv 0 \bmod q^{2(m-t)}$, but $2(m-t) \geq m+1$ because $m \geq 2t+1$.

The proof is completed similarly to the proof of 10.1.

COROLLARY 10.6. a) *The sequence below is exact*

$$K_1(A,q) \xrightarrow{H} KQ_1(A,\Lambda) \longrightarrow KQ_1(A/q,\Lambda/q) \longrightarrow 0 .$$

b) *More generally, if* $Y_0 = \{\sigma \epsilon K_1(A) \mid \sigma\bar{\sigma}^{-1} \epsilon Y\}$ *then the sequence below is exact*

$$Y_0 \xrightarrow{H} KQ_1(A,\Lambda)_Y \longrightarrow KQ_1(A/q,\Lambda/q)_{Y/q} \longrightarrow 0 .$$

c) *The canonical map below is an isomorphism*

$$WQ_1(A,\Lambda) \xrightarrow{\cong} WQ_1(A/q,\Lambda/q) .$$

COROLLARY 10.7. *If* t *satisfies the hypothesis in 10.3 then the canonical map below is surjective*

$$KH_1(A,\Lambda)_Y \to KH_1(A/q^{2t+1},\Lambda/q^{2t+1})_{Y/q^{2t+1}} \to 0 .$$

Proof of 10.6. a) The proof of 10.1 shows that if (M,B) and $(N,C) \epsilon Q(A,\Lambda)$ then any isomorphism $(M/qM,B) \to (N/qN,C)$ can be lifted to an isomorphism $(M,B) \to (N,C)$. The fact that the map $KQ_1(A,\Lambda) \to KQ_1(A/q,\Lambda/q)$ is surjective follows from the special case $(M,B) = H(A^n) = (N,C)$. Since the map $EQ(A,\Lambda) \to EQ(A/q,\Lambda/q)$ is surjective, it follows that the kernel $(KQ_1(A,\Lambda) \to KQ_1(A/q,\Lambda/q))$ is generated by all elements $\begin{pmatrix}\alpha & \beta\\ \gamma & \delta\end{pmatrix}$ in $GQ(A,\Lambda)$ such that $\begin{pmatrix}\alpha & \beta\\ \gamma & \delta\end{pmatrix} \equiv \begin{pmatrix}I & 0\\ 0 & I\end{pmatrix} \bmod q$. Since q is necessarily contained in the Jacobson radical of A, it follows that α is invertible. Thus, $\begin{pmatrix}\alpha & \beta\\ \gamma & \delta\end{pmatrix} = \begin{pmatrix}I & 0\\ \gamma\alpha^{-1} & I\end{pmatrix}\begin{pmatrix}I & \delta\bar{\alpha}\\ 0 & I\end{pmatrix}\begin{pmatrix}\alpha & 0\\ 0 & \bar{\alpha}^{-1}\end{pmatrix} \equiv (\bmod EQ(A,\Lambda))$ $\begin{pmatrix}\alpha & 0\\ 0 & \bar{\alpha}^{-1}\end{pmatrix}$. Thus, the sequence $K_1(A,q) \xrightarrow{H} KQ_1(A,\Lambda) \longrightarrow KQ_1(A/q,\Lambda/q)$ is exact.

b) is proved similarly to a). One replaces $Q(A,\Lambda)$ by $Q(A,\Lambda)_{\text{discr-based-}Y}$ and $Q(A/q,\Lambda/q)$ by $Q(A/q,\Lambda/q)_{\text{discr-based-}Y/q}$.

c) By a), we know that the map $(KQ_1(A,\Lambda)/H(K_1(A,\mathfrak{q}))) \xrightarrow{\cong} KQ_1(A/\mathfrak{q},\Lambda/\mathfrak{q})$ is an isomorphism. By a well-known fact [10, III 2.7], the map $K_1(A) \to K_1(A/\mathfrak{q})$ is surjective. Thus, $WQ_1(A,\Lambda) = \{\text{coker } \mathbf{H} : K_1(A) \to KQ_1(A,\Lambda)/H(K_1(A,\mathfrak{q}))\} \cong \{\text{coker } \mathbf{H} : K_1(A/\mathfrak{q}) \to KQ_1(A/\mathfrak{q},\Lambda/\mathfrak{q})\} = WQ_1(A/\mathfrak{q},\Lambda/\mathfrak{q})$.

Proof of 10.7. The based $\Lambda/\mathfrak{q}^{2t+1}$-metabolic planes $M(A/\mathfrak{q}^{2t+1},\alpha)_{\text{based}}$ $(\alpha \in \Lambda/\mathfrak{q}^{2t+1})$ are cofinal in $H(A/\mathfrak{q}^{2t+1})_{\text{discr-based-}Y/\mathfrak{q}^{2t+1}}$ by 2.10 d). If $\alpha_i \in \Lambda/\mathfrak{q}^{2t+1}$ and if $\tilde{\alpha}_i$ denotes a lifting to Λ then the proof of 10.3 shows that each isomorphism of $M(A/\mathfrak{q}^{2t+1},\alpha_1)_{\text{based}} \perp \cdots \perp M(A/\mathfrak{q}^{2t+1},\alpha_n)_{\text{based}}$ can be lifted to an isomorphism of $M(A,\tilde{\alpha}_1)_{\text{based}} \perp \cdots \perp M(A,\tilde{\alpha}_n)_{\text{based}}$. Thus, the map $KH_1(A,\Lambda)_Y \to KH_1(A/\mathfrak{q}^{2t+1}, \Lambda/\mathfrak{q}^{2t+1})_{Y/\mathfrak{q}^{2t+1}}$ is surjective.

§11. CHANGE OF FORM PARAMETER

In this chapter, we measure the change in the groups $KQ_i(A,\Lambda)$ and $KH_i(A,\Lambda)$ which takes place when one varies the form parameter Λ.

Throughout the chapter, we fix the following data: A denotes a ring with involution $a \mapsto \overline{a}$. λ denotes an element in center(A) such that $\lambda\overline{\lambda} = 1$. Λ and Γ denote form parameters defined with respect to λ. We shall assume that

$$\Lambda \subset \Gamma .$$

A. *The group* $S(\Gamma/\Lambda)$

A matrix $\begin{pmatrix} a & b \\ c & d \end{pmatrix}$ with coefficients in A defines a sesquilinear form B on $A \oplus A$ via the rule $B((v,w),(x,y)) = (\overline{v},\overline{w})\begin{pmatrix} a & b \\ c & d \end{pmatrix}\begin{pmatrix} x \\ y \end{pmatrix} = (\overline{v}a + \overline{w}c, \overline{v}b + \overline{w}d)\begin{pmatrix} x \\ y \end{pmatrix} = \overline{v}ax + \overline{w}cx + \overline{v}by + \overline{w}dy$. If we choose $\{1 \mid 1 \in A\}$ as a basis for A and if we choose the dual basis $\{f \mid f \in A^*, f(1) = 1\}$ for A^* then it is easy to check that the map $H(A) = (A \oplus A^*, B_A) \to (A \oplus A, \begin{pmatrix} 0 & 0 \\ 1 & 0 \end{pmatrix})$, $(1\cdot a, f\cdot b) \mapsto (a,b)$, is an isomorphism of quadratic modules for any form parameter. If one replaces $H(A)$ above by the based hyperbolic plane $H(A)_{\text{based}}$ and if one replaces $(A \oplus A, \begin{pmatrix} 0 & 0 \\ 1 & 0 \end{pmatrix})$ by the based quadratic module $(A \oplus A, \begin{pmatrix} 0 & 0 \\ 1 & 0 \end{pmatrix})_{\text{based}}$ where $A \oplus A$ is given the preferred basis $\{(1,0),(0,1)\}$ then the isomorphism above becomes one of based quadratic modules. Throughout the chapter, we shall use the isomorphisms above to identify

$$H(A) = (A \oplus A, \begin{pmatrix} 0 & 0 \\ 1 & 0 \end{pmatrix})$$

$$H(A)_{\text{based}} = (A \oplus A, \begin{pmatrix} 0 & 0 \\ 1 & 0 \end{pmatrix})_{\text{based}} .$$

DEFINITION 11.1. Let $a, b \in \Gamma$. Define

$$(a,b) = (A \oplus A, \begin{pmatrix} a & 0 \\ 1 & b \end{pmatrix})$$

and if $A \oplus A$ is given the preferred basis $\{(1,0),(0,1)\}$, define

$$(a,b)_{\text{based}} = (A \oplus A, \begin{pmatrix} a & 0 \\ 1 & b \end{pmatrix})_{\text{based}} .$$

The quadratic module (a,b) (resp. $(a,b)_{\text{based}}$) is called a *quasi hyperbolic plane* (resp. *based quasi hyperbolic plane*).

It is clear that $H(A) = (0,0)$ and $H(A)_{\text{based}} = (0,0)_{\text{based}}$. Furthermore, since the λ-hermitian form associated to $\begin{pmatrix} a & 0 \\ 1 & b \end{pmatrix}$ is $\begin{pmatrix} a & 0 \\ 1 & b \end{pmatrix} + \lambda\overline{\begin{pmatrix} a & 0 \\ 1 & b \end{pmatrix}} = \begin{pmatrix} 0 & \lambda \\ 1 & 0 \end{pmatrix}$, it follows that (a,b) and $(a,b)_{\text{based}}$ are nonsingular.

If $x \in A$ and $a \in \Gamma$ then the rule $a \mapsto xa\bar{x}$ (resp. $a \mapsto \bar{x}ax$) induces a left (resp. right) action of A on Γ/Λ. Define the group

$$S(\Gamma/\Lambda) = (\Gamma/\Lambda \otimes_A \Gamma/\Lambda)/\{a \otimes b - b \otimes a,\ a \otimes b - a \otimes ba\bar{b}\} .$$

The letter S is used in the definition of $S(\Gamma/\Lambda)$ to remind one that $S(\Gamma/\Lambda)$ is a quotient of the symmetric tensor product.

THEOREM 11.2. a) *There is an exact sequence*

$$KQ_1(A,\Lambda) \longrightarrow KQ_1(A,\Gamma) \xrightarrow{\partial} S(\Gamma/\Lambda) \xrightarrow{\rho} KQ_0(A,\Lambda) \longrightarrow KQ_0(A,\Gamma) \longrightarrow 0$$

where

$$\rho[a \otimes b] = [a,b] - [0,0]$$

$$\partial\begin{bmatrix} \alpha & \beta \\ \gamma & \delta \end{bmatrix} = \sum_{i=1}^{n} [a_i \otimes b_i] \quad \textit{where}$$

$a_1, \cdots, a_n$ *are the diagonal coefficients of* $\bar{\gamma}\alpha$

$b_1, \cdots, b_n$ *are the diagonal coefficients of* $\bar{\delta}\beta$.

Furthermore, the sequence is natural with respect to involution preserving, ring homomorphisms $f : A \to A'$ *such that*

$$f(\lambda) = \lambda', \quad f(\Lambda) \subset \Lambda', \quad \textit{and} \quad f(\Gamma) \subset \Gamma'.$$

b) *If* X *is an involution invariant subgroup of* $K_1(A)$ *satisfying the appropriate conditions (cf. §1B) then there is an exact sequence*

$$KQ_1(A,\Lambda)_X \to KQ_1(A,\Gamma)_X \to S(\Gamma/\Lambda) \to$$

$$\left\{\begin{matrix} KQ_0(A,\Lambda)_{based\text{-}X} & \to KQ_0(A,\Gamma)_{based\ X} \\ KQ_0(A,\Lambda)_{even\text{-}based\text{-}X} & \to KQ_0(A,\Gamma)_{even\text{-}based\text{-}X} \\ KQ_0(A,\Lambda)_{discr\text{-}based\text{-}X} & \to KQ_0(A,\Gamma)_{discr\text{-}based\text{-}X} \end{matrix}\right\} \to 0$$

where

$$\rho[a \otimes b] = [a,b]_{based} - [0,0]_{based}$$

$$\partial = \textit{as in part a)}.$$

Furthermore, the sequence is natural with respect to involution preserving, ring homomorphisms $f : A \to A'$ *such that* $f(\lambda) = \lambda'$, $f(\Lambda) \subset \Lambda'$, $f(\Gamma) \subset \Gamma'$, *and* $K_1(f)(X) \subset X'$.

For the proof of Theorem 11.2, we need a lemma. Recall that if $\mathbf{C}$ is a subcategory of a category $\mathbf{D}$ then $\mathbf{C}$ is called *full* if every morphism in $\mathbf{D}$ between objects in $\mathbf{C}$ is also a morphism in $\mathbf{C}$.

LEMMA 11.3. *Let* $\mathbf{D}$ *be a category with product. If* $\mathbf{C}$ *is a full, cofinal subcategory of* $\mathbf{D}$ *then the canonical map* $K_0\mathbf{C} \to K_0\mathbf{D}$ *is injective.*

Proof. If M is an object of $\mathbf{C}$ (resp. $\mathbf{D}$), let $[M]$ (resp. $\langle M \rangle$) denote its class in $K_0\mathbf{C}$ (resp. $K_0\mathbf{D}$). Let $[M]-[N] \in K_0\mathbf{C}$ such that $\langle M \rangle - \langle N \rangle = 0$. Then there is an object P in $\mathbf{D}$ such that $M \perp P \cong N \perp P$. Pick Q in $\mathbf{D}$ such that $P \perp Q$ is isomorphic to an object R in $\mathbf{C}$. Clearly, $M \perp R \cong N \perp R$. Thus, $[M] = [N]$.

Proof of Theorem 11.2. The proof of part b) is completely analogous to that of part a). So, we shall prove only part a).

Let F denote the canonical, product preserving functor $F : Q(A,\Lambda) \to Q(A,\Gamma)$. By Theorem 6.20, there is an exact sequence

$$KQ_1(A,\Lambda) \longrightarrow KQ_1(A,\Gamma) \xrightarrow{\partial_0} K_0F \xrightarrow{\rho_0} KQ_0(A,\Lambda) \longrightarrow KQ_0(A,\Gamma)$$

which has the naturality properties required in Theorem 11.2. Furthermore, from the definition of quadratic modules, it is clear that the map $KQ_0(A,\Lambda) \to KQ_0(A,\Gamma)$ is surjective. By (2.10), the functor F is cofinal. Thus, by (6.22), we can canonically identify $K_0'F = K_0F$. Below, we shall define an isomorphism $\theta : S(\Gamma/\Lambda) \to K_0'F$ such that $\theta^{-1}\partial_0 = \partial$ and $\rho_0\theta = \rho$. This will complete the proof of the theorem.

The plan of the rest of the proof is as follows. First, we define a group $KQ_0(A,\Lambda,\Gamma)$. Then we define homomorphisms $\theta_1 : KQ_0(A,\Lambda,\Gamma) \to K_0'F$ and $\theta_2 : S(\Gamma/\Lambda) \to KQ_0(A,\Lambda,\Gamma)$ and show that they are isomorphisms. Then, we set $\theta = \theta_1\theta_2$ and show that θ has the properties indicated above.

If (a_i, b_i) $(i = 1, \cdots, n)$ are quasi hyperbolic planes then there is an obvious isomorphism

$$(a_1,b_1) \perp \cdots \perp (a_n,b_n) \cong \left(A \oplus A, \left(\begin{array}{ccc|ccc} a_1 & & & & & \\ & \ddots & & & & \\ & & a_n & & & \\ \hline 1 & & & b_1 & & \\ & \ddots & & & \ddots & \\ & & 1 & & & b_n \end{array}\right)\right).$$

We shall use this isomorphism to identify the two sides above.

If (a,b) and (c,d) are regarded as objects of $Q(A,\Gamma)$ then the identity map $1 : A \oplus A \to A \oplus A$ defines an isomorphism $(a,b) \to (c,d)$. Let

$$Q(A,\Lambda,\Gamma)$$

denote the full subcategory with product of co F consisting of all objects $(\perp_{i=1}^{n}(a_i, b_i), 1, \perp_{i=1}^{n}(c_i, d_i))$. To simplify notation, we shall write $(\perp(a_i, b_i), \perp(c_i, d_i))$ in place of $(\perp_{i=1}^{n}(a_i, b_i), 1, \perp_{i=1}^{n}(c_i, d_i))$. Let

$$KQ_0(A, \Lambda, \Gamma) = K_0Q(A, \Lambda, \Gamma)/[M, N] + [N, P] = [M, P]$$

and let

$$\theta_1 : KQ_0(A, \Lambda, \Gamma) \to K_0{}'F$$

denote the canonical map.

We shall show that θ_1 is an isomorphism. Let $((M, B), \sigma(N, C)) \in$ **co** F. Since σ defines an isomorphism $M \to N$, we can identify M with N and assume that $\sigma = 1$. To simplify notation, we shall write $((M, B), (N, C))$ in place of $((M, B), 1, (N, C))$. The relations defining $K_0{}'F$ show that $[(N', C'), (N', C')] = 0$ for any Λ-quadratic module (N', C'). Given a Λ-quadratic module (N, C), there is by 2.10 a) a Λ-quadratic module (N', C') such that $(N, C) \perp (N', C') \cong \perp_{i=1}^{n}(0, 0)$ for some n. Thus, for some (N', C'), $((M, B), (N, C)) \perp ((N', C'), (N', C')) \cong ((M', B'), \perp_{i=1}^{n}(0, 0))$ for a suitable (M', B'). We shall show that $((M', B'), \perp_{i=0}^{n}(0, 0)) \cong (\perp_{i=1}^{n}(a_i, b_i), \perp_{i=1}^{n}(0, 0))$ for suitable quasi hyperbolic planes $(a_1, b_1), \cdots, (a_n, b_n)$. This will show that the map $KQ_0(A, \Lambda, \Gamma) \to K_0{}'F$ is surjective. Also, this will show that $Q(A, \Lambda, \Gamma)$ is cofinal in co F and thus, by Lemma 11.3, the map $KQ_0(A, \Lambda, \Gamma) \to K_0{}'F$ is injective as well. We need now a lemma.

LEMMA 11.4. *Let* F *be a free module of rank* n. *Let* B *be an* $n \times n$ *matrix. If* T *is an* $n \times n$ *matrix then the* Λ*-quadratic modules* (F, B) *and* $(F, B+T)$ *have the same associated* Λ*-quadratic and* λ*-hermitian forms* $\Longleftrightarrow$ $T = -\lambda\overline{T}$ *and the diagonal coefficients of* T *lie in* Λ.

The proof of 11.4 is straightforward. Try first the case $n = 2$.

We return now to the proof of 11.2. From the definition of $((M', B'), \perp_{i=1}^{n} (0,0))$, it follows that $M' = A^{2n}$. Recall the identification above of $\perp_{i=1}^{n} (0,0)$ with $\left(A^n \oplus A^n, \begin{pmatrix} 0 & 0 \\ I & 0 \end{pmatrix}\right)$. After making this identification, it follows that $M' = A^n \oplus A^n$. Let

$$I_- = \begin{pmatrix} 0 & 0 \\ I & 0 \end{pmatrix}.$$

Let D be a diagonal matrix, T^- a lower triangular matrix (i.e. all coefficients on and above the main diagonal are zero), and T^+ an upper triangular matrix such that with respect to the basis for M' given by the identification $M' = A^n \oplus A^n$,

$$B' = I_- + D + T^- + T^+ .$$

From the definition of $((M', B'), \perp_{i=1}^{n} (0,0))$, it follows that B' and I_- have the same associated λ-hermitian form $\begin{pmatrix} 0 & \lambda I \\ I & 0 \end{pmatrix}$, and the same associated Γ-quadratic form. Thus, by 11.4, $T^+ = -\lambda \overline{T}^-$ and the coefficients of D lie in Γ. By 11.4, the Λ-quadratic modules $(M', B') = (A^n \oplus A^n, I_- + D + T^- - \lambda \overline{T}^-)$ and $(A^n \oplus A^n, I_- + D)$ have the same associated λ-hermitian and Λ-quadratic forms. Thus, the identity map $A^n \oplus A^n \to A^n \oplus A^n$ defines an isomorphism $(M', B') \cong (A^n \oplus A^n, I_- + D)$. But, if

$$D = \begin{pmatrix} a_1 & & & & & \\ & \ddots & & & & \\ & & a_n & & & \\ & & & b_1 & & \\ & & & & \ddots & \\ & & & & & b_n \end{pmatrix}$$

then $(A^n \oplus A^n, I_- + D) = \perp_{i=1}^{n} (a_i, b_i)$. Thus, $((M', B'), \perp_{i=1}^{n} (0,0)) \cong (\perp_{i=1}^{n} (a_i, b_i), \perp_{i=1}^{n} (0,0))$.

Next, we show that the rule $[a \otimes b] \mapsto [(a,b),(0,0)]$ defines an isomorphism

$$\theta_2 : S(\Gamma/\Lambda) \to KQ_0(A, \Lambda, \Gamma).$$

To simplify notation, we shall write (a,b) in place of $((a,b),(0,0))$. To show that the rule above is a homomorphism, it suffices to show the following: (i) $[a+b, c] = [a,b]+[a,c]$, (ii) $[a, xb\bar{x}] = [\bar{x}ax, b]$, (iii) $[a,b] = [b,a]$, (iv) $[a,b] = [a, ba\bar{b}]$. We shall sometimes use Lemma 11.4 without explicit reference. By definition, a morphism $\perp(a_i, b_i) \to \perp(c_i, d_i)$ in $Q(A, \Lambda, \Gamma)$ is an isomorphism f of Λ-quadratic modules $\perp(A \oplus A, \begin{pmatrix} a_i & 0 \\ 1 & b_i \end{pmatrix}) \to \perp(A \oplus A, \begin{pmatrix} c_i & 0 \\ 1 & d_i \end{pmatrix})$ and an isomorphism g of Λ-quadratic modules $\perp(A \oplus A, \begin{pmatrix} 0 & 0 \\ 1 & 0 \end{pmatrix}) \to \perp(A \oplus A, \begin{pmatrix} 0 & 0 \\ 1 & 0 \end{pmatrix})$ such that the diagram

$$\begin{array}{ccc} A^n \oplus A^n & \xrightarrow{\;1\;} & A^n \oplus A^n \\ {\scriptstyle f}\downarrow & & \downarrow{\scriptstyle g} \\ A^n \oplus A^n & \xrightarrow[\;1\;]{} & A^n \oplus A^n \end{array}$$

commutes. Thus, a morphism, $\perp(a_i, b_i) \to \perp(c_i, d_i)$ in $Q(A, \Lambda, \Gamma)$ is an element $g \in \mathrm{Aut}(\perp^{n} (A \oplus A, \begin{pmatrix} 0 & 0 \\ 1 & 0 \end{pmatrix})) = GQ_{2n}(A, \Lambda)$ such that $g: A^n \oplus A^n \to A^n \oplus A^n$ defines also a morphism of the Λ-quadratic modules $\perp(A \oplus A, \begin{pmatrix} a_i & 0 \\ 1 & b_i \end{pmatrix}) \to \perp(A \oplus A, \begin{pmatrix} c_i & 0 \\ 1 & d_i \end{pmatrix})$. Thus, $g = \left(\begin{array}{cc|cc} 1 & \bar{x} & & \\ & 1 & & \\ \hline & & 1 & 0 \\ & & -x & 1 \end{array}\right)$ defines a morphism $(a,b) \perp (c,d) \to (a + \bar{x}cx, b) \perp (c, d + xb\bar{x})$ in $Q(A, \Lambda, \Gamma)$. Thus, in $KQ_0(A, \Lambda, \Gamma)$

$$(*) \qquad [a,b]+[c,d] = [a+\bar{x}cx, b]+[c, d+xb\bar{x}].$$

Setting $d = b$ and $x = 1$, we obtain that $[a,b]+[c,b] = [a+c,b]+[c,2b]$. To verify (i), it suffices to show that $[c,2b] = 0$. Since $2b \in \Lambda$, it follows from 11.4 that $[c,2b] = [c,0]$. We shall show that $[c,0] = 0$. By definition, $(c,0) = ((c,0),1,(0,0))$. The pair of maps $\begin{pmatrix}1 & -c\\ & 1\end{pmatrix}$ and $\begin{pmatrix}1 & 0\\ 0 & 1\end{pmatrix}$ define an isomorphism $((c,0),\begin{pmatrix}1 & 0\\ 0 & 1\end{pmatrix},(0,0)) \xrightarrow{\cong} ((0,0),\begin{pmatrix}1 & c\\ 0 & 1\end{pmatrix},(0,0))$ in $Q(A,\Lambda,\Gamma)$. We have noted above already that $((0,0),\begin{pmatrix}1 & c\\ 0 & 1\end{pmatrix},(0,0))$ and $((0,0),\begin{pmatrix}1 & c\\ 0 & 1\end{pmatrix},(0,0)) \perp (\overset{n}{\perp}(0,0), I, \overset{n}{\perp}(0,0)) = (\overset{n+1}{\perp}(0,0), \left(\begin{array}{cc|cc}1 & & c & \\ & I & & 0\\ \hline & & 1 & \\ & & & I\end{array}\right), \overset{n+1}{\perp}(0,0))$ determine the same element of $KQ_0(A,\Lambda,\Gamma)$.

Since $EQ(A,\Gamma)$ is by 3.9 the commutator subgroup of $GQ(A,\Gamma)$, it follows that for n sufficiently large ($n=3$ is okay), $\left(\begin{array}{cc|cc}1 & & c & \\ & I & & 0\\ \hline & & 1 & \\ & & & I\end{array}\right)$ can be written as a product of commutators $\prod_j [\alpha_j, \beta_j]$ such that $\alpha_j, \beta_j \in GQ_{2(n+1)}(A,\Gamma)$. But the relation $[M,\rho\sigma,M] = [M,\sigma,M]+[M,\rho,M]$ for $KQ_0(A,\Lambda,\Gamma)$ shows that $\left[\overset{n+1}{\perp}(0,0), \prod_j [\alpha_j,\beta_j], \overset{n+1}{\perp}(0,0)\right] = 0$. Thus, (i) is established. Next, we verify (ii). If in (*) above, we set $a = d = 0$ then we obtain that $-[\bar{x}cx, b] = [c, xb\bar{x}]$. But $-[\bar{x}cx,b] =$ (by 11.4) $-[\bar{x}cx - 2\bar{x}cx, b] =$ (by (i)) $-[\bar{x}cx,b]+2[\bar{x}cx,b] = [\bar{x}cx,b]$. Combining the two equations above, we obtain that $[\bar{x}cx,b] = [c,xb\bar{x}]$. Thus, (ii) is established. The map $\begin{pmatrix}0 & \lambda\\ 1 & 0\end{pmatrix} \in GQ_2(A,\Lambda)$ defines an isomorphism $(a,b) \to (b,a)$ in $Q(A,\Lambda,\Gamma)$. (iii) follows. Next, we verify (iv). The pair of maps $\begin{pmatrix} 1 & -b \\ 0 & 1 \end{pmatrix}$ and $\begin{pmatrix} 1 & -b \\ 0 & 1 \end{pmatrix}$ define an isomorphism $(b, a) = ((b, a), I, (0, 0)) \to ((ba\bar{b}, a), I, (-b, 0))$ in $Q(A, \Lambda, \Gamma)$. Above, we have seen that $(-b, 0) = ((-b, 0), I, (0, 0))$ vanishes in $KQ_0(A, \Lambda, \Gamma)$. Thus, $[b, a] = [(ba\bar{b}, a), I, (-b, 0)] + [(-b, 0), I, (0, 0)] = [(ba\bar{b}, a), I, (0, 0)] = [ba\bar{b}, a]$. (iv) follows now from (iii).

To show that $\theta_2 : S(\Gamma/\Lambda) \to KQ_0(A, \Lambda, \Gamma)$ is an isomorphism, we shall construct an inverse homomorphism. Let $(\perp_{i=1}^{n} (a_i, b_i), \sigma, \perp_{i=1}^{n} (c_i, d_i)) \in Q(A, \Lambda, \Gamma)$. From 11.4, we know that as Γ-quadratic modules, $(A \oplus A, \begin{pmatrix} a_i & 0 \\ 1 & b_i \end{pmatrix}) = (A \oplus A, \begin{pmatrix} c_i & 0 \\ 1 & d_i \end{pmatrix}) = (A \oplus A, \begin{pmatrix} 0 & 0 \\ 1 & 0 \end{pmatrix})$. Thus, σ defines an element of $GQ_{2n}(A, \Gamma)$. Conversely, if $\sigma \in GQ_{2n}(A, \Gamma)$ then by 11.4, the matrix

$$\sigma^{-1} \left(\begin{array}{ccc|ccc} a_1 & & & & & \\ & \ddots & & & & \\ & & a_n & & & \\ \hline 1 & & & b_1 & & \\ & \ddots & & & \ddots & \\ & & 1 & & & b_n \end{array} \right) \bar{\sigma}^{-1} = I_- + D + T - \lambda \bar{T}$$

for some matrix T and some diagonal matrix

$$D = \left(\begin{array}{ccc|ccc} a_1' & & & & & \\ & \ddots & & & & \\ & & a_n' & & & \\ \hline & & & b_1' & & \\ & & & & \ddots & \\ & & & & & b_n' \end{array} \right)$$

such that $a_i', b_i' \in \Gamma$. Since $I_- + D + T - \lambda\bar{T}$ and $I_- + D$ have the same associated Λ-quadratic and λ-hermitian forms, it follows that σ defines an isomorphism $\perp_{i=1}^{n} (A \oplus A, \begin{pmatrix} a_i & 0 \\ 1 & b_i \end{pmatrix}) \to \perp_{i=1}^{n} (A \oplus A, \begin{pmatrix} a_i' & 0 \\ 1 & b_i' \end{pmatrix})$ of Λ-quadratic modules. We shall show that the rule $[\perp_{i=1}^{n} (a_i, b_i), \sigma,$

$\overset{n}{\underset{i=1}{\perp}} (c_i, d_i)] \mapsto \sum_{i=1}^{n} [a'_i \otimes b'_i] - [c_i \otimes d_i]$ defines a homomorphism $KQ_0(A, \Lambda, \Gamma)$ $\to S(\Gamma/\Lambda)$. Once this has been done, it will be clear that this homomorphism and $\theta_2 : S(\Gamma/\Lambda) \to KQ_0(A, \Lambda, \Gamma)$ are mutually inverse. To show that the rule above defines a homomorphism, we must show the following: (i) if $(\overset{n}{\underset{i=1}{\perp}} (a_i, b_i), \sigma, \overset{n}{\underset{i=1}{\perp}} (c_i, d_i)) \cong (\overset{n}{\underset{i=1}{\perp}} (v_i, w_i), \rho, \overset{n}{\underset{i=1}{\perp}} (x_i, y_i))$ then $\sum_{i=1}^{n} [a_i' \otimes b_i'] - [c_i \otimes d_i] = \sum_{i=1}^{n} [v_i' \otimes w_i'] - [x_i \otimes y_i]$; (ii) the rule kills the relation $[M, \sigma, N] + [N, \rho, P] = [M, \rho\sigma, P]$.

We verify (i). The first step is to reduce to the case $\sigma = \rho = 1$ and $c_i = d_i = x_i = y_i = 0$ for all i. Clearly, one can replace $(\perp(a_i, b_i), \sigma, \perp(c_i, d_i))$ by $M = ((\perp(a_i, b_i)) \perp (\perp(-c_i, -d_i)),\ \sigma \perp 1, (\perp(c_i, d_i)) \perp (\perp(-c_i, -d_i))$ and $(\perp(v_i, w_i), \rho, \perp(x_i, y_i))$ by $N = ((\perp(v_i, w_i)) \perp (\perp(-x_i, -y_i)), \rho \perp 1, (\perp(x_i, y_i)) \perp (\perp(-x_i, -y_i)))$. Furthermore, since the matrix

$$\varepsilon = \left(\begin{array}{cc|cc} 1 & & 0 & \\ & 1 & & a \\ \hline & & 1 & \\ & & & 1 \end{array}\right)\left(\begin{array}{cc|cc} 1 & & & \\ & 1 & & \\ \hline -\bar{b} & & 1 & \\ & 0 & & 1 \end{array}\right)\left(\begin{array}{cc|cc} 1 & 1 & & \\ 0 & 1 & & \\ \hline & & 1 & 0 \\ & & -1 & 1 \end{array}\right) \in GQ_4(A, \Gamma)$$

defines an isomorphism

$$(A^2 \oplus A^2, \left(\begin{array}{cc|cc} a & & & \\ & -a & & \\ \hline 1 & & b & \\ & 1 & & -b \end{array}\right)) \to (A^2 \oplus A^2 \left(\begin{array}{cc|cc} 0 & & & \\ & 0 & & \\ \hline 1 & & 0 & \\ & 1 & & 0 \end{array}\right))$$

of Λ-quadratic modules, we can replace, for suitable τ and η, M by $(\perp(a_i, b_i) \perp (\perp(-c_i, -d_i)), \tau, \perp(0,0))$ and N by $(\perp(v_i, w_i) \perp (\perp(-x_i, -y_i)), \eta, \perp(0,0))$. Thus, we can assume at the outset that $c_i = d_i = x_i = y_i = 0$ for all i. Furthermore, if the pair of maps (f, g) defines the isomorphism $(\perp(a_i, b_i), \sigma, \perp(0,0)) \to (\perp(v_i, w_i), \rho, \perp(0,0))$ then $(\rho f \sigma^{-1}, g)$ defines an isomorphism $(\perp(a_i', b_i'), 1, \perp(0,0)) \to (\perp(v_i', w_i'), 1, \perp(0,0))$. Thus, one can assume that $\sigma = \rho = 1$. We note that with this assumption, $f = g \in GQ_{4n}(A, \Lambda)$.

Next, we show that we can reduce to the case $n = 1$. Let

$$\alpha = \begin{pmatrix} a_1 & & \\ & \ddots & \\ & & a_n \end{pmatrix}, \quad \beta = \begin{pmatrix} b_1 & & \\ & \ddots & \\ & & b_n \end{pmatrix}, \quad \nu = \begin{pmatrix} v_1 & & \\ & \ddots & \\ & & v_n \end{pmatrix} \text{ and } \omega = \begin{pmatrix} w_1 & & \\ & \ddots & \\ & & w_n \end{pmatrix}.$$

Let $M_n(A)$ denote the ring of $n \times n$ matrices with involution $(a_{ij}) \mapsto \text{transpose}(\overline{a_{ij}})$. Let $M_n(\Lambda)$ (resp. $M_n(\Gamma)$) denote the group of all $\tau \in M_n(A)$ such that $\tau = -\lambda\overline{\tau}$ and the diagonal coefficients of τ lie in Λ (resp. Γ). Thus, $M_n(\Lambda)$ and $M_n(\Gamma)$ are form parameters on $M_n(A)$. Consider the group $S(M_n(\Gamma)/M_n(\Lambda))$. One checks straightforward that the rule $[(a_{ij}) \otimes (b_{ij})] \mapsto \sum_{i=1}^{n} [a_{ii} \otimes b_{ii}]$ defines a homomorphism $\mathrm{tr}: S(M_n(\Gamma)/M_n(\Lambda)) \to S(\Gamma/\Lambda)$. (In fact the map is an isomorphism with inverse

$$[a \otimes b] \mapsto \left[\begin{pmatrix} a & 0 & \cdot & \cdot & 0 \\ 0 & 0 & \cdot & \cdot & 0 \\ \vdots & \vdots & & & \vdots \\ 0 & 0 & \cdot & \cdot & 0 \end{pmatrix} \otimes \begin{pmatrix} b & 0 & \cdot & \cdot & 0 \\ 0 & 0 & \cdot & \cdot & 0 \\ \vdots & \vdots & & & \vdots \\ 0 & 0 & \cdot & \cdot & 0 \end{pmatrix}\right].)$$

Clearly, $\mathrm{tr}\,[\alpha \otimes \beta] = \sum_{i=1}^{n} [a_i \otimes b_i]$ and $\mathrm{tr}\,[\nu \otimes \omega] = \sum_{i=1}^{n} [v_i \otimes w_i]$. Thus, to prove (i), it suffices to show that $[\alpha \otimes \beta] = [\nu \otimes \omega]$. But, this is the case $n = 1$ for the ring $M_n(A)$.

Recall above the map $g \in GQ_4(A, \Lambda)$. Write $g = \begin{pmatrix} M & N \\ P & Q \end{pmatrix}$. Let $a = a_1$, $b = b_1$, $v = v_1$, and $w = w_1$. Since g defines a morphism

$g: (A \oplus A, \begin{pmatrix} a & 0 \\ 1 & b \end{pmatrix}) \to (A \oplus A, \begin{pmatrix} v & 0 \\ 1 & w \end{pmatrix})$ of Λ-quadratic modules, it follows that the matrices $\begin{pmatrix} a & 0 \\ 1 & b \end{pmatrix}$ and

$$\overline{g}\begin{pmatrix} v & 0 \\ 1 & w \end{pmatrix} g = \begin{pmatrix} \overline{M}vM + \overline{P}M + \overline{P}wP & \overline{M}vN + \overline{P}N + \overline{P}wQ \\ \overline{N}vM + \overline{Q}M + \overline{Q}wP & \overline{N}vN + \overline{Q}N + \overline{Q}wQ \end{pmatrix}$$

have the same associated Λ-quadratic and λ-hermitian forms. Thus, by 11.4, the elements a and $\bar{M}vM + \bar{P}M + \bar{P}wP$ define the same element of Γ/Λ and the elements b and $\bar{N}vN + \bar{Q}N + \bar{Q}wQ$ define the same element of Γ/Λ. Thus, it suffices to show that $(\bar{M}vM + \bar{P}M + \bar{P}wP) \otimes (\bar{N}vN + \bar{Q}N + \bar{Q}wQ)$ and $v \otimes w$ define the same element of $S(\Gamma/\Lambda)$. From the definition 3.1 of $GQ(A, \Lambda)$, one knows that $\bar{P}M$ and $\bar{Q}N \in \Lambda$. Thus, one can drop these elements from the tensor product above. By 3.1, $\begin{pmatrix} M & N \\ P & Q \end{pmatrix}^{-1} = \begin{pmatrix} \bar{Q} & \lambda\bar{N} \\ \bar{\lambda}\bar{P} & \bar{M} \end{pmatrix}$ and thus, $M(\lambda\bar{N}) \in \Lambda$. Thus, $\lambda M \overline{\lambda\bar{N}} = N\bar{M} \in \Lambda$. Thus, $[\bar{M}vM \otimes \bar{N}vN] = [\bar{N}vN \otimes \bar{M}vM] = [v \otimes N\bar{M}v\overline{N\bar{M}}] = [v \otimes N\bar{M}] = 0$. Similarly, $[\bar{P}wP \otimes \bar{Q}wQ] = 0$. Thus, $[(\bar{M}vM + \bar{P}wP) \otimes (Nv\bar{N} + \bar{Q}wQ)] = [\bar{M}vM \otimes \bar{Q}wQ] + [\bar{P}wP \otimes \bar{N}vN] = [\bar{Q}wQ \otimes \bar{M}vM] + [\bar{P}wP \otimes \bar{N}vN] = [w \otimes Q\bar{M}v\overline{Q\bar{M}}] + [w \otimes P\bar{N}v\overline{P\bar{N}}] = [w \otimes (Q\bar{M} + P\bar{N})\, v\overline{(Q\bar{M} + P\bar{N})}]$. The equation $\begin{pmatrix} 1 & 0 \\ 0 & 1 \end{pmatrix} = \begin{pmatrix} M & N \\ P & Q \end{pmatrix}\begin{pmatrix} \bar{Q} & \lambda\bar{N} \\ \bar{\lambda}\bar{P} & \bar{M} \end{pmatrix} = \begin{pmatrix} * & * \\ * & \lambda P\bar{N} + Q\bar{M} \end{pmatrix}$ shows that $Q\bar{M} = 1 - \lambda P\bar{N}$. Thus, $[w \otimes (Q\bar{M} + P\bar{N}) v\overline{(Q\bar{M} + P\bar{N})}] = [w \otimes (1 - \lambda P\bar{N} + P\bar{N})\, v\overline{(1 - \lambda P\bar{N} + P\bar{N})}] = [w \otimes v] + [w \otimes \lambda P\bar{N} v \bar{\lambda}\overline{P\bar{N}}] + [w \otimes P\bar{N}\, v\, \overline{P\bar{N}}] = [w \otimes v] + 2[w \otimes P\bar{N}\, v\, \overline{P\bar{N}}] = [w \otimes v] = [v \otimes w]$.

Next, we prove (ii). The technik used to prove (i) can be applied in (ii) to reduce to case $\sigma = \rho = 1$. But here, it is clear that the rule kills the relations.

It remains to show that $\rho_0\theta = \rho$ and $\theta^{-1}\partial_0 = \partial$. If $[a \otimes b] \in S(\Gamma/\Lambda)$ then $\rho_0\theta[a \otimes b] = \rho_0[(a, b), 1, (0, 0)] = [a, b] - [0, 0]$. Thus, $\rho_0\theta = \rho$. Since Γ/Λ has exponent 2, it follows that $S(\Gamma/\Lambda)$ has exponent 2. Thus, $K_0'F$ has exponent 2. Thus, if $\begin{bmatrix} \alpha & \beta \\ \gamma & \delta \end{bmatrix} \in KQ_1(A, \Gamma)$ then $\partial_0\begin{bmatrix} \alpha & \beta \\ \gamma & \delta \end{bmatrix} = \partial_0\begin{bmatrix} \alpha & \beta \\ \gamma & \delta \end{bmatrix}^{-1} = [\perp(0, 0), \begin{pmatrix} \alpha & \beta \\ \gamma & \delta \end{pmatrix}^{-1}, \perp(0, 0)]$. The pair of maps $\begin{pmatrix} \alpha & \beta \\ \gamma & \delta \end{pmatrix}^{-1}$ and $\begin{pmatrix} I & 0 \\ 0 & I \end{pmatrix}$ define an isomorphism $(\perp(0, 0), \begin{pmatrix} \alpha & \beta \\ \gamma & \delta \end{pmatrix}^{-1}, \perp(0, 0)) \to (\perp(a_i, b_i), 1, \perp(0, 0))$ in co F where $a_1, \cdots, a_n$ (resp. $b_1, \cdots, b_n$) are

the diagonal coefficients of $\bar{\gamma}\alpha$ (resp. $\bar{\delta}\beta$). Thus, $\theta^{-1}\partial_0\begin{bmatrix}\alpha & \beta\\ \gamma & \delta\end{bmatrix} =$
$\theta^{-1}[\perp(a_i, b_i), 1, \perp(0,0)] = \sum_{i=1}^{n} [a_i \otimes b_i]$.

The next theorem provides a strong reduction result for $S(\Gamma/\Lambda)$.

THEOREM 11.4. a) *Suppose that* A *has a family of involution invariant ideals* $0 = \mathfrak{q}_0 \subseteq \mathfrak{q}_1 \subseteq \cdots \subseteq \mathfrak{q}_n \subseteq A$ *such that if* Λ_k *and* Γ_k *denote respectively the images of* Λ *and* Γ *in* $A_k = A/\mathfrak{q}_k$ *then for* $k = 0, \cdots, n-1$ *either* $\mathfrak{q}_{k+1}/\mathfrak{q}_k \subseteq \text{annihilator}_{A_k}(\Gamma_k/\Lambda_k)$ *or* A_k *is* $\mathfrak{q}_{k+1}/\mathfrak{q}_k$*-adically complete. Then the canonical map below is an isomorphism*

$$S(\Gamma/\Lambda) \xrightarrow{\cong} S(\Gamma_n/\Lambda_n) .$$

Furthermore, if A_n *is semisimple then in 11.2*

$$\partial = 0$$

and the sequences below are split exact

$$0 \to S(\Gamma/\Lambda) \xrightarrow{\rho} \begin{Bmatrix} KQ_0(A,\Lambda) \to KQ_0(A,\Gamma) \\ \\ WQ_0(A,\Lambda) \to WQ_0(A,\Gamma) \end{Bmatrix} \longrightarrow 0 .$$

b) *As in Theorem 11.2, any of the based versions of part a) are valid.*

The proof of Theorem 11.4 will be given after Lemma 11.6.

The next result gives a condition which guarantees that the hypotheses of 11.4 hold. Call a ring A with involution *trace noetherian* if A is a noetherian module over the subring generated additively by 1 and all $a + \bar{a}$ such that $a \in \text{center}(A)$. For example, any order of characteristic $\neq 2$ or, more generally, any ring A such that $A/2A$ is finite is trace noetherian.

LEMMA 11.5. *Trace noetherian rings satisfy the hypotheses of 10.4.*

Proof. Let k denote the subring of A generated additively by 1 and all $a+\bar{a}$ such that the $a \in$ center A. Let $\mathfrak{p}$ denote the ideal of k generated additively by all $a+\bar{a}$ above. Let $\mathfrak{q}_1 = \mathfrak{p}A$ and $\mathfrak{q}_2$ = inverse image in A of the Jacobson radical of $A/\mathfrak{q}_1$. Since $\mathfrak{p} \subset \mathrm{Ann}_A(\Gamma/\Lambda)$, it follows that $\mathfrak{q}_1 \subset \mathrm{Ann}_A(\Gamma/\Lambda)$. Since $k/\mathfrak{p} = 0$ or $\mathbf{Z}/2\mathbf{Z}$ it follows that $A_1 = A/\mathfrak{q}_1$ is finite. Thus, A_1 is $(\mathfrak{q}_2/\mathfrak{q}_1)$-adically complete and $A_2 = A/\mathfrak{q}_2$ is semisimple.

LEMMA 11.6. *If* A *has a product decomposition* $A = A_1 \times \cdots \times A_n$ *into involution invariant factors* A_i *then* Λ *has a corresponding product decomposition* $\Lambda = \Lambda_1 \times \cdots \times \Lambda_n$ *into form parameters* Λ_i *on* A_i.

Proof. Let $e_1, \cdots, e_n$ denote the system of idempotents corresponding to the decomposition of A; thus e_i is the identity element of $A_i = e_i A e_i$. If $\Lambda_i = e_i \Lambda \bar{e}_i$ then one can check easily that Λ_i is a form parameter on A_i and that $\Lambda = \Lambda_1 \times \cdots \times \Lambda_n$.

Proof of Theorem 11.4. Since the proofs of parts a) and b) are similar, we shall prove only part a). For the first assertion, it suffices to prove the case $n = 1$. Clearly, the map $S(\Gamma/\Lambda) \to S(\Gamma_1/\Lambda_1)$ is surjective. Suppose that $\mathfrak{q}_1 \subset \mathrm{Ann}_A(\Gamma/\Lambda)$. If $a, b \in \Gamma$ and $b \in \mathfrak{q}_1$ then the relation $a \otimes b = a \otimes ba\bar{b}$ shows that $a \otimes b$ represents 0 on $S(\Gamma/\Lambda)$. Thus, $S(\Gamma/\Lambda) \to S(\Gamma_1/\Lambda_1)$ is an isomorphism. Suppose that A is $\mathfrak{q}_1$-adically complete. Let $WQ_1(A,\Lambda) = KQ_1(A,\Lambda)/\{\begin{pmatrix} \alpha & 0 \\ 0 & \bar{\alpha}^{-1} \end{pmatrix} \mid \alpha \in GL(A)\}$. By 11.2, there are exact sequences

$$\begin{array}{ccccccccc}
WQ_1(A,\Lambda) & \longrightarrow & WQ_1(A,\Gamma) & \longrightarrow & S(\Gamma/\Lambda) & \longrightarrow & KQ_0(A,\Lambda) & \longrightarrow & KQ_0(A,\Gamma) \\
\downarrow f_1 & & \downarrow g_1 & & \downarrow & & \downarrow f_0 & & \downarrow g_0 \\
WQ_1(A_1,\Lambda_1) & \longrightarrow & WQ_1(A_1,\Gamma_1) & \longrightarrow & S(\Gamma_1/\Lambda_1) & \longrightarrow & KQ_0(A_1,\Lambda_1) & \longrightarrow & KQ_0(A_1,\Gamma_1)
\end{array}$$

By 10.2 and 10.6 c), f_j and g_j $(k=0,1)$ are isomorphisms. Thus, by the five lemma, $S(\Gamma/\Lambda) \to S(\Gamma_1/\Lambda_1)$ is an isomorphism.

We want to prove now the latter assertions in the theorem. We handle first the special case A is a division ring of characteristic 2. It suffices to show that if $K = \ker(KQ_0(A,\Lambda) \to KQ_0(A,\Gamma))$ and $W = \ker(WQ_0(A,\Lambda) \to WQ_0(A,\Gamma))$ then the canonical map $K \to W$ is an isomorphism and that in the exact sequence 11.2, $KQ_1(A,\Lambda) \longrightarrow KQ_1(A,\Gamma) \xrightarrow{\partial} S(\Gamma/\Lambda) \xrightarrow{\rho} KQ_0(A,\Lambda) \longrightarrow KQ_0(A,\Gamma)$, $\partial = 0$ and ρ has a retract. Since A is a division ring, all modules are free. Thus, $WQ_0(A,\Lambda) = KQ_0(A,\Lambda)/[0,0]$ and $WQ_0(A,\Gamma) = KQ_0(A,\Gamma)/[0,0]$. Since a typical element in K is a difference $\sum_{i=1}^{n}([a_i,b_i]-[0,0])$, it follows that K maps onto W. If $\sum_{i=1}^{n}([a_i,b_i]-[0,0])$ vanishes in W then there is an r such that $(a_1,b_1)\perp\cdots\perp(a_n,b_n)\perp(\overset{r}{\perp}(0,0)) \cong \overset{n+r}{\perp}(0,0)$. Thus, $\sum_{i=1}^{n}([a_i,b_i]-[0,0]) = \sum_{i=1}^{n}[a_i,b_i]+r[0,0] = (n+r)[0,0] = 0$. Thus, $K \to W$ is an isomorphism. Next, we show that $\partial = 0$. By standard stability arguments [1], [11], $KQ_1(A,\Gamma)$ is generated by $GQ_2(A,\Gamma)$. If $\begin{pmatrix} a & b \\ c & d\end{pmatrix} \in GQ_2(A,\Gamma)$ then a or $b \neq 0$. We want to be able to assume that $a \neq 0$. If $a = 0$ then $b \neq 0$, and after multiplying $\begin{pmatrix} a & b \\ c & d\end{pmatrix}$ by $\begin{pmatrix} 0 & b \\ \overline{\lambda b}^{-1} & 0\end{pmatrix}$, we can assume that $a \neq 0$. If $a \neq 0$ then $\begin{pmatrix} a & b \\ c & d\end{pmatrix} = \begin{pmatrix} a & \\ & \bar{a}^{-1}\end{pmatrix}\begin{pmatrix} 1 & \\ \bar{a}c & 1\end{pmatrix}\begin{pmatrix} 1 & a^{-1}b \\ & 1\end{pmatrix}$. An easy exercise using 3.1 shows that $a^{-1}b$ and $\overline{\bar{a}c} \in \Gamma$. Thus, the class of $\begin{pmatrix} a & b \\ c & d\end{pmatrix}$ in $KQ_1(A,\Gamma)$ is the same as that of $\begin{pmatrix} a & \\ & \bar{a}^{-1}\end{pmatrix}$. But $\begin{pmatrix} a & \\ & \bar{a}^{-1}\end{pmatrix}$ lies also in $GQ_2(A,\Lambda)$. Thus, $\partial = 0$. It follows now that the sequence $0 \longrightarrow S(\Gamma/\Lambda) \xrightarrow{\rho'} WQ_0(A,\Lambda) \to WQ_0(A,\Gamma) \to 0$ is exact. Clearly, ρ has a retract $\Longleftrightarrow$ ρ' has one. But, since the char $A = 2$, it follows from 2.9 that $WQ_0(A,\Lambda)$ has exponent 2. Thus, ρ' has a retract.

Next, we handle the general case. Since the map $S(\Gamma/\Lambda) \to S(\Gamma_n/\Lambda_n)$ is an isomorphism and A_n is semisimple and since the canonical map

$\ker(KQ_0(A,\Lambda) \to KQ_0(A,\Gamma)) \to \ker(WQ_0(A,\Lambda) \to WQ_0(A,\Gamma))$ is surjective (because the hyperbolic modules one factors out of $KQ_0(A,\Gamma)$ to get $WQ_0(A,\Gamma)$ come from $KQ_0(A,\Lambda)$), it follows that we can reduce to the case A is semisimple. Factor A as a product of rings $A = B_1 \times \cdots \times B_m$ such that either B_i is simple or B_i is a product $B_i = B'_i \times B'_i$ of simple rings B'_i interchanged by the involution. In the latter case B_i is called *hyperbolic*. By Lemma 11.6, Λ and Γ have corresponding decompositions $\Lambda = \Lambda_1 \times \cdots \times \Lambda_m$ and $\Gamma = \Gamma_1 \times \cdots \times \Gamma_m$. The K-theory and Witt groups have corresponding decompositions, so that we can reduce to the case $A = B_i$. If A is hyperbolic then $\Lambda = \Gamma$ and the assertions we want to prove are trivial. Suppose that A is simple. If the char $A \neq 2$ then $\Lambda = \Gamma$ and the assertions are again trivial. If the char $A = 2$ then we can reduce by Morita theory §9 to the case A is a division ring and note that this case was handled already above.

Next, we compute $S(\Gamma/\Lambda)$ when Theorem 11.4 allows us to reduce to the case that A_n is a semisimple, finite ring. Other computations of $S(\Gamma/\Lambda)$ can be found in [8.2].

THEOREM 11.7. *Assume the data of Theorem 11.4. Assume further that A_n is a semisimple, finite ring. Let* $A_n = \prod_{i=1}^{r} M_{n_i}(k_i)$ *be a product decomposition of A_n into involution invariant, matrix rings $M_{n_i}(k_i)$ over rings k_i such that k_i is either a field or a hyperbolic ring. If*

$$r_\Gamma \quad (\text{resp. } r_\Lambda)$$

denotes the number of fields k_i with characteristic 2 and trivial involution such that $k_i \subset$ image $(\Gamma \to A_n)$ (resp. image $(\Lambda \to A_n)$) then

$$S(\Gamma/\Lambda) \cong (\mathbf{Z}/2\mathbf{Z})^{(r_\Gamma - r_\Lambda)}.$$

Proof. It is clear that we can reduce to the case $n = 1$. By Lemma 11.6, we can reduce further to the case $r = 1$. Let $k = k_1$. If k is hyperbolic or the characteristic $k \neq 2$ or the involution on k is nontrivial then $\Lambda = \Gamma$ and the theorem follows. Suppose now that k has characteristic 2 and trivial involution. By Morita theory §9, $S(\Gamma/\Lambda) \cong S(k \cap \Gamma / k \cap \Lambda)$. Since k is finite, it follows that $k = \{c^2 \mid c \in k\}$. Thus, k has precisely two form parameters, namely 0 and k. If $k \cap \Lambda = k \cap \Gamma$ then it is clear that $S(k \cap \Gamma / k \cap \Lambda) = 0$. In the remaining case, $k \cap \Lambda = 0$ and $k \cap \Gamma = k$. Here, it is an easy exercise to show that $S(k/0) \cong k/\{c + c^2 \mid c \in k\} \cong \mathbf{Z}/2\mathbf{Z}$.

We close our discussion of $S(\Gamma/\Lambda)$ with a useful general observation.

LEMMA 11.8. *If Γ is a sum of form parameters $\Gamma_i (i \in I)$ such that $\Lambda \subset \Gamma_i$ and such that the canonical map $\coprod_{i \in I} \Gamma_i/\Lambda \xrightarrow{\cong} \Gamma/\Lambda$ is an isomorphism then the canonical map below is an isomorphism*

$$\coprod_{i \in I} S(\Gamma_i/\Lambda) \xrightarrow{\cong} S(\Gamma/\Lambda) .$$

Proof. Clear.

B. *The group* $T(\Gamma/\Lambda)$

To motivate the definition of the group $T(\Gamma/\Lambda)$, we develop a few facts concerning the group $GQ(A, \Lambda) \cap EQ(A, \Gamma)$, which will be needed later anyways.

Let

$$EQ(A, \Lambda, \Gamma) = GQ(A, \Lambda) \cap EQ(A, \Gamma) .$$

LEMMA 11.9. *Let π denote the $2n \times 2n$ matrix*

$$\pi = \begin{pmatrix} \begin{pmatrix} 0 & -\lambda \\ 1 & 0 \end{pmatrix} & & \\ & \ddots & \\ & & \begin{pmatrix} 0 & -\lambda \\ 1 & 0 \end{pmatrix} \end{pmatrix}$$

and let

$$\omega = \begin{pmatrix} 0 & \pi \\ -\pi^{-1} & 0 \end{pmatrix}.$$

Let β *and* γ *denote* $2n \times 2n$ *matrices such that*

$$\overline{\gamma}, \beta \text{ are } \Gamma\text{-hermitian,}$$

$$\beta + \overline{\beta}\,\overline{\gamma}\beta \text{ is } \Lambda\text{-hermitian.}$$

Let

$$(\gamma, \beta) = \begin{pmatrix} I & \\ \gamma & I \end{pmatrix} \begin{pmatrix} I & \beta \\ & I \end{pmatrix} \begin{pmatrix} I & \\ -\gamma & I \end{pmatrix}.$$

a) *Then* $(\gamma, \beta) \in EQ(A, \Lambda, \Gamma)$.

b) *Conversely, each element of* $EQ(A, \Lambda, \Gamma)$ *can be written as a product*

$$\omega(\gamma, \beta) \begin{pmatrix} I & \\ \gamma' & I \end{pmatrix} \begin{pmatrix} \varepsilon & \\ & \overline{\varepsilon}^{-1} \end{pmatrix}$$

for some n *and some* $2n \times 2n$, $\overline{\Gamma}$*-hermitian matrix* γ' *and some* $\varepsilon \in E_{2n}(A)$. *Furthermore, the class of the matrix above in* $EQ(A, \Lambda, \Gamma)/EQ(A, \Lambda)$ *is the same as that of* (γ, β).

Proof. a) It is clear that $(\gamma, \beta) \in EQ(A, \Gamma)$. Thus, we must show that $(\gamma, \beta) \in GQ(A, \Lambda)$. Multiplying out (γ, β), one obtains that

$$(\gamma, \beta) = \begin{pmatrix} I-\beta\gamma & \beta \\ \gamma\beta\gamma & I+\gamma\beta \end{pmatrix}.$$

By the defining equations (3.2) a) for $GQ(A, \Lambda)$, the matrix (γ, β) defines an element of $GQ(A, \Lambda) \Longleftrightarrow$ the matrices $\overline{(I+\gamma\beta)}\beta$ and $\overline{\gamma\beta\gamma}(I-\beta\gamma)$ are Λ-hermitian. The matrix $\overline{(I+\gamma\beta)}\beta = \beta + \overline{\beta}\,\overline{\gamma}\beta$ is Λ-hermitian, by definition of (γ, β). The matrix $-\overline{\gamma\beta\gamma}(I-\beta\gamma) = -\overline{\gamma}\,\overline{\beta}\,\overline{\gamma} - \overline{\gamma}\,\overline{\beta}\,\overline{\gamma} + \overline{\gamma}\,\overline{(I+\gamma\beta)}\beta\gamma =$ (because $\overline{\gamma} = -\lambda\gamma$) $-\overline{\gamma}\beta\gamma + \lambda\overline{\overline{\gamma}\beta\gamma} + \overline{\gamma}\,\overline{(I+\gamma\beta)}\beta\gamma$ is, thus, also Λ-hermitian.

b) By Theorem 3.10, every element of $EQ(A, \Lambda, \Gamma)$ can be written, for some n, as a product

$$\begin{pmatrix} I & U \\ & I \end{pmatrix} \omega \begin{pmatrix} I & B \\ & I \end{pmatrix} \begin{pmatrix} I & \\ L & I \end{pmatrix} \begin{pmatrix} E & \\ & \bar{E}^{-1} \end{pmatrix}$$

such that $E \in E_{2n}(A)$. If $\gamma = -\pi^{-1}U\pi^{-1}$ then we can rewrite the product above in the form

$$\omega \begin{pmatrix} I & \\ \gamma & I \end{pmatrix} \begin{pmatrix} I & B \\ & I \end{pmatrix} \begin{pmatrix} I & \\ L & I \end{pmatrix} \begin{pmatrix} E & \\ & \bar{E}^{-1} \end{pmatrix} .$$

To complete the proof, it suffices to show that $L = -\gamma + \gamma'$ for some $\bar{\Lambda}$-hermitian matrix γ'. Thus, it suffices to show that $\overline{L + \gamma}$ is Λ-hermitian. Since ω, $\begin{pmatrix} E & \\ & \bar{E}^{-1} \end{pmatrix} \in EQ(A, \Lambda, \Gamma)$, it follows that $\begin{pmatrix} I & \\ \gamma & I \end{pmatrix} \begin{pmatrix} I & B \\ & I \end{pmatrix} \begin{pmatrix} I & \\ L & I \end{pmatrix} \in EQ(A, \Lambda, \Gamma)$. The preceding product is the matrix $\begin{pmatrix} I+BL & B \\ \gamma+L+\gamma BL & I+\gamma B \end{pmatrix}$. From the defining equations (3.2) a) for $GQ(A, \Lambda)$, it follows that the matrices $\overline{(I+\gamma B)}\, B$ and $\overline{(\gamma+L+\gamma BL)}\,(I+BL)$ are Λ-hermitian. We rewrite the last matrix: $\overline{(\gamma+L+\gamma BL)}\,(I+BL) =$ $\overline{(\gamma+(I+\gamma B)\, L)}\,(I+BL) = \bar{\gamma} + \bar{\gamma} BL + \bar{L} + \bar{L}\,\bar{B}\,\bar{\gamma} + \bar{L}\overline{(I+\gamma B)}\, BL =$ (because $\bar{\gamma} = -\lambda\gamma$) $\overline{\gamma + L} + (\bar{\gamma}BL - \lambda(\overline{\bar{\gamma}BL})) + \bar{L}(\overline{(I+\gamma B)}\, B)L$. It follows that $\overline{\gamma + L}$ is a sum of Λ-hermitian matrices. Thus, $\overline{\gamma + L}$ is Λ-hermitian.

LEMMA 11.10. *If* $\alpha \in GL_n(A)$ *and* $\begin{pmatrix} \alpha & \beta \\ \gamma & \delta \end{pmatrix} \in GQ_{2n}(A, \Lambda)$ *then* $\bar{\alpha}^{-1}\gamma$ *and* $\alpha^{-1}\beta$ *are* Λ*-hermitian and*

$$\begin{pmatrix} \alpha & \beta \\ \gamma & \delta \end{pmatrix} = \begin{pmatrix} \alpha & \\ & \bar{\alpha}^{-1} \end{pmatrix} \begin{pmatrix} I & \\ \bar{\alpha}\gamma & I \end{pmatrix} \begin{pmatrix} I & \alpha^{-1}\beta \\ & I \end{pmatrix} .$$

Proof. Since $\begin{pmatrix} \alpha & \\ & \bar{\alpha}^{-1} \end{pmatrix} \in GQ_{2n}(A, \Lambda)$, it follows that

$$\begin{pmatrix} \alpha & \\ & \bar{\alpha}^{-1} \end{pmatrix}^{-1} \begin{pmatrix} \alpha & \beta \\ \gamma & \delta \end{pmatrix} = \begin{pmatrix} I & \alpha^{-1}\beta \\ \bar{\alpha}\gamma & \bar{\alpha}\delta \end{pmatrix} \in GQ_{2n}(A, \Lambda) .$$

The defining equations (3.2) a) for $GQ_{2n}(A, \Lambda)$ show that $\bar{\alpha}\gamma$ is

$\overline{\Lambda}$-hermitian. Thus, $\begin{pmatrix} I & 0 \\ \overline{a}\gamma & I \end{pmatrix}^{-1} \begin{pmatrix} I & a^{-1}\beta \\ \overline{a}\gamma & \overline{a}\delta \end{pmatrix} = \begin{pmatrix} I & a^{-1}\beta \\ 0 & \tau \end{pmatrix} \epsilon\ GQ_{2n}(A, \Lambda)$ where $\tau = \overline{a}\delta - (\overline{a}\gamma)(a^{-1}\beta)$. Next, we show that $\tau = I$. By Lemma 3.1 a), $\begin{pmatrix} I & a^{-1}\beta \\ 0 & \tau \end{pmatrix}^{-1} = \begin{pmatrix} \overline{\tau} & \lambda\overline{a^{-1}\beta} \\ 0 & I \end{pmatrix}$. Thus, $\begin{pmatrix} I & \\ & I \end{pmatrix} = \begin{pmatrix} I & a^{-1}\beta \\ & \tau \end{pmatrix}\begin{pmatrix} \overline{\tau} & \lambda\overline{a^{-1}\beta} \\ & I \end{pmatrix}$ $= \begin{pmatrix} \overline{\tau} & a^{-1}\beta + \lambda\overline{a^{-1}\beta} \\ & \tau \end{pmatrix}$. Thus, $\tau = I$. Now, the defining equations (3.2) a) applied to $\begin{pmatrix} I & a^{-1}\beta \\ & I \end{pmatrix}$ show that $a^{-1}\beta$ is Λ-hermitian.

Let $(W, C) \epsilon EQ(A, \Lambda, \Gamma)$ be defined as in (11.9). By *stabilizing* (W, C), we shall mean replacing (W, C) by either itself or $\left(\begin{pmatrix} W & 0 \\ 0 & 0 \end{pmatrix}, \begin{pmatrix} C & 0 \\ 0 & -\pi \end{pmatrix}\right)$, where π is defined as in (11.9). It is clear that the class of (W, C) in $EQ(A, \Lambda, \Gamma)/EQ(A, \Lambda)$ is not changed by stabilization.

LEMMA 11.11. *Let* $(W, C) \epsilon EQ(A, \Lambda, \Gamma)$. *Then* $(W, C) \epsilon EQ(A, \Lambda) \Longleftrightarrow$ *after stabilizing* (W, C), *there is* $\overline{\Lambda}$*-hermitian matrix* γ *such that the matrix*

$$\begin{pmatrix} C & \pi \\ \pi & -\pi(W-\gamma)\pi \end{pmatrix}$$

is invertible and the discriminant matrix

$$\begin{pmatrix} -\pi^{-1} & \\ & \pi^{-1} \end{pmatrix}\begin{pmatrix} C & \pi \\ \pi & -\pi(W-\gamma)\pi \end{pmatrix}$$

vanishes in $K_1(A)$.

Proof. $\Longrightarrow$) If $(W, C) \epsilon EQ(A, \Lambda)$ then by (3.10) for the form parameter Λ, there are $\begin{pmatrix} I & \\ \gamma & I \end{pmatrix}, \begin{pmatrix} I & \beta \\ & I \end{pmatrix}, \begin{pmatrix} I & \\ \gamma' & I \end{pmatrix} \epsilon EQ(A, \Lambda)$ and $\varepsilon \epsilon E(A)$ such that $\omega\begin{pmatrix} I & \\ W & I \end{pmatrix}\begin{pmatrix} I & C \\ & I \end{pmatrix}\begin{pmatrix} I & \\ -W & I \end{pmatrix} = \omega\begin{pmatrix} I & \\ \gamma & I \end{pmatrix}\begin{pmatrix} I & \beta \\ & I \end{pmatrix}\begin{pmatrix} I & \\ \gamma' & I \end{pmatrix}\begin{pmatrix} \varepsilon & \\ & \overline{\varepsilon}^{-1} \end{pmatrix}$. One can conjugate $\begin{pmatrix} \varepsilon & \\ & \bar{\varepsilon}^{-1} \end{pmatrix}$ past $\begin{pmatrix} I & \\ \gamma' & I \end{pmatrix}$ at the expense of replacing γ' by $\bar{\varepsilon}\gamma'\varepsilon$. After doing this and then transposing terms on the right hand side of the equation above to the left hand side, one can write

$$(\omega\begin{pmatrix} I & \\ \gamma & I \end{pmatrix}\begin{pmatrix} I & \beta \\ & I \end{pmatrix})^{-1} (\omega\begin{pmatrix} I & \\ W & I \end{pmatrix}\begin{pmatrix} I & C \\ & I \end{pmatrix})\begin{pmatrix} I & \\ -(W+\bar{\varepsilon}\gamma'\varepsilon) & I \end{pmatrix} = \begin{pmatrix} \varepsilon & \\ & \bar{\varepsilon}^{-1} \end{pmatrix}.$$

Now, from 3.16, it follows that there are E and F in E(A) and $\begin{pmatrix} I & \\ K & I \end{pmatrix}$ $\begin{pmatrix} I & \\ L & I \end{pmatrix} \in EQ(A,\Gamma)$ such that the left hand side of the equation above is the product

$$\begin{pmatrix} E & \\ & \bar{E}^{-1} \end{pmatrix} \omega \begin{pmatrix} I & \\ K & I \end{pmatrix} \left(\begin{array}{cc|cc} I & & C & \pi \\ & I & \pi & -\pi(W-\gamma)\pi \\ \hline & & I & \\ & & & I \end{array}\right) \begin{pmatrix} I & \\ L & I \end{pmatrix} \begin{pmatrix} F & \\ & \bar{F}^{-1} \end{pmatrix}.$$

Thus,

$$\omega \begin{pmatrix} I & \\ K & I \end{pmatrix} \left(\begin{array}{cc|cc} I & & C & \pi \\ & I & \pi & -\pi(W-\gamma)\pi \\ \hline & & I & \\ & & & I \end{array}\right) \begin{pmatrix} I & \\ L & I \end{pmatrix} = \begin{pmatrix} E^{-1}\varepsilon F^{-1} & \\ & \overline{E^{-1}\varepsilon F^{-1}}^{-1} \end{pmatrix}.$$

Multiplying out the left hand side above, one obtains that

$$\left(\begin{array}{c|c} * & * \\ \hline * & \begin{pmatrix} -\pi^{-1} & \\ & -\pi^{-1} \end{pmatrix} \begin{pmatrix} C & \pi \\ \pi & -\pi(W-\gamma)\pi \end{pmatrix} \end{array}\right) = \left(\begin{array}{c|c} E^{-1}\varepsilon F^{-1} & \\ \hline & \overline{E^{-1}\varepsilon F^{-1}}^{-1} \end{array}\right).$$

Thus, $\begin{pmatrix} C & \pi \\ \pi & -\pi(W-\gamma)\pi \end{pmatrix}$ is invertible and its discriminant matrix

$\begin{pmatrix} -\pi^{-1} & \\ & -\pi^{-1} \end{pmatrix} \begin{pmatrix} C & \pi \\ \pi & -\pi(W-\gamma)\pi \end{pmatrix}$ vanishes in $K_1(A)$.

$\Leftarrow$) The proof will be accomplished in two steps. If $x \in EQ(A,\Lambda,\Gamma)$, let [x] denote its class in $EQ(A,\Lambda,\Gamma)/EQ(A,\Lambda)$.

Step I. $[W, C] = \left[\begin{pmatrix} W-\gamma & 0 \\ 0 & 0 \end{pmatrix}, \begin{pmatrix} C & \pi \\ \pi & -\pi(W-\gamma)\pi \end{pmatrix}\right].$

Step II. If C is invertible and its discriminant matrix $-\pi^{-1}C$ vanishes in $K_1(A)$ then $[W, C] = 1$.

Step I. $[W, C] = \left[\omega\begin{pmatrix} I & \\ \gamma & I \end{pmatrix}\right]^{-1} \left[\omega\begin{pmatrix} I & \\ W & I \end{pmatrix}\begin{pmatrix} I & C \\ & I \end{pmatrix}\begin{pmatrix} I & \\ -W & I \end{pmatrix}\right] =$ (by (3.13), for suitable $E, F \in E_{4n}(A)$ and $\begin{pmatrix} I & \\ L & I \end{pmatrix} \in EQ_{8n}(A, \Gamma)$)

$$\begin{bmatrix} E & \\ & \bar{E}^{-1} \end{bmatrix} \left[\begin{array}{cc|cc} I & & & \\ & I & & \\ \hline W-\gamma & -\bar{\pi} & I & \\ -\bar{\pi} & & & I \end{array}\right] \left[\begin{array}{cc|cc} I & & C & \pi \\ & I & \pi & -\pi(W-\gamma)\pi \\ \hline & & I & \\ & & & I \end{array}\right] \begin{bmatrix} I & \\ L & I \end{bmatrix} \begin{bmatrix} F & \\ & \bar{F}^{-1} \end{bmatrix} =$$

$$\left[\begin{array}{cc|cc} I & & & \\ & I & & \\ \hline W-\gamma & & I & \\ & 0 & & I \end{array}\right] \left[\begin{array}{cc|cc} I & & C & \pi \\ & I & \pi & -\pi(W-\gamma)\pi \\ \hline & & I & \\ & & & I \end{array}\right] \begin{bmatrix} I & \\ L & I \end{bmatrix} = \textcircled{a}.$$

As in the proof of (11.9) b) where it was shown that $L + \gamma$ is $\bar{\Lambda}$-hermitian, one can show that $\left(L + \begin{pmatrix} W-\gamma & 0 \\ 0 & 0 \end{pmatrix}\right)$ is $\bar{\Lambda}$-hermitian. Thus,

$$\textcircled{a} = \textcircled{a} \left[\begin{array}{c|c} I & \\ \hline -\left(L + \begin{pmatrix} W-\gamma & 0 \\ 0 & 0 \end{pmatrix}\right) & I \end{array}\right] = \left[\begin{pmatrix} W-\gamma & 0 \\ 0 & 0 \end{pmatrix}, \begin{pmatrix} C & \pi \\ \pi & -\pi(W-\gamma)\pi \end{pmatrix}\right].$$

Step II. If $-\pi^{-1}C$ banishes in $K_1(A)$ then so does $-C\pi^{-1} = \pi(-\pi^{-1}C)\pi^{-1}$.

Since $[\omega] = 1$, it follows that $[W, C] = [W, C][\omega] = \begin{bmatrix} -C\pi^{-1} & -(I-CW)\pi \\ -(I+WC)\pi^{-1} & -WCW\pi \end{bmatrix}.$

But, by Lemma 11.10, the last matrix vanishes in $EQ(A, \Lambda, \Gamma)/EQ(A, \Lambda)$.

Define the category with product

$$T(\Gamma/\Lambda)$$

as follows: Its objects are pairs (γ, β) of $n \times n$ matrices, n even, such that

$$\bar{\gamma}, \beta \text{ are } \Gamma\text{-hermitian,}$$

$$\beta + \bar{\beta}\bar{\gamma}\beta \text{ is } \Lambda\text{-hermitian.}$$

We shall say that $(\gamma, \beta) = (\gamma', \beta')$ if $\beta = \beta'$ and $\gamma - \gamma'$ is $\bar{\Lambda}$-hermitian. If $\varepsilon \in E_n(A)$ then ε defines a morphism $\varepsilon: (\gamma, \beta) \to (\bar{\varepsilon}^{-1}\gamma\varepsilon^{-1}, \varepsilon\beta\bar{\varepsilon})$. The product is defined by $(\gamma, \beta) \oplus (\gamma', \beta') = \left(\begin{pmatrix}\gamma & \\ & \gamma'\end{pmatrix}, \begin{pmatrix}\beta & \\ & \beta'\end{pmatrix}\right)$. Let π be as in 11.9. By *stabilizing* (γ, β) we shall mean replacing (γ, β) by either itself or $(\gamma, \beta) \oplus (0, -\pi)$ for some π.

Let $(V, B) \in T(\Gamma/\Lambda)$ and suppose that V, B, and π are $n \times n$ matrices. If γ is an $n \times n$, $\bar{\Lambda}$-hermitian matrix such that the matrix

$$\begin{pmatrix} B & \pi \\ \pi & -\pi(V-\gamma)\pi \end{pmatrix}$$

is invertible and such that the discriminant matrix

$$\begin{pmatrix} -\pi^{-1} & \\ & -\pi^{-1} \end{pmatrix} \begin{pmatrix} B & \pi \\ \pi & -\pi(V-\gamma)\pi \end{pmatrix}$$

vanishes in $K_1(A)$, define

$$\Lambda_\gamma(V, B) = \left(\begin{pmatrix} V-\gamma & 0 \\ 0 & 0 \end{pmatrix}, \begin{pmatrix} B & \pi \\ \pi & -\pi(V-\gamma)\pi \end{pmatrix}\right).$$

Define the group

$$T(\Gamma/\Lambda) = K_0T(\Gamma/\Lambda)/\left\{\left[\begin{pmatrix} 0 & 0 \\ 0 & 0 \end{pmatrix}, \begin{pmatrix} 0 & 1 \\ -\lambda & 0 \end{pmatrix}\right] = 0,\ [V, B] = [\Lambda_\gamma(V, B)]\right\}.$$

THEOREM 11.12. *There is an exact sequence*

$$WH_0(A, \Lambda)_{\text{discr-based-0}} \longrightarrow$$

$$WH_0(A, \Gamma)_{\text{discr-based-0}} \xrightarrow{\partial} T(\Gamma/\Lambda) \xrightarrow{\rho} KQ_1(A, \Lambda) \longrightarrow KQ_1(A, \Gamma)$$

where

$$\partial[M, B] = [B^{-1}, B]$$

$$\rho[V, B] = \begin{bmatrix} I & \\ V & I \end{bmatrix}\begin{bmatrix} I & B \\ & I \end{bmatrix}\begin{bmatrix} I & \\ -V & I \end{bmatrix}.$$

Furthermore, the sequence is natural with respect to involution preserving, ring homomorphisms $f: A \to A'$ *such that* $f(\lambda) = \lambda'$, $f(\Lambda) \subset \Lambda'$, *and* $f(\Gamma) \subset \Gamma'$.

To prove the theorem, we shall have to establish a number of lemmas.

LEMMA 11.13. *Let* $(V, B) \in T(\Gamma/\Lambda)$. *If* B *is invertible then* $(V, B) = (B^{-1}, B)$.

Proof. We must show that the diagonal coefficients of $B^{-1} - V$ lie in $\bar{\Lambda}$. By definition, the diagonal coefficients of $B + \bar{B}\bar{V}B$ lie in Λ. Thus, the diagonal coefficients of $\bar{B}^{-1}(B + \bar{B}\bar{V}B)B^{-1} = \bar{B}^{-1} + \bar{V}$ lie in Λ. Thus, the diagonal coefficients of $B^{-1} + V$ lie in $\bar{\Lambda}$. Since the diagonal coefficients of $-2V$ lie in $\bar{\Lambda}$, it follows that the diagonal coefficients of $B^{-1} - V$ lie in $\bar{\Lambda}$.

LEMMA 11.14. *Let* $(V, B) \in T(\Gamma/\Lambda)$. *If* B *is an invertible,* Λ*-hermitian matrix then* (V, B) *vanishes in* $T(\Gamma/\Lambda)$.

Proof. By Lemma 11.13, $(V, B) = (B^{-1}, B)$. Since B is Λ-hermitian, it follows that B^{-1} is $\bar{\Lambda}$-hermitian. Thus, $(B^{-1}, B) = (0, B)$. To prove the lemma, it suffices to show that $(0, B)$ vanishes in $T(\Gamma/\Lambda)$. If $\varepsilon = \varepsilon_{21}(\pi B^{-1})$ then ε defines an isomorphism $(0, B) \oplus (0, B) \to \Lambda_{\pi^{-1}B\pi^{-1} + B^{-1}}(0, B)$. Thus, $2[0, B] = [0, B]$. Thus, $[0, B] = 0$.

LEMMA 11.15. *Let* $(V, B) \in T(\Gamma/\Lambda)$. *If* V *is* $\bar{\Lambda}$*-hermitian then* (V, B) *vanishes in* $T(\Gamma/\Lambda)$.

Proof. By definition, $(V, B) = (0, B)$. By definition, $B + \bar{B}0B$ is Λ-hermitian. Thus, B is Λ-hermitian. If B is an $n \times n$ matrix and if γ is the zero $n \times n$ matrix, then $\Lambda_\gamma(0, B) = \left(\begin{pmatrix} 0 & 0 \\ 0 & 0 \end{pmatrix}, \begin{pmatrix} B & \pi \\ \pi & 0 \end{pmatrix}\right)$. Since

$\begin{pmatrix} B & \pi \\ \pi & 0 \end{pmatrix}$ is invertible and Λ-hermitian, it follows from Lemma 11.14 that

$\left[\begin{pmatrix} 0 & 0 \\ 0 & 0 \end{pmatrix}, \begin{pmatrix} B & \pi \\ \pi & 0 \end{pmatrix}\right] = 0$. Thus, $[0, B] = 0$.

LEMMA 11.16. *Let* $(V, B) \epsilon T(\Gamma/\Lambda)$. *Assume that* V, B, β *and* a *are* $n \times n$ *matrices. If* β *and* $\overline{a}\overline{V}a$ *are* Λ*-hermitian then* $[V, B] = [V, B + a\beta\overline{a}]$.

Proof. We consider first the special case that B is invertible. By Lemma 11.15, $[0, -\beta] = 0$. Thus, $[V, B+a\beta\overline{a}] = [V, B+a\beta\overline{a}] + [0, -\beta]$. If $\varepsilon = \varepsilon_{21}(-\beta\overline{a}B^{-1})\varepsilon_{12}(a)$ then ε defines an isomorphism $(V \oplus 0, (B+a\beta\overline{a}) \oplus -\beta) \xrightarrow{\simeq} (V \oplus 0, B \oplus -(\beta + \beta\overline{a}B^{-1}a\beta))$. Thus, $[V, B+a\beta\overline{a}] = [V, B] + [0, -(\beta+\beta\overline{a}B^{-1}a\beta] =$ (by 11.15) $[V, B]$.

Suppose now that B is arbitrary. Let $\langle V, B\rangle$ denote the class of the matrix $\begin{pmatrix} I & \\ V & I \end{pmatrix}\begin{pmatrix} I & B \\ & I \end{pmatrix}\begin{pmatrix} I & \\ -V & I \end{pmatrix}$ in $EQ(A, \Lambda, \Gamma)/EQ(A, \Lambda)$. By Lemma 11.9 b), $-\langle V, B\rangle = \langle W, C\rangle$ for some $(W, C) \epsilon T(\Gamma/\Lambda)$. To prove the lemma, it suffices to show that $[V \oplus W, B \oplus C] = [V \oplus W, (B+a\beta\overline{a}) \oplus C]$. We leave it as an exercise to show that $\langle V, B\rangle = \langle V, B+a\beta\overline{a}\rangle$. Thus, $1 = \langle V\oplus W, B\oplus C\rangle = \langle V \oplus W, (B+a\beta\overline{a})\oplus C\rangle$. Thus, by Lemma 11.11, after stabilizing (W, C) if necessary, there are $\overline{\Lambda}$-hermitian matrices ρ and τ such that the matrices

$$\left(\begin{array}{cc|cc} B & & \pi & \\ & C & & \pi \\ \hline \pi & & & \\ & \pi & \multicolumn{2}{c}{-\pi\left(\begin{pmatrix} V & \\ & W \end{pmatrix} - \rho\right)\pi} \end{array}\right), \left(\begin{array}{cc|cc} B+a\beta\overline{a} & & \pi & \\ & C & & \pi \\ \hline \pi & & & \\ & \pi & \multicolumn{2}{c}{\pi\left(\begin{pmatrix} V & \\ & W \end{pmatrix} - \tau\right)\pi} \end{array}\right)$$

are invertible and their discriminants vanish in $K_1(A)$. Thus, $[V \oplus W, B \oplus C] = [\Lambda_\rho(V\oplus W, B \oplus C)]$ and $[V \oplus W, (B+a\beta\overline{a})\oplus C] = [\Lambda_\tau(V \oplus W, (B+a\beta\overline{a})\oplus C)]$. But, the right hand sides of both equations are equal by the special case in the paragraph above.

LEMMA 11.17. *Let* $(V,B) \in T(\Gamma/\Lambda)$. *Assume that* V, B, *and* γ *are* $n\times n$ *matrices. If* γ *is* $\bar{\Lambda}$*-hermitian then* $[V,B] = \left[\begin{pmatrix} V & 0\\ 0 & 0\end{pmatrix}, \begin{pmatrix} B & \pi\\ \pi & -\pi(V-\gamma)\pi\end{pmatrix}\right]$.

Proof. Let $<V,B>$ denote the class of the matrix $\begin{pmatrix} I & \\ V & I\end{pmatrix}\begin{pmatrix} I & B\\ & I\end{pmatrix}\begin{pmatrix} I & \\ -V & I\end{pmatrix}$ in $EQ(A,\Lambda,\Gamma)/EQ(A,\Lambda)$. One deduces directly from Corollary 3.13 (with $\beta = 0$) that $<V,B> = \left\langle\begin{pmatrix} V-\gamma & -\bar{\pi}\\ -\bar{\pi} & 0\end{pmatrix}, \begin{pmatrix} B & \pi\\ \pi & -\pi(V-\gamma)\pi\end{pmatrix}\right\rangle$. Since $\begin{pmatrix} -\gamma & -\bar{\pi}\\ -\bar{\pi} & 0\end{pmatrix}$ is $\bar{\Lambda}$-hermitian, it follows that $<V,B> = \left\langle\begin{pmatrix} V & 0\\ 0 & 0\end{pmatrix}, \begin{pmatrix} B & \pi\\ \pi & -\pi(V-\gamma)\pi\end{pmatrix}\right\rangle$.

By Lemma 11.9 b), $-<V,B> = <W,C>$ for some $(W,C) \in T(\Gamma/\Lambda)$. Since

$$1 = \left\langle\begin{pmatrix} V & \\ & W\end{pmatrix}, \begin{pmatrix} B & \\ & C\end{pmatrix}\right\rangle = \left\langle\begin{pmatrix} V & 0 & \\ 0 & 0 & \\ & & W\end{pmatrix}, \begin{pmatrix} B & \pi & \\ \pi & -\pi(V-\gamma)\pi & \\ & & C\end{pmatrix}\right\rangle,$$

it follows from Lemma 11.11 that after stabilizing (W,C) if necessary, there are $\bar{\Lambda}$-hermitian matrices ρ and τ such that the matrices

$$\left(\begin{array}{cc|c} B & & \pi \\ & C & \pi \\ \hline \pi & & \\ & \pi & -\pi\left(\begin{pmatrix} V & \\ & W\end{pmatrix}-\rho\right)\pi \end{array}\right), \left(\begin{array}{ccc|c} B & \pi & & \pi \\ \pi & -\pi(V-\gamma)\pi & & \pi \\ & & C & \pi \\ \hline \pi & & & \\ & \pi & & \\ & & \pi & -\pi\left(\begin{pmatrix} V & 0 & \\ 0 & 0 & \\ & & W\end{pmatrix}-\tau\right)\pi \end{array}\right)$$

are invertible and their discriminants vanish in $K_1(A)$. Thus,

(1) $[V\oplus W, B\oplus C] = [\Lambda_\rho(V\oplus W, B\oplus C)]$, and

(2) $\left[\begin{pmatrix} V & 0\\ 0 & 0\end{pmatrix}\oplus W, \begin{pmatrix} B & \pi\\ \pi & -\pi(V-\gamma)\pi\end{pmatrix}\oplus C\right] = \left[\Lambda_\tau\left(\begin{pmatrix} V & 0\\ 0 & 0\end{pmatrix}\oplus W, \begin{pmatrix} B & \pi\\ \pi & -\pi(V-\gamma)\pi\end{pmatrix}\oplus C\right)\right]$.

To complete the proof, it suffices to show that the right-hand side of (1) is equal to the right-hand side of (2).

By Lemma 11.16, we know that the right-hand sides of (1) and (2) are unchanged if we set $\rho = \tau = 0$. So, we set $\rho = \tau = 0$. If $\varepsilon = \varepsilon_{15}(-I)$ then ε defines an isomorphism from the right-hand side of (2) to

$$\left[\left(\begin{array}{ccc|ccc} V & 0 & & & & \\ 0 & 0 & & & & \\ & & W & & & \\ \hline & & & 0 & & \\ & & & & V & \\ & & & & & 0 \end{array}\right), \left(\begin{array}{ccc|c} B & 0 & & \pi \quad \\ 0 & -\pi(V-\gamma)\pi & & \quad \pi \\ & & C & \qquad \pi \\ \hline \pi & & & \\ & \pi & & -\pi\begin{pmatrix} V & 0 & \\ 0 & 0 & \\ & & W \end{pmatrix}\pi \\ & & \pi & \end{array}\right)\right] =$$

$$\left[\left(\begin{array}{cc|cc} V & & & \\ & W & & \\ \hline & & 0 & \\ & & & 0 \end{array}\right), \left(\begin{array}{cc|c} B & & \pi \quad \\ & C & \quad \pi \\ \hline \pi & & -\pi\begin{pmatrix} V & \\ & W \end{pmatrix}\pi \\ & \pi & \end{array}\right)\right] + \left[\begin{pmatrix} 0 & \\ & V \end{pmatrix}, \begin{pmatrix} -\pi(V-\gamma)\pi & \pi \\ \pi & 0 \end{pmatrix}\right].$$

To complete the proof, we must show that $[\begin{pmatrix} 0 & \\ & V \end{pmatrix}, \varphi] = 0$ when φ is invertible. By Lemma 11.13, $[\begin{pmatrix} 0 & \\ & V \end{pmatrix}, \varphi] = [\varphi^{-1}, \varphi]$. It is clear that the map $WH_0(A,\Gamma)_{\text{discr-based-0}} \to T(\Gamma/\Lambda)$, $[\psi] \mapsto [\psi^{-1}, \psi]$, is a homomorphism and that $[\varphi^{-1}, \varphi]$ lies in the image of this homomorphism. But, by Lemma 2.13 and Corollary 2.12, $[\varphi] = 0$.

LEMMA 11.18. *If* $(V,B), (\begin{pmatrix} V & 0 \\ 0 & 0 \end{pmatrix}, \begin{pmatrix} B & C \\ -\lambda\bar{C} & D \end{pmatrix}) \in T(\Gamma/\Lambda)$ *then* $[V,B] = [\begin{pmatrix} V & 0 \\ 0 & 0 \end{pmatrix}, \begin{pmatrix} B & C \\ -\lambda\bar{C} & D \end{pmatrix}]$.

Proof. By Lemma 11.17, $[\begin{pmatrix} V & 0 \\ 0 & 0 \end{pmatrix}, \begin{pmatrix} B & C \\ -\lambda\bar{C} & D \end{pmatrix}] =$

$$\left[\left(\begin{array}{cc|cc} V & & & \\ & 0 & & \\ \hline & & 0 & \\ & & & 0 \end{array}\right), \left(\begin{array}{cc|cc} B & C & \pi & \\ -\lambda\bar{C} & D & & \pi \\ \hline \pi & & V & \\ & \pi & & 0 \end{array}\right)\right].$$

If $\varepsilon = \varepsilon_{14}(-C\pi^{-1})$ and if $W = \bar{\pi}^{-1}\bar{C}\,V\,C\,\pi^{-1}$ then ε defines an isomorphism from the right-hand side of the equation above to

$$\left[\left(\begin{array}{cc|cc} V & & & \\ & 0 & & \\ \hline & & 0 & \\ & & & W \end{array}\right), \left(\begin{array}{cc|cc} B & & \pi & \\ & D & & \pi \\ \hline \pi & & V & \\ & \pi & & 0 \end{array}\right)\right] = \left[\begin{pmatrix} V & \\ & 0 \end{pmatrix}, \begin{pmatrix} B & \pi \\ \pi & V \end{pmatrix}\right] +$$

$$\left[\begin{pmatrix} 0 & \\ & W \end{pmatrix}, \begin{pmatrix} D & \pi \\ \pi & 0 \end{pmatrix}\right] = \text{(by Lemma 11.17) } [V, B] + \left[\begin{pmatrix} 0 & \\ & W \end{pmatrix}, \begin{pmatrix} D & \pi \\ \pi & 0 \end{pmatrix}\right].$$

If $\varphi = \begin{pmatrix} D & \pi \\ \pi & 0 \end{pmatrix}$ then as in the proof of Lemma 11.17, one can show that $\left[\begin{pmatrix} 0 & \\ & W \end{pmatrix}, \varphi\right] = 0$.

LEMMA 11.19. *If* $(V, B), (V, C) \in T(\Gamma/\Lambda)$ *then* $[V, B+C] = [V,B]+[V,C]$.

Proof. It suffices to show that $[V, B+C]+[V, -B] = [V, C]$. If $\varepsilon = \varepsilon_{12}(I)$ then $[V, B+C]+[V, -B] = [V \oplus V, (B+C) \oplus -B] = [\bar{\varepsilon}^{-1}(V \oplus V)\varepsilon^{-1}$, $\varepsilon((B+C) \oplus -B)\bar{\varepsilon}] = \left[V \oplus 0, \begin{pmatrix} C & -B \\ -B & -B \end{pmatrix}\right] =$ (by Lemma 11.18) $[V, C]$.

Let

$$\tilde{T}(\Gamma/\Lambda)$$

denote the free abelian group generated by the objects of $T(\Gamma/\Lambda)$ modulo the following relations:

(i) $(V, B)+(W, C) = (V \oplus W, B \oplus C)$.

(ii) $(V, B) = \Lambda_\gamma(V, B)$, providing as usual that the matrix $\begin{pmatrix} B & \pi \\ \pi & -\pi(V-\gamma)\pi \end{pmatrix}$ is invertible and its discriminant matrix vanishes in $K_1(A)$.

(iii) $(V, B) = (\varepsilon V \bar{\varepsilon}, \overline{\varepsilon^{-1}} B \varepsilon^{-1})$, providing ε vanishes in $K_1(A)$ and providing B is invertible and its discriminant matrix $-\pi^{-1}B$ vanishes in $K_1(A)$.

LEMMA 11.20. *The canonical map below is an isomorphism*

$$\tilde{T}(\Gamma/\Lambda) \xrightarrow{\ \simeq\ } T(\Gamma/\Lambda).$$

Proof. The map is clearly surjective. To show that it is an isomorphism, it suffices to show that if $(V,B)\in T(\Gamma/\Lambda)$ and $\varepsilon\in E_n(A)$ then (V,B) and $(\varepsilon V\bar{\varepsilon},\bar{\varepsilon}^{-1}B\varepsilon^{-1})$ determine the same element of $\tilde{T}(\Gamma/\Lambda)$. By replacing (V,B) by $(V,B)\oplus(V,-B)$ and by replacing $(\varepsilon V\bar{\varepsilon},\bar{\varepsilon}^{-1}B\varepsilon^{-1})$ by $(\varepsilon V\bar{\varepsilon},\bar{\varepsilon}^{-1}B\varepsilon^{-1})\oplus(V,-B)$, we can reduce to the case that (V,B) vanishes in $KQ_1(A,\Lambda)$. By Lemma 11.11, there is after stabilizing if necessary, a $\bar{\Lambda}$-hermitian matrix γ such that the matrix $\begin{pmatrix} B & \pi \\ \pi & -\pi(V-\gamma)\pi \end{pmatrix}$ is invertible and its discriminant matrix vanishes in $K_1(A)$. The matrix $\begin{pmatrix} \bar{\varepsilon} & \\ & \pi^{-1}\varepsilon^{-1}\pi \end{pmatrix}$ defines an isomorphism $\Lambda_\gamma(V,B)\to\Lambda_{\varepsilon\gamma\bar{\varepsilon}}(\varepsilon V\bar{\varepsilon},\bar{\varepsilon}^{-1}B\varepsilon^{-1})$. Thus, if $[V,B]$ denotes the class of (V,B) in $\tilde{T}(\Gamma/\Lambda)$ then $[V,B]=$ (by (ii)) $[\Lambda_\gamma(V,B)]=$ (by (iii)) $[\Lambda_{\varepsilon\gamma\bar{\varepsilon}}(\varepsilon V\bar{\varepsilon},\bar{\varepsilon}^{-1}B\varepsilon^{-1})]=$ (by (ii)) $[\varepsilon V\bar{\varepsilon},\bar{\varepsilon}^{-1}B\varepsilon^{-1}]$.

Let Ω be a form parameter on A. Define the category with product

$$\hat{H}(A,\Omega)$$

as follows: Its objects are invertible Ω-hermitian matrices B such that B has even rank and the discriminant matrix $-\pi^{-1}B$, where $\operatorname{rank}(\pi)=\operatorname{rank} B$, vanishes in $K_1(A)$. The product is defined by $B\oplus C=\begin{pmatrix} B & 0 \\ 0 & C\end{pmatrix}$.

A morphism $p: B\to C$ is defined when there are decompositions $B=B_1\oplus\cdots\oplus B_n$ and $C=C_1\oplus\cdots\oplus C_n$ such that the set $\{B_1,\cdots,B_n\}=\{C_1,\cdots,C_n\}$ and p is the obvious permutation matrix such that $pB\bar{p}=C$. For example, if $n=2$, $B_1=C_2=\begin{pmatrix} a & 0 \\ 0 & a\end{pmatrix}$ and $B_2=C_1=\begin{pmatrix} b & 0 \\ 0 & b\end{pmatrix}$ then we take $p=\left(\begin{array}{cc|cc} & & 1 & \\ & & & 1 \\ \hline 1 & & & \\ & 1 & & \end{array}\right)$ and we don't consider possibilities such as

$\left(\begin{array}{cc|cc} & & & 1 \\ & & 1 & \\ \hline & 1 & & \\ 1 & & & \end{array}\right)$. Call B *indecomposable* if it is not possible to write $B=B_1\oplus B_2$. Clearly, each B has a unique up-to-order decomposition into indecomposables. Clearly, two objects are isomorphic (and by a unique isomorphism) $\Longleftrightarrow$ their indecomposable decompositions are the same up to order. Let

$$E(\Omega) \subset K_0\hat{H}(A,\Omega)$$

denote the subgroup of $K_0\hat{H}(A,\Omega)$ generated by $\begin{bmatrix} 0 & 1 \\ -\lambda & 0 \end{bmatrix}$ and by all $[B] - \overline{[\epsilon^{-1}B\epsilon^{-1}]}$ such that $B \in \hat{H}(A,\Omega)$ and $\epsilon \in GL_{\mathrm{rank}(B)}(A) \cap E(A)$. Clearly, the map below is an isomorphism

$$KH_0(A,\Omega)_{\text{discr-based-0}} \xrightarrow{\cong} K_0\hat{H}(A,\Omega)/E(\Omega)\,.$$
$$[M,B] \longmapsto [B]\,.$$

Define the category with product

$$\hat{T}(\Gamma/\Lambda)$$

as follows: Its objects and product are those of $T(\Gamma/\Lambda)$. A morphism $p:(V,B) \to (W,C)$ is given when there are decompositions $(V,B) = (V_1,B_1)\oplus\cdots\oplus(V_n,B_n)$ and $(W,C) = (W_1,C_1)\oplus\cdots\oplus(W_n,C_n)$ such that the set $\{(V_i,B_i)|i=1,\cdots,n\} = \{(W_i,C_i)|i=1,\cdots,n\}$ and such that p is the obvious permutation matrix (as above) such that $pB\overline{p} = C$ and $pV\overline{p} = W$. From the definition of equality in $\hat{T}(\Gamma/\Lambda)$, it follows that one can set the off diagonal coefficients of V equal to zero. Thus, given a decomposition $B = B_1 \oplus B_2$, there are matrices V_1 and V_2 such that $(V,B) = (V_1,B_1)\oplus(V_2,B_2)$. Call (V,B) *indecomposable* if one cannot write $B = B_1 \oplus B_2$. Clearly, two objects are isomorphic (and by a unique isomorphism) $\Longleftrightarrow$ their indecomposable decompositions are the same up to order.

Define

$$\hat{T}(\Gamma/\Lambda) = K_0\hat{T}(\Gamma/\Lambda)/\{[V,B] = [\Lambda_\gamma(V,B)]\}$$

where $\Lambda_\gamma(V,B)$ is defined as usual and it is assumed as usual that the right-hand matrix in $\Lambda_\gamma(V,B)$ is invertible and its discriminant matrix vanishes in $K_1(A)$. Let

$$E(\Gamma/\Lambda) \subset \hat{T}(\Gamma/\Lambda)$$

denote the subgroup of $\hat{T}(\Gamma/\Lambda)$ generated by $\left[\begin{pmatrix} 0 & 0 \\ 0 & 0 \end{pmatrix}, \begin{pmatrix} 0 & 1 \\ -\lambda & 0 \end{pmatrix}\right]$ and by all $[V,B] - [\bar{\varepsilon}^{-1} V \varepsilon^{-1}, \varepsilon B \bar{\varepsilon}]$ such that B is invertible and its discriminant matrix $-\pi^{-1}B$ vanishes in $K_1(A)$ and $\varepsilon \in GL_{\mathrm{rank}(B)}(A) \cap E(A)$. By Lemma 11.20, the canonical map below is an isomorphism

$$\hat{T}(\Gamma/\Lambda)/E(\Gamma/\Lambda) \xrightarrow{\cong} T(\Gamma/\Lambda)\,.$$

If (V,B) and $(W,C) \in \hat{T}(\Gamma/\Lambda)$ such that $(V,B) \cong (V_1,B_1) \oplus \cdots \oplus (V_n,B_n)$ and $(W,C) \cong (V_1,B_1) \oplus \cdots \oplus (V_{n-1},B_{n-1}) \oplus \Lambda_\gamma(V_n,B_n)$ then (W,C) is called a *pop up* of (V,B) and (V,B) is called a *pop down* of (W,C). If $(W,C) \cong (V,B)$ then (W,C) and (V,B) are called *trivial* pop ups and pop downs of one another. If there is a chain of pop ups from (V,B) to (W,C) then we shall also say that (W,C) is a *pop up* of (V,B) and that (V,B) is a *pop down* of (W,C). (V,B) and (W,C) are called *related* if one can get from (V,B) to (W,C) by a chain of pop ups and pop downs. (V,B) is called *minimal* if it is not a nontrivial pop up. (V,B) is called a *root* of (W,C) if (V,B) is minimal and (W,C) is a pop up of (V,B).

KEY LEMMA 11.21. *Let* $(V,B),(W,C) \in \hat{T}(\Gamma/\Lambda)$. *Suppose that* (V,B) *and* (W,C) *are related. If* (V,B) *is minimal then* (W,C) *is a pop up of* (V,B).

Proof. Let (V,B) and $(W,C) \in \hat{T}(\Gamma/\Lambda)$. If $(V,B) \cong (V_1,B_1) \oplus \cdots \oplus (V_n,B_n)$ and $(W,C) \cong (V_1,B_1) \oplus \cdots \oplus (V_{n-1},B_{n-1}) \oplus \Lambda_\gamma(V_n,B_n)$ then we shall write $\Lambda_\gamma^{+}(V,B) = (W,C)$ and $\Lambda_\gamma^{-}(W,C) = (V,B)$. Suppose now that (V,B) and $(W,C) \in \hat{T}(\Gamma/\Lambda)$ such that (V,B) is minimal and (V,B) is related to (W,C). We can suppose that $(V,B) \not\cong (W,C)$. Thus, there are matrices $\gamma_1,\cdots,\gamma_r$ such that if $f_i = \Lambda_{\gamma_i}^{\pm}$ then $f_r \cdots f_1(V,B) = (W,C)$. If $\gamma_1,\cdots,\gamma_r$

are chosen such that r is minimal then we claim that the case $f_i = \Lambda_{\gamma_i}^-$ doesn't occur. This will prove that (W, C) is a pop up of (V, B). The claim is proved by contradiction. Suppose that for some $i, f_i = \Lambda_{\gamma_i}^-$. Let q be the smallest integer such that a minus occurs. Clearly, $q > 1$, because (V, B) is minimal. We shall show that there is an integer p such that $1 \leq p < q$ and such that $f_q \cdots f_1(V, B) \cong f_{q-1} \cdots f_{p+1} f_{p-1} \cdots f_1(V,B)$. This will contradict the minimality of r. Let $(V^0, B^0) = (V, B)$ and let $(V^m, B^m) = f_m \cdots f_1(V, B)$. The effect of f_q on (V^{q-1}, B^{q-1}) is to replace an indecomposable factor $(X, D) = \Lambda_{\gamma_q}^+(X', D')$ of (V^{q-1}, B^{q-1}) by (X', D'). Let p be the largest integer $\leq q-1$ such that (X, D) is a factor of (V^p, B^p) but not of (V^{p-1}, B^{p-1}). Since (V, B) is minimal, it follows that $1 \leq p$. For $p \leq m \leq q-1$, we can write $(V^m, B^m) = (V_1^m, B_1^m) \oplus (X, D)$. Since f_m is a $(+)$-operation, it follows that for $p+1 \leq m \leq q-1$, $(V^m, B^m) \cong f_m(V^{m-1}, B^{m-1}) \cong f_m(V_1^{m-1}, B_1^{m-1}) \oplus (X, D)$. The effect of f_p on (V^{p-1}, B^{p-1}) is to replace a factor (X'', D'') of (V^{p-1}, B^{p-1}) by the indecomposable factor $(X, D) = \Lambda_{\gamma_p}(X'', D'')$. Since $\Lambda_{\gamma_q}(X', D') = (X, D) = \Lambda_{\gamma_p}(X'', D'')$, it follows that $(X', D') = (X'', D'')$. If $(V^{p-1}, B^{p-1}) \cong (V_1^{p-1}, B_1^{p-1}) \oplus (X'', D'')$ then $(V_1^{p-1}, B_1^{p-1}) \cong (V_1^p, B_1^p)$. Also, if $(V^q, B^q) \cong (V_1^q, B_1^q) \oplus (X', B')$ then $(V_1^q, B_1^q) \cong (V_1^{q-1}, B_1^{q-1})$. Thus, $f_{q-1} \cdots f_{p+1} f_{p-1} \cdots f_1(V, B) = f_{q-1} \cdots f_{p+1}(V^{p-1}, B^{p-1}) = f_{q-1} \cdots f_{p+1}((V_1^{p-1}, B_1^{p-1}) \oplus (X'', D'')) \cong f_{q-1} \cdots f_{p+1}((V_1^p, V_1^p) \oplus (X'', D'')) = (f_{q-1} \cdots f_{p+1}(V_1^p, B_1^p)) \oplus (X'', D'') \cong (V_1^{q-1}, B_1^{q-1}) \oplus (X'', D'') \cong (V_1^q, B_1^q) \oplus (X', D') \cong (V^q, B^q)$. Thus, $f_{q-1} \cdots f_{p+1} f_{p-1} \cdots f_1(V, B) = f_q \cdots f_1(V, B)$.

COROLLARY 11.22. a) *Each* $(V, B) \in \hat{T}(\Gamma/\Lambda)$ *has a root. Moreover, any two roots are isomorphic.*

b) *Let* (V, B) *and* $(W, C) \in \hat{T}(\Gamma/\Lambda)$. *Let* $[V, B]$ *and* $[W, C]$ *denote respectively their classes in* $\hat{T}(\Gamma/\Lambda)$ *and let* $\sqrt{(V, B)}$ *and* $\sqrt{(W, C)}$ *denote respectively a root for each. Then* $[V, B] = [W, C] \Longleftrightarrow \sqrt{(V, B)} \cong \sqrt{(W, C)}$.

c) *Adopt the data in part b). Then* $\sqrt{(V,B)\oplus(W,C)} \cong \sqrt{(V,B)}\oplus\sqrt{(W,C)}$.

Proof. a) By dimension counting, it is clear that each (V,B) has a root. Suppose that (W,C) and (W',C') are roots for (V,B). Thus, (W,C) and (W',C') are related. Since they are both minimal, it follows from Lemma 11.21 that they are pop ups of one another. But this can happen only if they are trivial pop ups of one another. Thus, $(W,C) \cong (W',C')$.

b) $[V,B] = [W,C] \iff$ there is an $(X,D) \in \hat{T}(\Gamma/\Lambda)$ such that $(V,B)\oplus(X,D)$ is related to $(W,C)\oplus(X,D) \iff$ there is an $(X,D) \in \hat{T}(\Gamma/\Lambda)$ such that $\sqrt{(V,B)}\oplus\sqrt{(X,D)}$ is related to $\sqrt{(W,C)}\oplus\sqrt{(X,D)}$. Since $\sqrt{(V,D)}$, $\sqrt{(X,D)}$, and $\sqrt{(W,C)}$ are minimal, it follows that $\sqrt{(V,B)}\oplus\sqrt{(X,D)}$ and $\sqrt{(W,C)}\oplus\sqrt{(X,D)}$ are minimal. Thus, by Lemma 11.21, $\sqrt{(V,B)}\oplus\sqrt{(X,D)} \cong \sqrt{(W,C)}\oplus\sqrt{(X,D)}$. Since cancellation holds in $\hat{T}(\Gamma/\Lambda)$, it follows that $\sqrt{(V,B)} \cong \sqrt{(W,C)}$.

c) follows from b).

COROLLARY 11.23. *Let*

$$\hat{P}(\Gamma/\Lambda) \subset K_0\hat{H}(A,\Gamma)$$

denote the subgroup of $K_0\hat{H}(A,\Gamma)$ *generated by all* $[C]-[C_1]$ *such that* $\sqrt{(C^{-1},C)} \cong \sqrt{(C_1^{-1},C_1)}$. *Then the sequence below is exact*

$$0 \to \hat{P}(\Gamma/\Lambda) \to K_0\hat{H}(A,\Gamma) \to \hat{T}(\Gamma/\Lambda)$$
$$[B] \mapsto [B^{-1},B]\,.$$

Proof. From the definition of $\hat{T}(\Gamma/\Lambda)$, it is clear that the composite mapping at $K_0\hat{H}(A,\Gamma)$ is trivial. On the other hand, let $[B]-[C] \in K_0H(A,\Gamma)$. If $[B]-[C]$ vanishes in $\hat{T}(\Gamma/\Lambda)$ then by Lemma 11.21, $\sqrt{(B^{-1},B)} \cong \sqrt{(C^{-1},C)}$. Thus, $[B]-[C] \in \hat{P}(\Gamma/\Lambda)$.

Proof of Theorem 11.12. That ρ is a well-defined homomorphism follows from the quadratic Whitehead Lemma 3.8, Lemma 11.9, and Lemma 11.11. That ∂ is a well-defined homomorphism follows from the relations which define $T(\Gamma/\Lambda)$. Exactness at $KQ_1(A,\Lambda)$ follows directly from Lemma 11.9 b). Exactness at $T(\Gamma/\Lambda)$ follows from Step II in the proof of Lemma 11.11 and from Lemma 11.11. That the composite mapping at $KH_0(A,\Gamma)_{\text{discr-based-0}}$ is trivial follows from Lemma 11.14. Consider now the exact sequence $\hat{P}(\Gamma/\Lambda) \to K_0\hat{H}(A,\Gamma) \to \hat{T}(\Gamma/\Lambda)$ in Corollary 11.23. Since $K_0\hat{H}(A,\Gamma)/E(\Gamma) = KH_0(A,\Gamma)_{\text{discr-based-0}}, \hat{T}(\Gamma/\Lambda)/E(\Gamma/\Lambda) = T(\Gamma/\Lambda)$ and $E(\Gamma)$ maps onto $E(\Gamma/\Lambda)$, it follows that there is an exact sequence $\hat{P}(\Gamma/\Lambda) \xrightarrow{p} KH_0(A,\Gamma)_{\text{discr-based-0}} \to T(\Gamma/\Lambda)$. Let $h: KH_0(A,\Lambda)_{\text{discr-based-0}} \to KH_0(A,\Gamma)_{\text{discr-based-0}}$. To complete the proof, it suffices to show that the image $p \subset$ image h. Let $(V,B) \in T(\Gamma/\Lambda)$ such that we may write $\Lambda_\gamma(V,B)$ and $\Lambda_\delta(V,B)$. Let $[\quad]$ denote taking a class in $KH_0(A,\Gamma)_{\text{discr-based-0}}$. By Corollary 11.22, it suffices to show that the following holds:

(i) If B is invertible then $[B] - \begin{bmatrix} B & \pi \\ \pi & -\pi(V-\gamma)\pi \end{bmatrix} \in$ image h.

(ii) $\begin{bmatrix} B & \pi \\ \pi & -\pi(V-\gamma)\pi \end{bmatrix} - \begin{bmatrix} B & \pi \\ \pi & -\pi(V-\delta)\pi \end{bmatrix} \in$ image h.

(i) follows from the fact that the matrix $\varepsilon_{21}(-\pi B^{-1})$ defines an isomorphism

$$\begin{pmatrix} B & \pi \\ \pi & -\pi(B^{-1}-\gamma)\pi \end{pmatrix} \xrightarrow{\cong} \begin{pmatrix} B & \\ & -\pi(2B^{-1}-\gamma)\pi \end{pmatrix}$$

of (nonsingular) Γ-hermitian forms and the fact that $-\pi(2B^{-1}-\gamma)\pi$ is Λ-hermitian.

We prove now (ii). Let $\sigma = \begin{pmatrix} B & \pi \\ \pi & -\pi(V-\delta)\pi \end{pmatrix}$ and let $\begin{pmatrix} * & * \\ * & C \end{pmatrix} = \sigma^{-1}$. If

$$\varepsilon = \left(\begin{array}{c|c} \begin{matrix} I & \\ & I \end{matrix} & \\ \hline -\begin{pmatrix} 0 & \pi \\ 0 & 0 \end{pmatrix}\sigma^{-1} & \begin{matrix} I & \\ & I \end{matrix} \end{array}\right) \varepsilon_{43}((-\pi(\gamma-\delta)\pi)\pi^{-1})\,\varepsilon_{24}(I)$$

then ε defines an isomorphism

$$\left(\left(\begin{array}{c|c} V & \\ \quad 0 & \\ \hline & 0 \\ & \quad 0 \end{array}\right), \left(\begin{array}{cc|cc} B & \pi & & \\ \pi & -\pi(V-\gamma)\pi & & \\ \hline & & 0 & \pi \\ & & \pi & -\pi(\gamma-\delta)\pi \end{array}\right)\right) \xrightarrow{\cong} \left(\left(\begin{array}{c|c} V & \\ \quad 0 & \\ \hline & 0 \\ & \quad 0 \end{array}\right),\right.$$

$$\left.\left(\begin{array}{cc|cc} B & \pi & & \\ \pi & -\pi(V-\delta)\pi & & \\ \hline & & -\pi C\pi & \pi \\ & & \pi & -\pi(\gamma-\delta)\pi \end{array}\right)\right).$$

Thus, $\begin{pmatrix} -\pi C\pi & \pi \\ \pi & -\pi(\gamma-\delta)\pi \end{pmatrix}$ is nonsingular. If (v, b) denotes the object on the right in the isomorphism above then the lowest, right-hand, 2×2 block of the matrix $b+\bar{b}\bar{v}b$ is the matrix $\begin{pmatrix} -\pi C\pi & \pi \\ \pi & -\pi(\gamma-\delta)\pi \end{pmatrix}$. Thus, $\begin{pmatrix} -\pi C\pi & \pi \\ \pi & -\pi(\gamma-\delta)\pi \end{pmatrix}$ is Λ-hermitian. Clearly, $\begin{pmatrix} 0 & \pi \\ \pi & -\pi(\gamma-\delta)\pi \end{pmatrix}$ is also Λ-hermitian. (ii) follows, because ε defines also an isomorphism

$$\begin{pmatrix} B & \pi \\ \pi & -\pi(V-\gamma)\pi \end{pmatrix} \perp \begin{pmatrix} 0 & \pi \\ \pi & -\pi(\gamma-\delta)\pi \end{pmatrix} \xrightarrow{\cong} \begin{pmatrix} B & \pi \\ \pi & -\pi(V-\delta)\pi \end{pmatrix} \perp \begin{pmatrix} -\pi C\pi & \pi \\ \pi & -\pi(\gamma-\delta)\pi \end{pmatrix}$$

of hermitian forms.

The naturality assertions of the theorem are obvious.

LEMMA 11.24. *Let* $\mathfrak{q}$ *be an involution invariant ideal of* A. *If* $\mathfrak{q}\cap\Gamma\subset\Lambda$ *then the canonical homomorphism below is surjective*

$$T(\Gamma/\Lambda) \to T((\Gamma/\mathfrak{q}\cap\Gamma)/(\Lambda/\mathfrak{q}\cap\Lambda)).$$

REMARK 11.25. Let $x \in \text{center}(A)$. If $s = x+\bar{x}$ *is a nonzero divisor of* A *then* $\mathfrak{q} = sA$ *satisfies the hypotheses of the lemma above. In particular, if* 2 *is a nonzero divisor of* A *then* $\mathfrak{q} = 2A$ *satisfies the hypotheses of the lemma.*

Proof of Lemma 11.24. Let $(\gamma, \beta) \in T((\Gamma/\mathfrak{q}\cap\Gamma)/(\Lambda/\mathfrak{q}\cap\Lambda))$. Let $\bar{V}$ and B be Γ-hermitian matrices which lift respectively $\bar{\gamma}$ and β. To prove the lemma, it suffices to show that (V, B) is an object of $T(\Gamma/\Lambda)$. Thus, we must show that $B+\overline{B}\overline{V}B$ is Λ-hermitian. Clearly, $B+\overline{B}\overline{V}B \equiv \beta + \overline{\beta}\overline{\gamma}\beta \mod \mathfrak{q}$. Since $\beta+\overline{\beta}\overline{\gamma}\beta$ is $\Lambda/\mathfrak{q}\cap\Lambda$-hermitian, it follows that each diagonal coefficient b of $B+\overline{B}\overline{V}B$ can be written as a sum $b = x+q$ for some $x \in \Lambda$ and $q \in \mathfrak{q}$. Since $q = b-x \in \mathfrak{q}\cap\Gamma\subset\Lambda$, it follows that $b \in \Lambda$. Thus, $B+\overline{B}\overline{V}B$ is Λ-hermitian.

The next theorem provides us with a strong reduction result for $T(\Gamma/\Lambda)$.

THEOREM 11.26. *Let* $\mathfrak{q}$ *be an involution invariant ideal of* A. *If*

$$\mathfrak{q} \subset \{\sum_i a_i \mid a_i \in \text{annihilator}_A(\Gamma/\Lambda)\},$$

$$\mathfrak{q}\cap\Gamma \subset \{\sum_i a_i\Lambda\overline{a}_i \mid a_i \in \text{annihilator}_A(\Gamma/\Lambda)\}$$

then the canonical homomorphism below is an isomorphism

$$T(\Gamma/\Lambda) \xrightarrow{\cong} T(\Gamma/\mathfrak{q}\cap\Gamma, \Lambda/\mathfrak{q}\cap\Lambda).$$

REMARK 11.27. *Let* $s,t \in \text{center}(A)$. *If* s,t *are nonzero divisors in* A *such that* $s=\overline{s}$, $s\Gamma\subset\Lambda$, *and* $t \in \text{annihilator}_A(\Gamma/\Lambda)$ *then the ideal* $\mathfrak{q} = st\overline{t}A$ *satisfies the hypotheses of the theorem above. In particular, if* 2 *is a nonzero divisor in* A *then* $\mathfrak{q} = 8A$ *satisfies the hypotheses of the theorem.*

Proof of Theorem 11.26. By Lemma 11.24, the canonical homomorphism $T(\Gamma/\Lambda) \to T((\Gamma/\mathfrak{q}\cap\Gamma)/(\Lambda/\mathfrak{q}\cap\Lambda))$ is surjective. We shall construct an inverse to this homomorphism. Let $(\gamma,\beta) \in T((\Gamma/\mathfrak{q}\cap\Gamma)/(\Lambda/\mathfrak{q}\cap\Lambda))$. Let $(V,B),(W,C) \in T(\Gamma/\Lambda)$ be such that $V\equiv W\equiv\gamma \mod \mathfrak{q}$ and $B\equiv C\equiv\beta \mod \mathfrak{q}$. We shall show first that $[V,B]=[W,C]$. Since $V\equiv W \mod \mathfrak{q}$ and $\mathfrak{q}\cap\Gamma\subset\Lambda$, it follows that $(W,C)=(V,C)$. Write $C = B+D$. Thus, $D\equiv 0 \mod \mathfrak{q}$. The conditions on $\mathfrak{q}$ quarantee that D can be written as a sum $D = \sum_i a_i D_i \overline{a}_i$ such that each D_i is Λ-hermitian and $a_i \in \text{annihilator}_A(\Gamma/\Lambda)$. Thus, $\overline{a}_i V a_i$ is $\overline{\Lambda}$-hermitian. Thus, by Lemmas

11.16 and 11.19, $[V, C] = [V, B]$. We construct now the inverse promised above. Let $(\gamma, \beta) \in T((\Gamma/q \cap \Gamma)/(\Lambda/q \cap \Lambda))$. If $\bar{V}$ and B are Γ-hermitian matrices covering respectively $\bar{\gamma}$ and β then in the proof of Lemma 11.24, we showed that $(V, B) \in T(\Gamma/\Lambda)$. Furthermore, above we showed that the class $[V, B]$ is independent of the choice of V and B. The rule $[\gamma, \beta] \mapsto [V, B]$ clearly preserves the relation $[0, \pi] = 0$. That the rule preserves the relation $[\gamma, \beta] = [\bar{\varepsilon}^{-1}\gamma\varepsilon^{-1}, \varepsilon\beta\bar{\varepsilon}]$ follows from the fact that the canonical map $E_n(A) \to E_n(A/q)$ is surjective. By Lemma 11.18, the rule preserves the relation $[\gamma, \beta] = [\Lambda_{\gamma'}(\gamma, \beta)]$. Thus, the rule induces a homomorphism and the homomorphism is clearly inverse to the canonical map in the other direction.

The following general observation is useful.

LEMMA 11.28. *Let* $\{\Gamma_i \mid i \in I\}$ *be a set of form parameters on* A *such that for each* i, $\Lambda \subset \Gamma_i \subset \Gamma$. *If the canonical homomorphism* $\coprod_i \Gamma_i/\Lambda \to \Gamma/\Lambda$ *is bijective then the canonical homomorphism below is an isomorphism*

$$\coprod_i T(\Gamma_i/\Lambda) \to T(\Gamma/\Lambda) .$$

Proof. Clear.

To close the section, we want to connect the results in the current section with those in the previous section.

LEMMA 11.29. $WQ_2(A, \Lambda) =$ *center* $(StQ(A, \Lambda)/H(K_2(A)) =$ *ker* $(StQ(A, \Lambda)/H(K_2(A)) \to EQ(A, \Lambda))$.

Proof. By 3.17, $WQ_2(A, \Lambda) \subseteq$ center $StQ(A, \Lambda)/H(K_2(A))$. Conversely, let p denote the canonical map $StQ(A, \Lambda) \to EQ(A, \Lambda)$. If $x \in$ center $StQ(A, \Lambda)/H(K_2(A))$, let s be a representative in $StQ(A, \Lambda)$ for x. Since p is surjective, it is clear that $p(s) \in$ center $(EQ(A, \Lambda))$. But, the center $(EQ(A, \Lambda)) = 1$ (although, it is possible that the center $(EQ_{2n}(A, \Lambda)) \neq 1$). Thus, $s \in KQ_2(A, \Lambda)$. Thus, $x \in WQ_2(A, \Lambda)$. Clearly,

$WQ_2(A, \Lambda) \subset \ker(StQ(A, \Lambda)/H(K_2(A)) \to EQ(A, \Lambda))$. Conversely, if $x \in \ker(StQ(A, \Lambda)/H(K_2(A)) \to EQ(A, \Lambda))$, let s be a representative in $StQ(A, \Lambda)$ for x. Then $s \in \ker p = KQ_2(A, \Lambda)$. Thus, $x \in WQ_2(A, \Lambda)$.

Let $(M, B) \in H(A, \Lambda)_{\text{discr-based-0}}$. By *stabilizing* (M, B), we shall mean replacing it by $(M, B) \perp H(A)_{\text{based}} \perp \cdots \perp H(A)_{\text{based}}$ for a suitable number of modules $H(A)_{\text{based}}$. Using the preferred basis $e_1, \cdots, e_{2n}$ on M, we can associate to B a $2n \times 2n$ matrix $(B(e_i, e_j))$. We shall denote this matrix also by B. Let π denote the $2n \times 2n$ matrix $\begin{pmatrix} 0 & -1 \\ \lambda & 0 \end{pmatrix} \oplus \cdots \oplus \begin{pmatrix} 0 & -1 \\ \lambda & 0 \end{pmatrix}$. By definition, the discriminant matrix $\operatorname{discr}(M, B) = -\pi^{-1}B$ vanishes in $K_1(A)$. Let $\begin{pmatrix} 0 & \pi \\ -\pi^{-1} & 0 \end{pmatrix}$ denote the element of $StQ(A, \Lambda)$ or $StQ(A, \Lambda)/H(K_2(A))$ defined by

$$\begin{pmatrix} 0 & \pi \\ -\pi^{-1} & 0 \end{pmatrix} = \begin{pmatrix} I & 0 \\ -\pi^{\ 1} & I \end{pmatrix}\begin{pmatrix} I & \pi \\ 0 & 0 \end{pmatrix}\begin{pmatrix} I & 0 \\ -\pi^{-1} & I \end{pmatrix}.$$

If $x(B)$ is the element of $StQ(A, \Lambda)/H(K_2(A))$ defined by

$$x(B) = \begin{pmatrix} 0 & \pi \\ -\pi^{-1} & 0 \end{pmatrix}\begin{pmatrix} I & 0 \\ -B^{-1} & I \end{pmatrix}\begin{pmatrix} I & B \\ 0 & I \end{pmatrix}\begin{pmatrix} I & 0 \\ -B^{-1} & I \end{pmatrix}\begin{pmatrix} -\pi B^{-1} & \\ & \overline{(-\pi B^{-1})}^{-1} \end{pmatrix}^{-1}$$

then clearly $x(B) \in WQ_2(A, \Lambda)$, because the image of $x(B)$ in $EQ(A, \Lambda)$ is trivial.

LEMMA 11.30. *The rule*

$$(M, B) \mapsto x(B)$$

defines a surjective homomorphism of groups

$$WH_0(A, \Lambda)_{\text{discr-based-0}} \to WQ_2(A, \Lambda)\ .$$

Proof. Let r and s be two integers such that $r < s$. Let

$\tau = \begin{pmatrix} 0 & -1 \\ \lambda & 0 \end{pmatrix} \oplus \cdots \oplus \begin{pmatrix} 0 & -1 \\ \lambda & 0 \end{pmatrix}$ (r times) and $\rho = \begin{pmatrix} 0 & -1 \\ \lambda & 0 \end{pmatrix} \oplus \cdots \oplus \begin{pmatrix} 0 & -1 \\ \lambda & 0 \end{pmatrix}$

(s times). If M has rank 2n and if the discriminant matrix $\operatorname{discr}(B\oplus-\tau)\in E_{2(n+r)}(A)$ then we shall show that $x(B\oplus-\tau)=x(B\oplus-\rho)$. This will show that x is independent of stabilization and that one can assume that the $\operatorname{discr}(B)\in E_{2n}(A)$. Let $\pi=\begin{pmatrix}0&-1\\\lambda&0\end{pmatrix}\oplus\cdots\oplus\begin{pmatrix}0&-1\\\lambda&0\end{pmatrix}$ (n–times) and let $\sigma=\begin{pmatrix}0&-1\\\lambda&0\end{pmatrix}\oplus\cdots\oplus\begin{pmatrix}0&-1\\\lambda&0\end{pmatrix}$ (s–r times). Note that $\begin{pmatrix}0&\sigma\\-\sigma^{-1}&0\end{pmatrix}\begin{pmatrix}I&0\\\sigma^{-1}&I\end{pmatrix}\begin{pmatrix}I&-\sigma\\0&I\end{pmatrix}\begin{pmatrix}I&0\\\sigma^{-1}&I\end{pmatrix}=\begin{pmatrix}I&0\\0&I\end{pmatrix}$ in $StQ(A,\Lambda)$. Thus,

$$x(B\oplus-\tau)=\left(\begin{array}{cc|cc}&&\pi&\\&&&\tau\\\hline-\pi^{-1}&&&\\&-\tau^{-1}&&\end{array}\right)\left(\begin{array}{cc|cc}I&&&\\&I&&\\\hline-B^{-1}&&I&\\&\tau^{-1}&&I\end{array}\right)\left(\begin{array}{cc|cc}I&&B&\\&I&&-\tau\\\hline&&I&\\&&&I\end{array}\right)$$

$$\left(\begin{array}{cc|cc}I&&&\\&I&&\\\hline-B^{-1}&&I&\\&\tau^{-1}&&I\end{array}\right)\left(\begin{array}{cc|cc}I&&B&\\&I&&-\tau\\\hline&&I&\\&&&I\end{array}\right)\left(\begin{array}{cc|cc}-\pi B^{-1}&&&\\&(-\tau)(-\tau)^{-1}&&\\\hline&&\overline{(-\pi B^{-1})}^{-1}&\\&&&I\end{array}\right)=$$

$$\left(\begin{array}{ccc|ccc}&&&\pi&&\\&&&&\tau&\\&&&&&\sigma\\\hline-\pi^{-1}&&&&&\\&-\tau^{-1}&&&&\\&&-\sigma^{-1}&&&\end{array}\right)\left(\begin{array}{ccc|ccc}I&&&&&\\&I&&&&\\&&I&&&\\\hline-B^{-1}&&&I&&\\&\tau^{-1}&&&I&\\&&\sigma^{-1}&&&I\end{array}\right)$$

$$\left(\begin{array}{ccc|ccc}I&&&B&&\\&I&&&-\tau&\\&&I&&&-\sigma\\\hline&&&I&&\\&&&&I&\\&&&&&I\end{array}\right)\left(\begin{array}{ccc|ccc}I&&&&&\\&I&&&&\\&&I&&&\\\hline-B^{-1}&&&I&&\\&\tau^{-1}&&&I&\\&&\sigma^{-1}&&&I\end{array}\right)$$

$$\left(\begin{array}{ccc|ccc}-\pi B^{-1}&&&&&\\&(-\tau)(-\tau)^{-1}&&&&\\&&(-\sigma)(\sigma)^{-1}&&&\\\hline&&&\overline{(-\pi B^{-1})}^{-1}&&\\&&&&I&\\&&&&&I\end{array}\right)^{-1}=x(B\oplus-\rho)\,.$$

If ε is an isomorphism $(M, B) \to (N, C)$ in $H(A, \Lambda)_{\text{discr-based-0}}$ then

$$C = \bar{\varepsilon}^{-1} B \varepsilon^{-1} \quad \text{and} \quad x(C) = \begin{pmatrix} & \pi \\ -\pi^{-1} & \end{pmatrix} \begin{pmatrix} I & \\ -\varepsilon B^{-1}\bar{\varepsilon} & I \end{pmatrix} \begin{pmatrix} I & \bar{\varepsilon}^{-1} B \varepsilon^{-1} \\ & I \end{pmatrix}$$

$$\begin{pmatrix} I & \\ -\varepsilon B^{-1}\bar{\varepsilon} & I \end{pmatrix} \begin{pmatrix} -\pi\varepsilon B^{-1}\bar{\varepsilon} & \\ & \overline{(-\pi\varepsilon B^{-1}\bar{\varepsilon})}^{-1} \end{pmatrix} = \begin{pmatrix} & \pi \\ -\pi^{-1} & \end{pmatrix} \begin{pmatrix} \bar{\varepsilon}^{-1} & \\ & \varepsilon \end{pmatrix}$$

$$\begin{pmatrix} I & \\ -B^{-1} & I \end{pmatrix} \begin{pmatrix} I & B \\ & I \end{pmatrix} \begin{pmatrix} I & \\ -B^{-1} & I \end{pmatrix} \begin{pmatrix} \bar{\varepsilon} & \\ & \varepsilon^{-1} \end{pmatrix} \begin{pmatrix} -\pi\varepsilon B^{-1}\bar{\varepsilon} & \\ & \overline{(-\pi\varepsilon B^{-1}\bar{\varepsilon})}^{-1} \end{pmatrix} =$$

$$\begin{pmatrix} & \pi \\ -\pi^{-1} & \end{pmatrix} \begin{pmatrix} \bar{\varepsilon}^{-1} & \\ & \varepsilon \end{pmatrix} \begin{pmatrix} & \pi \\ -\pi^{-1} & \end{pmatrix}^{-1} \begin{pmatrix} -\pi B^{-1} & \\ & \overline{(-\pi B^{-1})}^{-1} \end{pmatrix} \begin{pmatrix} \bar{\varepsilon} & \\ & \varepsilon^{-1} \end{pmatrix}$$

$$\begin{pmatrix} -\pi\varepsilon B^{-1}\bar{\varepsilon} & \\ & \overline{(-\pi\varepsilon B^{-1}\bar{\varepsilon})}^{-1} \end{pmatrix}^{-1} = \begin{pmatrix} \pi\varepsilon\pi^{-1} & \\ & \overline{\pi\varepsilon\pi^{-1}}^{-1} \end{pmatrix} x(B) \begin{pmatrix} \pi\varepsilon\pi^{-1} & \\ & \overline{\pi\varepsilon\pi^{-1}}^{-1} \end{pmatrix}^{-1}$$

$= x(B)$.

Let (M, B) and $(N, C) \in H(A, \Lambda)_{\text{discr-based-0}}$. If M has rank $2m$ and N rank $2n$, let I_{2m} (resp. I_{2n}) denote the identity matrix of rank $2m$ (resp. $2n$). Then

$$x(B)\,x(C) = x(B) \left(\begin{array}{cc|cc} & I_{2m} & & \\ I_{2n} & & & \\ \hline & & & I_{2m} \\ & & I_{2n} & \end{array}\right) x(C) \left(\begin{array}{cc|cc} & I_{2m} & & \\ I_{2n} & & & \\ \hline & & & I_{2m} \\ & & I_{2n} & \end{array}\right) = x(B \oplus C).$$

The above establishes that x is a homomorphism. The fact that x is surjective follows from 3.23.

REMARK 11.31. *By a theorem of R. Sharpe* [26], *the homomorphism in (11.30) is an isomorphism provided* $\Lambda = \min$.

Let

$$StQ(A, \Lambda, \Gamma)$$

denote the inverse image in $StQ(A, \Gamma)$ of the group $EQ(A, \Lambda, \Gamma)$ defined in §A. Consider the commutative square

$$\begin{array}{ccc} \mathrm{StQ}(A,\Lambda,\Gamma) & \longrightarrow & \mathrm{StQ}(A,\Gamma) \\ \downarrow & & \downarrow \\ \mathrm{EQ}(A,\Lambda,\Gamma) & \longrightarrow & \mathrm{EQ}(A,\Gamma)\,. \end{array}$$

Since the bottom map is injective, it follows that the square is a fibre product square. Thus, by (6.15), the image $(\mathrm{StQ}(A,\Lambda) \to \mathrm{StQ}(A,\Lambda,\Gamma))$ is the commutator subgroup of $\mathrm{StQ}(A,\Lambda,\Gamma)$ and by (2.10) and (6.24), the canonical map $\mathrm{coker}(\mathrm{StQ}(A,\Lambda) \to \mathrm{StQ}(A,\Lambda,\Gamma)) \to KQ_1((A,\Lambda),(A,\Gamma))$ is an isomorphism where $KQ_1((A,\Lambda),(A,\Gamma))$ is the relative group corresponding to the functor $F : \mathcal{Q}(A,\Lambda) \to \mathcal{Q}(A,\Gamma)$. Define

$$\mathrm{StQ}(\Gamma/\Lambda) = \mathrm{coker}(\mathrm{StQ}(A,\Lambda) \to \mathrm{StQ}(A,\Lambda,\Gamma))\,.$$

LEMMA 11.32. *The rule*

$$[V,B] \mapsto \begin{bmatrix} & \pi \\ -\pi^{-1} & \end{bmatrix}\begin{bmatrix} I & \\ V & I \end{bmatrix}\begin{bmatrix} I & B \\ & I \end{bmatrix}\begin{bmatrix} I & \\ -V & I \end{bmatrix}$$

defines a surjective homomorphism

$$T(\Gamma/\Lambda) \to \mathrm{StQ}(\Gamma/\Lambda)\,.$$

Proof. The proof is similar to that of Lemma 11.30, except for the verification that the relation $[V,B] = [\Lambda_\gamma(V,B)]$ is preserved. But this can be handled by lifting the arguments in the proof of (11.11), up to the quadratic Steinberg group. In doing this, the reference (3.13) for the elementary group should be replaced by the corresponding reference (3.22) for the Steinberg group. The details will be left to the reader.

The following theorem connects in the terms of a commutative diagram the results of the current section with those in the previous section.

THEOREM 11.33. *Let* $WH_0(A,\Omega) = WH_0(A,\Omega)_{discr\text{-}based\text{-}0}$. *Then the diagram below is commutative, its rows are exact, and its vertical maps are surjective.*

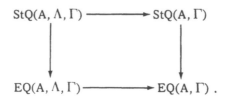

$$\begin{array}{ccccccccc}
WH_0(A,\Lambda) & \longrightarrow & WH_0(A,\Gamma) & \longrightarrow & T(\Gamma/\Lambda) & \longrightarrow & KQ_1(A,\Lambda) & \longrightarrow & KQ_1(A,\Gamma) \\
\downarrow & & \downarrow & & \downarrow & & \| & & \| \\
WQ_2(A,\Lambda) & \longrightarrow & WQ_2(A,\Gamma) & \longrightarrow & StQ(\Gamma/\Lambda) & \longrightarrow & KQ_1(A,\Lambda) & \longrightarrow & KQ_1(\Lambda,\Gamma) \longrightarrow
\end{array}$$

$$S(\Gamma/\Lambda) \longrightarrow KQ_0(A,\Lambda) \longrightarrow KQ_0(A,\Gamma) \longrightarrow 0\,.$$

Proof. The theorem follows from (11.2), (11.12), (11.30), (11.32), and the relative K-theory exact sequence (6.20).

§12. INDUCTION THEORY

The chapter is divided into two parts. In the first part, it is shown how one adapts induction theory to our setting. The main goal is to give the groups $KF_i(R\pi, \Lambda)$ ($F = Q$ or H) (π a finite group) a Frobenius module structure over $GU_0(R, \pi)$. In the second part, we summarize in a convenient form the basic induction results in the literature. These results show how one can exploit the Frobenius module structure on the groups $KF_i(R\pi, \Lambda)$ to reduce questions concerning the torsion subgroup of $KF_i(R\pi, \Lambda)$ to the special case that π is cyclic or hyperelementary. The approach taken is based on a paper [14] of Andreas Dress. I would like to thank Dress for several informative conversations related to his paper.

A. *Frobenius modules* $KF_i(R\pi, \Lambda_\pi)$ *and* $WF_i(R\pi, \Lambda_\pi)$

Let R be a commutative ring with involution and let $\lambda \in R$ such that $\lambda\bar{\lambda} = 1$. Let π be a finite group and let $R\pi$ denote the group ring of π over R. Give $R\pi$ the involution which inverts each element of π and which agrees with the involution already on R. We shall assume that all modules are right modules. Recall that $P(R\pi)$ is the category with product of finitely generated, projective $R\pi$-modules. We begin by recalling a couple facts concerning modules over group rings. They do not require that R has an involution.

LEMMA 12.1. *Let* $P \in P(R\pi)$ *and let* M *be an* $R\pi$ *module such that* $M \in P(R)$. *Then* $P \otimes_R M$ *with diagonal* π*-action, i.e.* $(p \otimes m)\sigma = p\sigma \otimes m\sigma$, *lies in* $P(R\pi)$.

Proof. $P \otimes_R M \in P(R\pi)$ if and only if it is a direct summand of a finitely generated, free $R\pi$-module. Thus, after choosing an $R\pi$-module N such that $M \oplus N$ is a finitely generated, free R-module and after choosing $Q \in P(R\pi)$ such that $P \oplus Q$ is a finitely generated, free $R\pi$-module, we can reduce to the case that $P = R\pi$ and M is finitely generated and free as an R-module. Let $x_1, \cdots, x_n$ be an R-basis for M. Clearly, $\{\sigma \otimes x_i \mid \sigma \in \pi,\ i = 1, \cdots, n\}$ is an R-basis for $R\pi \otimes_R M$. Hence, $\{1 \otimes x_i \mid i = 1, \cdots, n\}$ is an $R\pi$ basis for $R\pi \otimes_R M$.

LEMMA 12.2. *Let* P *and* M *be as in 12.1. Let* V *be an* $R\pi$*-submodule of* M *which is an* R*-direct summand. Then* $P \otimes_R V$ *is an* $R\pi$*-direct summand of* $P \otimes_R M$.

Proof. Since the sequence $0 \to V \to M \to M/V \to 0$ is split exact over R, we can tensor it with P and obtain an exact $R\pi$-sequence $0 \to P \otimes_R V \to P \otimes_R M \to P \otimes_R (M/V) \to 0$. Since the R-module $M/V \in P(R)$, it follows from 12.1 that $P \otimes_R (M/V) \in P(R\pi)$. Hence, the sequence above splits over $R\pi$.

For the rest of the section, R will have an involution. If A is any ring with involution and M is a right A-module, we make $M^* = \mathrm{Hom}_A(M,A)$ into a right A-module by defining $(fa)m = \bar{a} f(m)$.

LEMMA 12.3. *Every* R*-projective* $R\pi$*-module is* $R\pi$*-reflexive.*

Proof. Let $tr: R\pi \to R$, $\sum_{\sigma \in \pi} a_\sigma \sigma \mapsto a_1$. Let M be an R-projective $R\pi$-module. Let $M^* = \mathrm{Hom}_{R\pi}(M, R\pi)$ and $M^\times = \mathrm{Hom}_R(M, R)$. Define

$$tr: M^* \to M^\times, \quad g \mapsto tr\, g$$

$$sr: M^\times \to M^*, \quad f \mapsto \sum_{\sigma \in \pi} f(\_\sigma^{-1})\sigma .$$

sr and tr are inverse R-isomorphisms. Define

$$tr_{**}: M^{**} \to M^{\times\times}, \quad k \mapsto (f \mapsto tr(k(sr(f)))$$

$$sr_{\times\times}: M^{\times\times} \to M^{**}, \quad \ell \mapsto (g \mapsto \sum_{\sigma \in \pi} \sigma \ell(\sigma^{-1}(tr(g))) .$$

$sr_{\times\times}$ and tr_{**} are inverse R-isomorphisms. Furthermore, the diagram

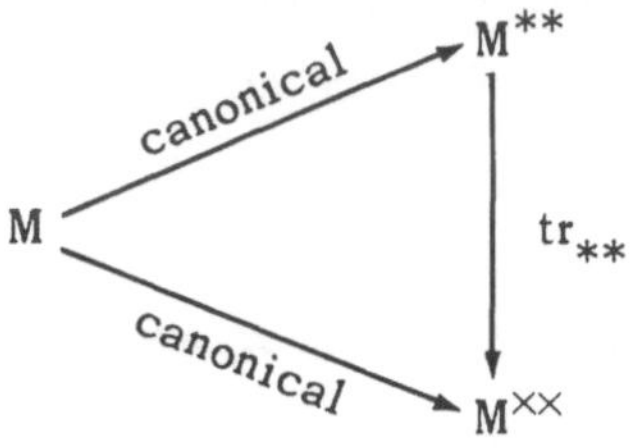

commutes (the canonical map is defined by $m \mapsto (f \mapsto \overline{f(m)})$). But $M \to M^{\times\times}$ is an isomorphism because M is R-projective. Hence, $M \to M^{**}$ is an isomorphism.

Let

$$H(R, \pi)$$

be the category with product whose objects are pairs (M, B) where M is an $R\pi$-module such that $M \in P(R)$ and B is an 1-hermitian form on M such that $M \to \mathrm{Hom}_{R\pi}(M, R\pi)$, $m \mapsto B(m, \ )$, is an isomorphism. A morphism $(M, B) \to (N, C)$ is an $R\pi$-isomorphism $M \to N$ which preserves the hermitian forms. $H(R, \pi)$ has a natural product defined by $(M, B) \perp (N, C) = (M \oplus N, B \oplus C)$. (M, B) is called *hyperbolic* if $(M, B) = (P \oplus \mathrm{Hom}_{R\pi}(P, R\pi), B_P)$ where P is an $R\pi$-module such that $P \in P(R)$ and $B_P((p, f), (q, g)) = f(q) + \overline{g(p)}$. Since P is reflexive by 12.3, the map $M \to M^*$, $m \mapsto B_P(m, \ )$, is an isomorphism. We let

$$KH_0(R, \pi) = K_0(H(R, \pi))$$
$$WH_0(R, \pi) = KH_0(R, \pi)/\text{hyperbolic modules.}$$

Next we give an alternative description of $H(R, \pi)$. Let

$$H_{\pi\text{-inv}}(R)$$

be the subcategory of $H^1(R, \max)$ (= category of nonsingular, 1-hermitian forms on finitely generated, projective R-modules) of all (M, B) such that M has a π-action which commutes with the action of R and leaves B invariant, i.e. $B(m\sigma, n\sigma) = B(m,n)$ for all $\sigma \in \pi$. A morphism in $H_{\pi\text{-inv}}(R)$ is a π-invariant morphism in $H^1(R, \max)$. $H_{\pi\text{-inv}}(R)$ inherits a natural product from $H^1(R, \max)$. An object of $H_{\pi\text{-inv}}(R)$ is called *hyperbolic* if it is a hyperbolic object $H(P) = (P \oplus \mathrm{Hom}_R(P, R), B_P)$ of $H^1(R, \max)$ such that π leaves P and $\mathrm{Hom}_R(P, R)$ invariant and such that the action of π on the latter is defined by $(f\sigma)(p) = f(p\sigma^{-1})$. In this case, B_P is automatically π-invariant.

To see that $H(R, \pi)$ and $H_{\pi\text{-inv}}(R)$ are equivalent we need a lemma. If A is a ring with involution, let $\mathrm{Sesq}(A)$ denote the category with product of all pairs (M, B) where M is an A-module and $B \in \mathrm{Sesq}_A(M)$. A morphism $(M, B) \to (N, C)$ is an A-linear map $M \to N$ which preserves the sesquilinear forms. $\mathrm{Sesq}(A)$ has a natural product defined by $(M, B) \perp (N, C) = (M \oplus N, B \oplus C)$. We let $\mathrm{Sesq}_{\pi\text{-inv}}(R)$ be the subcategory of $\mathrm{Sesq}(R)$ of all (M, B) such that M has a π-action which commutes with the action of R and leaves B invariant. A morphism in $\mathrm{Sesq}_{\pi\text{-inv}}(R)$ is a π-invariant morphism in $\mathrm{Sesq}(R)$. $\mathrm{Sesq}_{\pi\text{-inv}}(R)$ inherits a natural product from $\mathrm{Sesq}(R)$. If $(M, B) \in \mathrm{Sesq}_{\pi\text{-inv}}(R)$, define $(M, \mathrm{sr}\, B) \in \mathrm{Sesq}(R\pi)$ by $\mathrm{sr}\, B(m, n) = \sum_{\sigma \in \pi} B(m, n\sigma^{-1})\sigma$. Define $\mathrm{tr}: R\pi \to R$, $\sum_{\sigma \in \pi} a_\sigma \sigma \mapsto a_1$.

LEMMA 12.4 (Fröhlich and McEvett [16], p. 24). *The functors*

$$\mathrm{Sesq}(R\pi) \to \mathrm{Sesq}_{\pi\text{-inv}}(R),\ B \mapsto \mathrm{tr}\, B$$
$$\mathrm{Sesq}_{\pi\text{-inv}}(R) \to \mathrm{Sesq}(R\pi),\ B \mapsto \mathrm{sr}\, B$$

are inverse isomorphisms which preserve products, nonsingularity, λ-hermitian forms, even λ-hermitian forms, hyperbolic forms, and totally isotropic, $R\pi$-invariant subspaces.

Proof. $\mathrm{sr}\,\mathrm{tr}\, B = B$ and $\mathrm{tr}\,\mathrm{sr}\, B = B$. The details are straightforward.

The equivalence of $H(R, \pi)$ and $H_{\pi\text{-inv}}(R)$ follows immediately from 12.4.

Next, we introduce into $KH_0(R, \pi)$ and $WH_0(R, \pi)$ a relation used by Quillen [21, §5]. Define a *Quillen pair* (M, U) to be a pair consisting of an object $M \in H(R, \pi)$ and an R-direct, $R\pi$-subspace U of M such that $U = U^{\perp}$. Define

$$GU_0(R, \pi) = KH_0(R, \pi)/\{[M] - [H(U)] \mid (M, U) \text{ Quillen pair}\}$$

$$GW_0(R, \pi) = WH_0(R, \pi)/\{[M] \mid (M, U) \text{ Quillen pair}\}.$$

The plan of the rest of the section is as follows. We shall introduce a multiplication on $KH_0(R, \pi)$ and show that $KH_0(R, \pi)$, $WH_0(R, \pi)$, $GU_0(R, \pi)$ and $GW_0(R, \pi)$ are commutative rings with identity. Then we shall establish that all of these rings satisfy the Frobenius reciprocity law. Let

$$F = H \text{ or } Q.$$

F stands for form, hermitian or quadratic. We shall define a ring action of $GU_0(R, \pi)$ (resp. $GW_0(R, \pi)$) on $KF_i(R\pi, \Lambda)$ (resp. $WF_i(R\pi, \Lambda)$) $(i = 0, 1)$ and then establish that the resulting modules are Frobenius modules.

We define a commutative multiplication on $KH_0(R, \pi)$ by the rule

$$[N, C][M, B] = [N \otimes_R M, C \otimes_R B]$$

such that $N \otimes_R M$ has the diagonal π-action and $C \otimes_R B(n \otimes m, n_1 \otimes m_1) = \sum_{\sigma \in \pi} c_\sigma b_\sigma \sigma$ where $C(n, n_1) = \sum_\sigma c_\sigma \sigma$ and $B(m, m_1) = \sum_\sigma b_\sigma$. The multiplicative identity is (R, B) where R has the trivial π-action and $B(r, r_1) = \bar{r} r_1$. If we use 12.4 to identify $KH_0(R, \pi)$ with $K_0(H_{\pi\text{-inv}}(R))$ then on the latter group, multiplication looks the same except that $C \otimes_R B$ is defined by $C \otimes_R B(n \otimes m, n_1 \otimes m_1) = C(n, n_1) B(m, m_1)$. We shall show now that the subgroup of $KH_0(R, \pi)$ generated by the Quillen relations is an ideal. It suffices to show that if (M, U) is a Quillen pair then $(N \otimes_R M, N \otimes_R U)$ is a Quillen pair. Think of M as an object of $H_{\pi\text{-inv}}(R)$ and let B and C be the hermitian forms on M and N respectively. It follows from 2.8 that we can find an R-direct complement V to U such that $V \cong \mathrm{Hom}_R(U, R)$ and $B(f, u) = f(u)$ for all $f \in V$ and $u \in U$. Write $N \otimes_R M = N \otimes_R U \oplus N \otimes_R V$. We must show that $N \otimes_R U = (N \otimes_R U)^{\perp}$. Clearly, $N \otimes_R U \subset (N \otimes_R U)^{\perp}$. Thus, it suffices to show that given $y \in N \otimes_R V$, there is an $x \in N \otimes_R U$ such that $C \otimes_R B(y,x) \neq 0$. If we choose an $R\pi$-module U_1 such that $U \oplus U_1$ is R-free and if we add $H(U_1)$ to M then we can reduce to the case U is R-free. Let $e_1, \cdots, e_m$ be an R-basis for U and let $f_1, \cdots, f_m$ be its dual basis in V. Each element of $N \otimes_R V$ can be written as a sum $\sum_{i=1}^{m} n_i \otimes f_i$ for suitable $n_i \in N$. If $n_j \neq 0$, we can find an n'_j such that $C(n_j, n'_j) \neq 0$. Thus, $C \otimes_R B(\sum_{i=1}^{m} n_i \otimes f_i, n'_j \otimes e_j) = C(n_j, n'_j) \neq 0$. Thus, $(N \otimes_R M, N \otimes_R U)$ is a Quillen pair and $[N]([M]-[H(U)]) = [N \otimes_R M] - [H(N \otimes_R U)]$ is a Quillen relation.

The paragraph above shows that $KH_0(R,\pi)$ and $GU_0(R,\pi)$ are commutative rings. It is clear that the subgroup of $KH_0(R,\pi)$ generated by all hyperbolic spaces is an ideal and thus, $WH_0(R,\pi)$ is also a commutative ring. Moreover, the paragraph above shows that the subgroup of $WH_0(R,\pi)$ generated by all Quillen pairs is an ideal. Thus, $GW_0(R,\pi)$ is also a commutative ring.

We define a right $KH_0(R,\pi)$-action on $KF_0(R\pi,\Lambda)$ by the rule

$$[P,C][M,B] = [P \otimes_R M, C \otimes_R B]$$

such that $N \otimes_R M$ has the diagonal π-action and $C \otimes_R B(p \otimes m, p_1 \otimes m_1) = \sum_{\sigma \in \pi} c_\sigma b_\sigma \sigma$ where $C(p,p_1) = \sum_\sigma c_\sigma \sigma$ and $B(m,m_1) = \sum_\sigma b_\sigma \sigma$. We shall show that this action kills the Quillen relations. It suffices to show that if (M,U) is a Quillen pair then $[P \otimes_R M] = [H(P \otimes_R U)]$. Clearly, $P \otimes_R U$ is a totally isotropic subspace of $P \otimes_R M$. Moreover, in the proof above that the Quillen relations generate (additively) an ideal of $KH_0(R,\pi)$, we saw that $P \otimes_R U = (P \otimes_R U)^\perp$, and by 12.2 we know that $P \otimes_R U$ is an $R\pi$-direct summand of $P \otimes_R M$. Hence, by 2.8, $P \otimes_R M \cong H(P \otimes_R U)$ if $F = Q$, and $P \otimes_R M \cong \Lambda - M(P \otimes_R U)$ if $F = H$. Furthermore, in the latter case, it follows from 2.11 that $[\Lambda - M(P \otimes_R U)] = [H(P \otimes_R U)]$.

The paragraph above shows that $KF_0(R\pi,\Lambda)$ is a $KH_0(R,\pi)$-module and a $GU_0(R,\pi)$-module. It is clear that the $KH_0(R,\pi)$-action on $KF_0(R\pi,\Lambda)$ induces a $WH_0(R,\pi)$-action on $WF_0(R\pi,\Lambda)$. Moreover, the paragraph above shows that the latter action factors through $GW_0(R,\pi)$. Thus, $WF_0(R,\pi)$ is also a $GW_0(R,\pi)$-module.

Let $\pi' \subset \pi$ be a subgroup of π. Let $tr_{\pi/\pi'}: R\pi \to R\pi'$, $\sum_{\sigma \in \pi} a_\sigma \sigma \mapsto \sum_{\sigma \in \pi'} a_\sigma \sigma$. We define induction and restriction maps

$$i^* : GU_0(R, \pi') \to GU_0(R, \pi) \quad \text{(group homomorphism)}$$

$$[M, B] \mapsto [M \otimes_{R\pi'} R\pi, B \text{ extended linearly}]$$

$$i_* : GU_0(R, \pi) \to GU_0(R, \pi') \quad \text{(ring homomorphism)}$$

$$[M, B] \mapsto [M, tr_{\pi/\pi'} B] .$$

To avoid confusion, we write sometimes $i_{*_{\pi' \to \pi}}$ in place of i_*, and $i^{*\pi \to \pi'}$ in place of i^*. There are of course analogous induction and restriction maps on KH_0, WH_0 and GW_0.

THEOREM 12.5 (Frobenius reciprocity). *If* $x \in GU_0(R, \pi')$ *and if* $y \in GU_0(R, \pi)$ *then*

$$i^*(x(i_*(y))) = (i^*x)(y) .$$

There are of course analogous results for KH_0, WH_0, *and* GW_0.

The proof of 12.5 is routine and similar to that of the analogous result [29, 2.2] for modules.

Let $\pi' \subset \pi$ be a subgroup of π. If ${}^\lambda(R\pi', \Lambda_{\pi'})$ and ${}^\lambda(R\pi, \Lambda_\pi)$ are form rings such that $\Lambda_{\pi'} \subset \Lambda_\pi$ and $tr_{\pi/\pi'}(\Lambda_\pi) \subset \Lambda_{\pi'}$ (e.g. $\Lambda_{\pi'} = \min$ (resp. max) and $\Lambda_\pi = \min$ (resp. max)) then we define induction and restriction maps

$$i^* : KF_0(R\pi', \Lambda_{\pi'}) \to KF_0(R\pi, \Lambda_\pi)$$

$$[M, B] \mapsto [M \otimes_{R\pi'} R\pi, B \text{ extended linearly}]$$

$$i_* : KF_0(R\pi, \Lambda_\pi) \to KF_0(R\pi', \Lambda_{\pi'})$$

$$[M, B] \mapsto [M, tr_{\pi/\pi'} B] .$$

There are of course analogous induction and restriction maps on the Witt groups.

We shall refer to the following situation by a

$*$.

C is a category of finite groups and monomorphisms. To each group π in C, we associate a form parameter Λ_π in $R\pi$ defined with respect to λ. We assume that if $\pi' \hookrightarrow \pi$ is a morphism in C then the associated embedding $R\pi' \subset R\pi$ of group rings has the property that $\Lambda_{\pi'} \subset \Lambda_\pi$ and $tr_{\pi/\pi'} \Lambda_\pi \subset \Lambda_{\pi'}$.

The definition of a *Frobenius module* which is needed in the next theorem is given prior to Theorem 12.13 in the next section.

THEOREM 12.6. *Assume* $*$. *Then* $KF_0(R\pi, \Lambda_\pi)$ *is a* $GU_0(R, \pi)$-*Frobenius module and* $WF_0(R\pi, \Lambda_\pi)$ *is a* $GW_0(R, \pi)$-*Frobenius module.*

The proof of 12.6 is routine and similar to that of the analogous result [29, 2.4] for modules.

If C is a category, we recall the construction [10, I1.3] of the *automorphism category* ΣC of C. The objects of ΣC are pairs (P, α) where $P \in C$ and $\alpha \in \mathrm{Aut}_C(P)$. A morphism $(P, \alpha) \to (Q, \beta)$ is a morphism $f: P \to Q$ in C such that $f\alpha = \beta f$. If C is a category with product $\perp$ then ΣC has a natural product defined by $(P, \alpha) \perp (Q, \beta) = (P \perp Q, \alpha \perp \beta)$. By definition, $K_1(C) = K_0'(\Sigma C)$ [10, VII (1.4)]. Next we adapt a theorem of Bass [10, XI 8.1] to our situation. We define a $KH_0(R, \pi)$-action on $KF_1(R\pi, \Lambda)$ by

$$[(P, C), \alpha][M, B] = [(P \otimes_R M, C \otimes_R B), \alpha \otimes 1_B] .$$

Remember that $C \otimes_R B(p \otimes m, p_1 \otimes m_1) = \sum_{\sigma \in \pi} c_\sigma b_\sigma \sigma$ where $C(p, p_1) = \sum_\sigma c_\sigma \sigma$ and $B(m, m_1) = \sum_\sigma b_\sigma \sigma$. One shows that the $KH_0(R, \pi)$-action above kills the Quillen relations in the same way that one showed that the $KH_0(R, \pi)$-action on $KF_0(R\pi, \Lambda)$ kills the Quillen relations. Similarly, we have a $WH_0(R, \pi)$-action on $WF_1(R\pi, \Lambda)$ which kills the Quillen relations. Thus, $KF_1(R\pi, \Lambda)$ (resp. $WF_1(R\pi, \Lambda)$) is a $GU_0(R, \pi)$-module (resp. $GW_0(R, \pi)$-module).

Let $\pi' \subset \pi$ be a subgroup of π. If ${}^{\lambda}(R\pi', \Lambda_{\pi'})$ and ${}^{\lambda}(R\pi, \Lambda_{\pi})$ are form rings such that $\Lambda_{\pi'} \subset \Lambda_{\pi}$ and $\mathrm{tr}_{\pi/\pi'} \Lambda_{\pi} \subset \Lambda_{\pi'}$ then we define induction and restriction maps

$$i^* : KF_1(R\pi', \Lambda_{\pi'}) \to KF_1(R\pi, \Lambda_{\pi})$$
$$[(P, C), a] \mapsto [(P \otimes_{R\pi'} R\pi, C \text{ extended linearly}), a \otimes 1]$$

$$i_* : KF_1(R\pi, \Lambda_{\pi}) \to KF_1(R\pi', \Lambda_{\pi'})$$
$$[(P, C), a] \mapsto [(P, \mathrm{tr}_{\pi/\pi'} C), a] .$$

There are of course analogous induction and restriction maps on the Witt groups.

THEOREM 12.7. *Suppose* $*$. *Then* $KF_1(R\pi, \Lambda_{\pi})$ *is a* $GU_0(R, \pi)$-*Frobenius module and* $WF_1(R\pi, \Lambda_{\pi})$ *is a* $GW_0(R, \pi)$-*Frobenius module.*

The proof of 12.7 is similar to that of 12.6.

Next we decorate the Frobenius modules $KF_i(R\pi, \Lambda)$ and $WF_i(R\pi, \Lambda)$ with the subscripts X and based-Y where, as in §9,

based-Y

denotes any of the based subscripts introduced in §1B and C.

Let $P(R, \pi)$ be the category of finitely generated, R-projective $R\pi$-modules. Let

$$G_0(R, \pi)$$

be the abelian group with generators $[M]$ where $M \in P(R, \pi)$ and relations $[M] = [M'] + [M'']$ for each exact sequence $0 \to M' \to M \to M'' \to 0$. One defines a multiplication on $G_0(R, \pi)$ by the rule

$$[N][M] = [N \otimes_R M]$$

where $N \otimes_R M$ has the diagonal π-action. There is a canonical map $GU_0(R, \pi) \to G_0(R, \pi)$. The groups $K_0(R\pi)$ and $K_1(R\pi)$ are $G_0(R, \pi)$-modules [10, XI (1.8)] and via the homomorphism above, one can consider them as $GU_0(R,\pi)$-modules. Let X and Y be involution invariant subgroups of $K_0(R\pi)$ and $K_1(R\pi)$ respectively satisfying the conditions imposed in §1B. We assume further that

$$(1) \qquad \begin{array}{l} X \text{ is a } GU_0(R,\pi)\text{-submodule} \\ Y \text{ is a } GU_0(R,\pi)\text{-submodule.} \end{array}$$

We define a $GU_0(R,\pi)$-action on $KF_0(R\pi,\Lambda)_X$ and $KF_1(R\pi,\Lambda)_{\text{based-}Y}$ in the same way we defined a $GU_0(R,\pi)$-action on these groups when $X = K_0(A)$ and $Y = K_1(A)$. The $GU_0(R,\pi)$-actions above induce $GU_0(R,\pi)$-actions on the corresponding Witt groups. However, in order that the latter actions induce $GW_0(R,\pi)$-actions on the Witt groups, we must impose a further restriction on X and Y:

$$(2) \qquad \begin{array}{l} X \text{ is a } G_0(R,\pi)\text{-module} \\ Y \text{ is a } G_0(R,\pi)\text{-module.} \end{array}$$

Next, we define a $GU_0(R,\pi)$-action on $KF_0(R\pi,\Lambda)_{\text{based-}Y}$. Drop restrictions (1) and (2) and add the restrictions

$$(3) \qquad \begin{array}{l} \text{every finitely generated, projective } R\text{-module is free} \\ Y \text{ contains } K_1(R)\,. \end{array}$$

Then we define the action by

$$[P,C]\,[M,B] = [P \otimes_R M, C \otimes_R B]$$

such that $P \otimes_R M$ has the preferred basis $\{f_j \otimes e_i \mid 1 \le j \le n, 1 \le i \le m\}$ where $f_1, \cdots, f_n$ is a preferred $R\pi$-basis for P and $e_1, \cdots, e_m$ is any R-basis for M. This action induces a $GW_0(R,\pi)$-action on the Witt groups.

We shall refer to the following situation by a

**

Assume $*$. To each group $\pi \in \mathbf{C}$, we associate an involution invariant subgroup $X_\pi \subset K_0(R\pi)$ and an involution invariant subgroup $Y_\pi \subset K_1(R\pi)$ satisfying the conditions imposed in §1B. We assume that if $Z = X$ or Y and if $\pi' \to \pi$ is a morphism in $\mathbf{C}$ then $i^{*\pi' \to \pi} Z_{\pi'} \subset Z_\pi$ and $i_{*_{\pi \to \pi'}} Z_\pi \subset Z_{\pi'}$. We assume also one of the following:

(i) each Z_π is a $GU_0(R, \pi)$-submodule

(ii) each Z_π is a $G_0(R, \pi)$-submodule

(iii) every finitely generated, projective R-module is free, and each Y_π contains $K_1(R)$.

The following theorem generalizes 12.6 and 12.7.

THEOREM 12.8. *Suppose* $**$.

a) *If (i) is satisfied then* $KF_0(R\pi, \Lambda_\pi)_{X_\pi}$ *and* $KF_1(R\pi, \Lambda_\pi)_{\text{based-}Y_\pi}$ *are* $GU_0(R, \pi)$*-Frobenius modules.*

b) *If (ii) is satisfied then the corresponding Witt groups are* $GW_0(R, \pi)$*-Frobenius modules.*

c) *If (iii) is satisfied then* $KF_0(R\pi, \Lambda_\pi)_{\text{based-}Y_\pi}$ *is a* $GU_0(R\pi)$*-Frobenius module, and the corresponding Witt group is a* $GW_0(R, \pi)$*-Frobenius module.*

The proof of 12.8 is similar to that of 12.6 and 12.7.

B. *Induction machine*

The section is organized as follows. First, we give just enough notation to state the main induction results 12.9 - 12.12 which interest us. The results are conclusions of an induction machine which is described in the rest of the section. Proofs which are short are given in detail. The remaining proofs are outlined such that details are referenced to the literature.

We make the assumption that our modules are *left modules* in order to be consistent with the literature on induction.

Let p be a prime integer. A finite group π is called p-*hyperelementary* if it has a cyclic normal subgroup γ such that π/γ has exponent a power of p, i.e. π/γ is a p-group. π is called *hyperelementary* if it is p-hyperelementary for some p. If S is a nonempty (resp. empty) set of primes then π is called *S-hyperelementary* if π is p-hyperelementary for some $p \in S$ (resp. π is cyclic). Let $H_S(\pi)$ denote the set of all S-hyperelementary subgroups of π. For completeness, we mention that π is called p-*elementary* if it can be written as a direct product $\pi = \gamma \times \tau$ such that γ is cyclic and τ is a p-group. An equivalent definition that π is p-hyperelementary is that it is a semidirect product $\pi = \gamma \rtimes \tau = \{(a,x) \mid a \in \gamma,\ x \in \tau;\ (a,x)(b,y) = (a\,{}^{x}b, xy),\ {}^{x}b$ denotes the action of x on b$\}$ of a p-group τ acting on a cyclic group γ whose order is prime to p. (If π is p-hyperelementary in the first sense then the largest cyclic subgroup γ' of γ whose order is prime to p is a product of normal Sylow subgroups of π. The quotient π/γ' is a p-group, and the exact sequence $1 \to \gamma' \to \pi \to \pi/\gamma' \to 1$ splits by a theorem of Sylow.)

Let **Ab** denote the category of abelian groups. Let **C** denote a subcategory of **Ab** whose objects form a set. Define the *direct limit group* $\varinjlim \mathbf{C} = \coprod_{A \in \mathrm{Obj}(\mathbf{C})} A/\{x_A - fx_A \mid x_A \in A,$ f is a morphism $f: A \to B$ in $\mathbf{C}\}$, and define the *inverse limit group* $\varprojlim \mathbf{C} = \{\prod x_A \in \prod_{A \in \mathrm{Obj}(\mathbf{C})} A \mid$ if $f: A \to B$ is a morphism in $\mathbf{C}$ then $fx_A = x_B\}$. It is important not to overlook above the endomorphisms $f: A \to A$ in $\mathbf{C}$.

Let **G** denote the category whose objects are finite groups and whose morphisms are injective homomorphisms of finite groups. A *bifunctor* $F: \mathbf{G} \to \mathbf{Ab}$ is a pair (F^*, F_*) of functors, F^* covariant, F_* contravariant, such that for all $\pi \in \mathbf{G}$, $F^*(\pi) = F(\pi) = F_*(\pi)$ and the following rules are satisfied:

(1) If $f: \gamma \to \pi$ is an isomorphism then F^*f and F_*f are mutual inverse isomorphisms.

(2) If $f: \pi \to \pi$, $x \mapsto gxg^{-1} (g \in \pi)$, then $F^*f = F_*f = 1$.

We make $H_S(\pi)$ into a category by defining a morphism to be a map $\gamma \to \gamma'$, $x \mapsto gxg^{-1}$, such that $g \in \pi$ and $g\gamma g^{-1} \subset \gamma'$. Define $\varinjlim FH_S(\pi) = \varinjlim F^*H_S(\pi)$ and

$\varprojlim FH_S(\pi) = \varprojlim F_*H(\pi)$. If $i: \gamma \to \pi$, $x \mapsto x$, then the maps $F^*i: F\gamma \to F\pi$ and $F_*i: F\pi \to F\gamma$ induce homomorphisms $\varinjlim FH_S(\pi) \to F(\pi)$ and $F(\pi) \to \varprojlim FH_S(\pi)$.

Let Primes(π) = set of all primes which divide the order of π. Let $S' \subseteq$ Primes (π) and let S = Primes $(\pi) - S'$. Let $Z[S'^{-1}]$ denote the ring of fractions $\frac{a}{s'}$, such that $a \in Z$ and s' is a product of primes from S'. Call F *H-computable* if the canonical maps $Z[S'^{-1}] \otimes \varinjlim FH_S(\pi) \to Z[S'^{-1}] \otimes F(\pi)$ and $Z[S'^{-1}] \otimes F(\pi) \to Z[S'^{-1}] \otimes \varprojlim FH_S(\pi)$ are isomorphisms for all $S \subseteq$ Primes (π). It should be pointed out that the above definition differs somewhat from that given in Dress [14]. The following theorem can be attributed to the combined efforts of Frobenius, Artin, Brauer, Witt, Berman, Mackey, Swan, Lam, Green, and Dress.

THEOREM 12.9. *The functors below are H-computable bifunctors.*

(1) *The functors* $KF_i(R\pi, \Lambda_\pi)$ *and* $WF_i(R\pi, \Lambda_\pi)$ *defined in the first half of the chapter and any of their based versions. One can take* min *for each* Λ_π *or* max *for each* Λ_π.

(2) *The functors* $K_i(R\pi)$.

(3) *Kernels and cokernels of 'canonically defined' homomorphisms between the groups above and quotients thereof, for example the cohomology groups* $H^n(Z/2Z, K_i(R\pi))$ *of the* $Z/2Z$*-action on* $K_i(R\pi)$ *defined by the involution on* $R\pi$.

After making some useful observations below, we shall devote the rest of the section to outlining a proof of Theorem 12.9. The 'coup de grace' will be provided in the last paragraph of the section.

Some useful observations are the following.

12.10. Let $S' \subseteq$ Primes (π) denote a family of primes whose action via left multiplication on $F(\gamma)$ is invertible for all $\gamma \subseteq \pi$. Let S = Primes $(\pi) - S'$. If F is H-computable then $\varinjlim FH_S(\pi) = F(\pi) = \varprojlim FH_S(\pi)$. In particular, *if the exponent of* π *and* $F(\gamma)$ *are relatively prime for all* $\gamma \subseteq \pi$ *then* F *is H-computable* $\Rightarrow \varinjlim FH_\emptyset(\pi) = F(\pi) = \varprojlim FH_\emptyset(\pi)$.

12.10 follows from the fact that $\mathbf{Z}[S'^{-1}] \otimes F(\gamma) = F(\gamma)$.

Let $F_t(\pi)$ = subgroup of $F(\pi)$ of all torsion elements. Let $F/F_t(\pi) = F(\pi)/F_t(\pi)$. If F is a bifunctor then clearly so are F_t and F/F_t.

12.11. *If* F *is* $\mathbf{H}$-*computable and if* $G = F_t$ *or* F/F_t *then* $\mathbf{Z}[S'^{-1}] \otimes G(\pi) = \mathbf{Z}[S'^{-1}] \otimes \varprojlim GH_S(\pi)$ *for all* $S \subseteq$ *Primes* (π).

12.11 follows from the fact that the taking of torsion commutes with *finite* inverse limits, e.g. $(\varprojlim FH_S(\pi))_{\text{torsion}} = \varprojlim F_t H_S(\pi)$. Finite direct limits do not always have this property.

12.12. If F is $\mathbf{H}$-computable and if $F/F_t(\gamma)$ is finitely generated for all $\gamma \subseteq \pi$ then the ranks of all maximal, torsion free subgroups of $\varinjlim FH_\emptyset(\pi)$, $F(\pi)$, and $\varprojlim FH_\emptyset(\pi)$ are equal. In particular, *if the exponent of* π *is relatively prime to the exponent of each element of* $F_t(\gamma)$ *for all* $\gamma \subseteq \pi$ *then it follows from 12.10 and 12.11 that* $F(\pi)$ *and* $\varprojlim FH_\emptyset(\pi)$ *are isomorphic but not necessarily by the canonical maps.*

The observations above can be applied in the following way to the groups $KF_i(R\pi, \Lambda_\pi)$, $WF_i(R\pi, \Lambda_\pi)$, and their based versions, when R is a ring of integers in a global field, for example $R = \mathbf{Z}$. The torsion subgroups of the groups above are then known to have exponent a power of 2. Thus, if π has odd order (resp. has abelian 2-hyperelementary subgroups) (resp. is arbitrary finite) then the computation of the groups above is reduced to the case π is cyclic (resp. π is abelian) (resp. π is 2-hyperelementary).

The rest of the section is devoted to outlining a proof of Theorem 12.9. We begin with some definitions which appear in the literature.

Let $F : \mathbf{G} \to \mathbf{Ab}$ be a bifunctor. Let $\gamma, \gamma' \subseteq \pi$ be subgroups of π. Let $g_1, \cdots, g_n \in \pi$ be a full set of double coset representatives for γ and γ' in π; thus, π = the disjoint union $\dot{\bigcup}_{i=1}^{n} \gamma g_i \gamma'$. F is called a *Mackey functor* [14] if the square

$$\begin{array}{ccc} F(\gamma) & \xrightarrow{F^*} & F(\pi) \\ \Big\downarrow F_* & & \Big\downarrow F_* \\ \sum_{i=1}^{n} F(\gamma \cap g_i \gamma' g_i^{-1}) & \xrightarrow[F^*]{} & F(\gamma') \end{array}$$

commutes for all pairs $\gamma, \gamma' \subseteq \pi$. The vertical maps and the top horizontal map are induced from the natural inclusions. The map $F(\gamma \cap g_i \gamma' g_i^{-1}) \to F(\gamma')$ is F^* applied to $\gamma \cap g_i \gamma' g_i^{-1} \to \gamma'$, $x \mapsto g_i^{-1} x g_i$. The condition above is an axiomatization of the Mackey subgroup theorem. Let **CR** denote the category of commutative rings, and consider **CR** as a subcategory of **Ab**. A bifunctor $F : \mathbf{G} \to \mathbf{CR}$ is called a *Frobenius functor* [18] if F_* is a contravariant functor $\mathbf{G} \to \mathbf{CR}$ (no analogous assumption is made for F^*) and the Frobenius reciprocity law holds, namely if $f : \gamma \to \pi$ denotes a morphism in $\mathbf{G}$, and if $x \in F(\pi)$ and $y \in F(\gamma)$, and if we write F^* in place of F^*f and F_* in place of F_*f then

$$x F^*(y) = F^*(F_*(x) \cdot y) .$$

A bifunctor $F : \mathbf{G} \to \mathbf{CR}$ is called a *Green functor* [14] if it is both a Mackey functor and a Frobenius functor. Thus, we obtain the following convenient picture

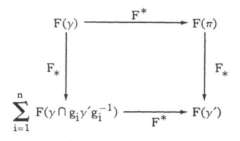

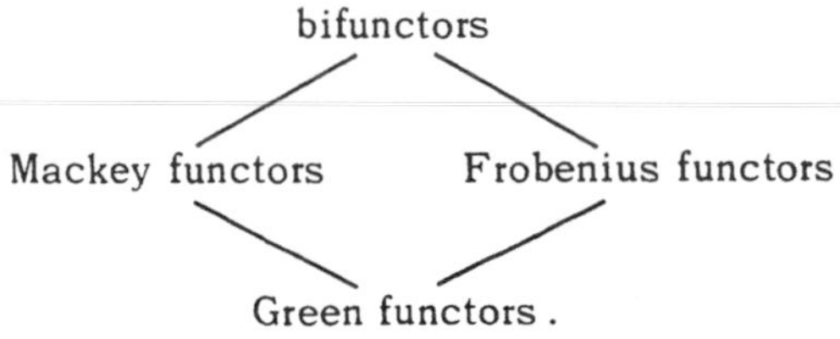

Let $F : \mathbf{G} \to \mathbf{CR}$ be a Frobenius functor. A bifunctor $M : \mathbf{G} \to \mathbf{Ab}$ is called a *Frobenius module over* F [18] if each $M(\pi)$ is an $F(\pi)$-module so that the following rules hold. Let $\gamma \to \pi$ be a morphism in $\mathbf{G}$. Let $x \in F(\pi)$, $m \in M(\pi)$, and let $y \in F(\gamma)$, $n \in M(\gamma)$. Write F^* (resp. M^*) in place of F^*f (resp. M^*f) and write F_* (resp. M_*) in place of F_*f (resp. M_*f). Then

$$(1)\quad M_*(xm) = F_*(x)M_*(m)$$

$$(2)\quad xM^*(n) = M^*(F_*(x)n)$$

$$(3)\quad (F^*y)m = F^*(yM_*(m)).$$

Clearly, a Frobenius functor is a Frobenius module over itself. Let F be a Green functor. A Mackey functor $M: \mathbf{G} \to \mathbf{Ab}$ is called a *Green module over* F [14] (or simply F-module) if M is a Frobenius module over F.

The main result below for Green functors and their modules is due to Dress [14, 1.1′, 1.2, 1.7].

THEOREM 12.13. *Let* F *be a Green functor and let* M *be an* F*-module. Let* $\mathrm{C}(\pi)$ *be a family of subgroups of* π *closed under the taking of subgroups, intersections, and conjugation by elements of* π. *Recall how* $\mathrm{H}(\pi)$ *was made into a category and make analogously* $\mathrm{C}(\pi)$ *into a category. Define analogously* $\varinjlim \mathrm{MC}(\pi)$ *and* $\varprojlim \mathrm{MC}(\pi)$.

a) *If the canonical map* $F^*: \coprod_{\gamma\in \mathrm{C}(\pi)} F(\gamma) \to F(\pi)$ *induced from the natural map* $\gamma \to \pi$, $x \mapsto x$, *is surjective then* $\varinjlim \mathrm{MC}(\pi) = M(\pi) = \varprojlim \mathrm{MC}(\pi)$.

b) *Suppose that any torsion element in* $F(\pi)$ *is nilpotent, i.e.* $nx = 0 \Longrightarrow x^m = 0$ *for some* m. *Let* $S' \subseteq$ *Primes* (π) *and let* $S =$ *Primes* $(\pi) - S'$. *Let* $\mathrm{C}^S(\pi) = \{\gamma' \subseteq \pi \mid \gamma'$ *contains a normal subgroup* $\gamma \in \mathrm{C}(\pi)$ *such that* γ'/γ *is a* p*-group with* $p \in S\}$ *(e.g. if* $\mathrm{C}(\pi) =$ *family of cyclic subgroups of* π *then* $\mathrm{C}^S(\pi) = \mathrm{H}_S(\pi)$). *If the canonical map* $\mathbf{Q}\otimes\coprod_{\gamma\in\mathrm{C}(\pi)} F(\gamma) \to \mathbf{Q}\otimes F(\pi)$ *is surjective then the canonical map* $\mathbf{Z}[S'^{-1}]\otimes\coprod_{\gamma\in\mathrm{C}^S(\pi)} F(\gamma) \to \mathbf{Z}[S'^{-1}]\otimes F(\pi)$ *is surjective for all* $S \subseteq$ *Primes* (π). *(Note that* $S =$ *Primes* $(\pi) \Longrightarrow \mathbf{Z}[S'^{-1}] = \mathbf{Z}$.)

COROLLARY 12.14. *Let* F *be a Green functor such that for all* $\pi \in \mathbf{G}$ *the torsion elements of* $F(\pi)$ *are nilpotent. Let* $\mathrm{C}(\pi) =$ *family of cyclic subgroups of* π. *Then any Green module* M *over* F *is* **H**-*computable* $\Longleftrightarrow$ *the canonical map* $\mathbf{Q}\otimes\coprod_{\gamma\in C(\pi)} F(\gamma) \to \mathbf{Q}\otimes F(\pi)$ *is surjective for all* $\pi \in \mathbf{G}$.

The corollary is an immediate consequence of the theorem.

Proof of 12.13. a) Let $i: \gamma \to \pi$, $x \mapsto x$, denote the natural embedding. Let $f^{\gamma} = F^{*}(i)$, $\theta^{\gamma} = M^{*}(i)$, and let $f_{\gamma} = F_{*}(i)$, $\theta_{\gamma} = M_{*}(i)$. By hypotheses, there are elements $a_{\gamma} \in F(\gamma)$ such that $1_{\pi} = \sum_{\gamma\in C(\pi)} f^{\gamma}(a_{\gamma})$.

Let $\theta: M(\pi) \to \varprojlim MC(\pi)$ denote the canonical map. We prove the injectivity of θ. Let $m \in \ker\theta$. Then $m = \sum f^{\gamma}(a_{\gamma})m =$ (Frobenius reciprocity) $\sum \theta^{\gamma}(a_{\gamma}\theta_{\gamma}(m)) = 0$, because $\theta_{\gamma}(m) = 0$. The surjectivity of θ is more difficult to prove. We refer the reader to Dress [13, 3.2]. A proof in the language of π-sets (sometimes called G-sets) is found in Dress [14, 1.2].

Now let θ denote the canonical map $\varinjlim MC(\pi) \to M(\pi)$. We prove the surjectivity of θ. Property (2) in the definition of a Frobenius module shows that the image θ^{γ} is an $F(\pi)$-submodule. Thus, the image θ is an $F(\pi)$-submodule and hence, $M(\pi)/\text{image } \theta$ is an $F(\pi)$-module. Property (3) in the definition of a Frobenius module shows that the (image f^{γ})$\cdot$ $M(\pi) \subseteq \text{image } \theta^{\gamma}$. Thus, the action of $F(\pi)$ on $M(\pi)/\text{image } \theta$ factors through $F(\pi)/\sum_{\gamma\in C(\pi)} \text{image } f^{\gamma}$. But, $F(\pi) = \sum \text{image } f^{\gamma}$ by hypothesis. Thus, $M(\pi) = \text{image } \theta$. The injectivity of θ is more difficult to show. We shall restrict our effort to translating two results of Dress. A π-*set* is simply a finite set with an action of π. Let $\hat{\pi}$ denote the category whose objects are π-sets and whose morphisms are functions which preserve the action of π. In [14, §0], Dress associates to any contravariant functor $L: \mathbf{G} \to \mathbf{Ab}$ a functor $\hat{L}: \hat{\pi} \to \mathbf{Ab}$. Let π/γ denote the π-set of left cosets $g\gamma$ of γ in π. Let $S = $ disjoint union $\dot{\bigcup}_{\gamma\in C(\pi)} \pi/\gamma$. [14, 1.2] says that if $\coprod_{\gamma\in C(\pi)} F(\gamma) \to F(\pi)$ is surjective then the functor $\hat{M}$ has a property called S-injectivity. [14, 1.1′] says that if $\hat{M}$ is S-injective then there is an exact sequence $\cdots \longrightarrow \hat{M}(S\times S) \xrightarrow{s_2} \hat{M}(S) \xrightarrow{s_1} \hat{M}(\pi/\pi) \longrightarrow 1$. But from the definition of $\hat{\ }$, one can deduce easily that $\hat{M}(\pi/\pi) = M(\pi)$ and that $\hat{M}(S) = \prod_{\gamma\in C(\pi)} M(\pi)$. With a bit more work, one can deduce that $\hat{M}(S\times S) = \prod_{\gamma\in C(\pi)} \prod_{g\in K_\gamma} M(\gamma \cap g\gamma g^{-1})$ where K_γ is a set of double coset representatives for γ in π. Computing the map s_2, one determines that the coker $s_2 = \varinjlim MC(\pi)$. This shows that θ is an isomorphism.

b) Let p be a prime. If A is an abelian group, let A_p denote the ring of all $\frac{a}{t}$ such that $a \in A$ and t is an integer not divisible by p. By a well-known lemma, $A = 0 \iff A_p = 0$ for all primes p. Let $C^p(\pi) = \{\gamma' \subseteq \pi \mid \gamma'$ contains a normal subgroup $\gamma \in C(\pi)$ such that γ'/γ is a p-group$\}$. Let S and S′ be as in 12.13 b). If $G = Z[S'^{-1}] \otimes F$ then we must show that $\theta: \prod_{\gamma\in C^S(\pi)} G(\gamma) \to G(\pi)$ is surjective. It suffices to show that for each p, the map $\theta_p: \prod_{\gamma\in C^S(\pi)} G(\gamma)_p \to G(\pi)_p$ is surjective. If $p \notin S'$ then one sees easily that $C^p(\pi) \subseteq C^S(\pi)$ and $F(\ )_p = G(\ )_p$. But, [14, 1.7] says that if $Q \otimes \prod_{\gamma\in C(\pi)} F(\gamma) \to Q \otimes F(\pi)$ is surjective then $\prod_{\gamma\in C^p(\pi)} F(\gamma)_p \to F(\pi)_p$ is surjective. Thus, θ_p is surjective. If $p \in S'$ then $G(\ )_p = Q \otimes F$. Thus, θ_p is surjective by hypothesis.

Next we give some examples of Green functors which will be relevant to the proof of 12.9. The first is the functor $GU_0(R,\pi)$ which was constructed in the first half of the chapter. We did not verify that $GU_0(R,\pi)$ satisfies the Mackey subgroup property and so we leave this as an exercise to the interested reader. $GW_0(R,\pi)$ is, of course, also a Green functor, as well as the functor $G_0(R,\pi)$.

Next we record some examples of Green modules. In the first half of the chapter, we showed how to pick form parameters Λ_π on $R\pi$ such that $KF_i(R\pi, \Lambda_\pi)$ (resp. $WF_i(R\pi,\Lambda_\pi)$) became a Green module over

$GU_0(R, \pi)$ (resp. $GW_0(R, \pi)$). For example, one can take each $\Lambda_\pi = \min$ or each $\Lambda_\pi = \max$. We did not verify the Mackey subgroup property for $KF_i(R\pi, \Lambda_\pi)$ and $WF_i(R\pi, \Lambda_\pi)$ and so we leave this again as an exercise. The based versions of $KF_i(R\pi, \Lambda_\pi)$ and $WF_i(R\pi, \Lambda_\pi)$ are, of course, also Green modules. In [29], it is verified (modulo the Mackey subgroup property) that $K_0(R\pi)$ and $K_1(R\pi)$ are Green modules over $G_0(R, \pi)$.

Proof of 12.9. Each functor in 12.9 is a Green module over either the Green functor $G_0(\mathbf{Z}, \pi)$ or the Green functor $GU_0(\mathbf{Z}, \pi)$. Let F denote G_0 or GU_0. Let $C(\pi)$ denote the family of cyclic subgroups of π. If the torsion elements of $F(\pi)$ are nilpotent for every finite group π then according to 12.13, it is enough to show that the map $\mathbf{Q} \otimes \coprod_{\gamma \epsilon C(\pi)} F(\gamma) \to F(\pi)$ is surjective for all π. The condition that torsion elements are nilpotent is satisfied according to a lemma of G. Segal [14, p. 295] if $F(\pi)$ is a λ-ring. But it is well known [19, XVI §8] that exterior products induce a λ-ring structure on $G_0(\mathbf{Z}, \pi)$, and a similar result [14, §3] shows that $GU_0(\mathbf{Z}, \pi)$ is a λ-ring. By Swan's application [29, 2.12] of Artin's induction theorem [29, 2.19], the map $\mathbf{Q} \otimes \coprod_{\gamma \epsilon C(\pi)} G_0(\mathbf{Z}, \gamma) \to \mathbf{Q} \otimes G_0(\mathbf{Z}, \pi)$ is surjective, and by Dress' theorem [14, Theorem 3] the map $\mathbf{Q} \otimes \coprod_{\gamma \epsilon C(\pi)} GU_0(\mathbf{Z}, \gamma) \to \mathbf{Q} \otimes GU_0(\mathbf{Z}, \pi)$ is surjective.

§13. ALTERNATE DEFINITIONS OF QUADRATIC MODULES

In this section, we compare several definitions of quadratic modules. Let ${}^{\lambda}(A, \Lambda)$ be a form ring. Let

$$\Lambda\text{-}\mathbf{quad}$$

be the category of all Λ-quadratic modules discussed in §1B. If M is a right A-module, let Λ-Herm(M) be the subgroup of Sesq(M) of all λ-hermitian forms B on M such that $B(m, m) \in \Lambda$ for all $m \in M$. Let

$$\Lambda\text{-}\mathbf{quad}'$$

be the category of all pairs (M, [B]) where M is a right A-module, $B \in \mathrm{Sesq}(M)$, and [B] is the class of B in $\mathrm{Sesq}(M)/\Lambda\text{-Herm}(M)$. A morphism $(M, [B]) \to (N, [C])$ is an A-linear map $f: M \to N$ which preserves the classes of B and C, i.e. $[B] = [C(f\_\_, f\_\_)]$. Λ-**quad**′ has a natural product defined by $(M, [B]) \perp (N, [C]) = (M \oplus N, [B \oplus C])$. Let

$$\Lambda\text{-}\mathbf{quad}''$$

be the category of all triples $(M, q, \langle\ ,\ \rangle)$ where M is a right A-module, $q: M \to A/\Lambda$, and $\langle\ ,\ \rangle$ is a λ-hermitian form on M such that

13.1 a) $q(ma) = \bar{a}q(m)a \quad (a \in A,\ m \in M)$

b) $q(m+n) - q(m) - q(n) \equiv \langle m, n\rangle \bmod \Lambda$

c) $\tilde{q}(m) + \lambda\overline{\tilde{q}(m)} = \langle m, m\rangle$ for any lifting $\tilde{q}(m)$ of $q(m)$.

q is called a *quadratic form.* A morphism $(M, q, \langle\ ,\ \rangle) \to (M', q', \langle\ ,\ \rangle')$ is a linear map $M \to M'$ which preserves the quadratic and hermitian forms. Λ-**quad**″ has a natural product defined by $\langle M, q, \langle\ ,\ \rangle) \perp (N, q', \langle\ ,\ \rangle') = (M \oplus N, q \oplus q', \langle\ ,\ \rangle \oplus \langle\ ,\ \rangle')$.

LEMMA 13.2. *If* $(M, B) \in \Lambda$-**quad** *and* $(M, [B]) \in \Lambda$-**quad**$'$ *then* $q_B = \langle\ ,\ \rangle_B = 0 \Longleftrightarrow [B] = 0$.

Proof. Clear.

COROLLARY 13.3. *The canonical functors below are product preserving, inverse equivalences*

$$\Lambda\text{-}\mathbf{quad} \to \Lambda\text{-}\mathbf{quad}', \quad (M, B) \mapsto (M, [B])$$

$$\Lambda\text{-}\mathbf{quad}' \to \Lambda\text{-}\mathbf{quad}, \quad (M, [B]) \mapsto (M, B).$$

Proof. Clear.

LEMMA 13.4. *If* $(M, B) \in \Lambda$-**quad** *then* q_B *and* $\langle\ ,\ \rangle_B$ *satisfy 13.1.*

Proof. Clear.

Hence, there is a canonical, product preserving functor Λ-**quad** $\to$ Λ-**quad**$''$, $(M, B) \mapsto (M, q_B, \langle\ ,\ \rangle_B)$. Let

$$\Lambda\text{-}\mathbf{quad}_{\text{proj}}$$

$$\Lambda\text{-}\mathbf{quad}''_{\text{proj}}$$

be respectively the full subcategories of Λ-**quad** and Λ-**quad**$''$ of all objects such that the underlying module is projective.

LEMMA 13.5. *The canonical functor below is a product preserving equivalence*

$$\Lambda\text{-}\mathbf{quad}_{\text{proj}} \to \Lambda\text{-}\mathbf{quad}''_{\text{proj}}.$$

Proof. It suffices to show that if $(M, q, \langle\ ,\ \rangle) \in \Lambda$-**quad**$''$ then there is an $(M, B) \in \Lambda$-**quad** such that $q_B = q$ and $\langle\ ,\ \rangle_B = \langle\ ,\ \rangle$. Reduce to the case M is free, e.g. pick M_1 such that $M' = M \oplus M_1$ is free and extend q and $\langle\ ,\ \rangle$ to M' via $q' = q \oplus 0_{M_1}$ and $\langle\ ,\ \rangle' = \langle\ ,\ \rangle \oplus 0_{M_1}$. Let $x_1, \cdots, x_m$ be a basis for M. Define B on M by $B(x_i, x_j) = \langle x_i, x_j \rangle$ for $i < j$, $B(x_i, x_j) = 0$ for $i > j$, and $B(x_i, x_i) = \tilde{q}(x_i)$ where

$\tilde{q}(x_i)$ is any lifting of $q(x_i)$ to A. One checks routinely that B has the desired properties.

If ${}^{-\lambda}(A, \Gamma)$ is a form ring let

$$\Gamma\text{-}\mathbf{herm}$$

be the category of all Γ-hermitian modules described in §1C. Let

$$\lambda\text{-}\mathbf{herm}$$
$$\text{min-}\mathbf{herm}$$

correspond to the maximum and minimum choices of Γ. Let

$$\text{even-}\mathbf{herm}$$

be the full subcategory of min-**herm** of all (M, B) such that B is even, i.e. $B = C + \lambda\bar{C}$. Let Γ-**herm**$_{\text{proj}}$, etc., be the full subcategory of Γ-**herm** of all (M, B) such that M is projective.

LEMMA 13.6. Even-**herm**$_{\text{proj}} \xrightarrow{\cong}$ min-**herm**$_{\text{proj}}$.

Proof. It suffices to show that if $(M, B) \in$ min-**herm**$_{\text{proj}}$ then $B = C + \lambda\bar{C}$ for some C. As in the proof of 13.4, we can reduce to the case M is free, say with basis $x_1, \cdots, x_n$. Suppose $B(x_i, x_i) = a_i + \lambda\bar{a}_i$. Define C on M by $C(x_i, x_i) = a_i$, $C(x_i, x_j) = B(x_i, x_j)$ for $i < j$, and $C(x_i, x_j) = 0$ for $i > j$. Then $B = C + \lambda\bar{C}$.

The idea to think of a quadratic module as a sesquilinear form modulo even skew-hermitian forms appears already in a paper of Klingenberg and Witt [17] and in a paper of Springer [27]. However, it was Tits who first jelled the idea into a definition. We give next Tits' definition [30, p. 21].

Let

$$\text{Tits-quad}$$

be the category of all pairs $(M, [B])$ where M is an A-module,

$B \in \mathrm{Sesq}(M)$, and $[B]$ is the class of B in $\mathrm{Sesq}(M)/\mathrm{even}(-\lambda)\text{-}\mathrm{Herm}(M)$. Morphisms and products are defined similarly to those in Λ**-quad′**. There is a canonical, product preserving functor **Tits-quad** $\to$ **min-quad**.

LEMMA 13.7. **Tits-quad**$_{\mathrm{proj}} \xrightarrow{\cong}$ **min-quad**$_{\mathrm{proj}}$.

Proof. The functor above can be factored as the composite **Tits-quad**$_{\mathrm{proj}}$ $\longrightarrow$ **min-quad′**$_{\mathrm{proj}} \xrightarrow{\cong}$ **min-quad**$_{\mathrm{proj}}$. The proof of 13.6 shows that even $(-\lambda)$-$\mathrm{Herm}(M) = \text{min-Herm}(M)$ for M projective. Hence, **Tits-quad**$_{\mathrm{proj}} \to$ **min-quad′**$_{\mathrm{proj}}$ is an equivalence.

LEMMA 13.8. *If* 1 *is a trace from the center of* A, *i.e.* $1 = a + \bar{a}$ *for some* $a \in \mathrm{center}(A)$, *then the maximum and minimum choices of a form parameter coincide.*

Proof. If $x = -\lambda\bar{x}$ then $x = (a+\bar{a})x = ax - \lambda\overline{ax}$. Hence, max = min.

REMARK. If 1 is a trace from the center of A then our refinements of the concepts of quadratic and hermitian form disappear. Furthermore, if we insist that the underlying modules are projective then there is no distinction between quadratic and hermitian forms (**min-quad** = **max-quad** $\overset{(1.1)}{=}$ **even-herm**, and **even-herm**$_{\mathrm{proj}} \overset{(13.6)}{=}$ **min-herm**$_{\mathrm{proj}}$ = **max-herm**$_{\mathrm{proj}}$). Thus, we have a conceptual explanation of the well-known fact that forms on projective modules over rings with $1/2$ behave better than other kinds of forms.

LEMMA 13.9. *Let* (M, B) *and* (N, C) *be* Λ*-quadratic modules. If* Λ *contains only the trivial* 2*-sided ideal then a linear map* $f: M \to N$ *is a morphism of* Λ*-quadratic modules* $\Longleftrightarrow$ f *preserves the associated* Λ*-quadratic forms.*

Proof. The implication from left to right is clear. Conversely, let $q = q_B - q_C(f\_\_)$ and $\langle\ ,\ \rangle = \langle\ ,\ \rangle_B - \langle f\_\_, f\_\_\rangle_C$. We must show $\langle\ ,\ \rangle$ is trivial. Let $\mathfrak{a} = \{\langle m, n\rangle \mid m, n \in M\}$. $\mathfrak{a}$ is clearly a 2-sided ideal of A. Since q is trivial, it follows from 13.1 b) that $\mathfrak{a} \subset \Lambda$. Hence, $\mathfrak{a} = 0$.

Wall generalizes in [32] the usual concept of an involution on a ring. Let A be a ring. Then an involution on A is a pair $(\lambda, -)$ where λ is a unit in A and $-$ is an antisomorphism of A such that $\bar{\bar{a}} = \lambda a \bar{\lambda}$. It follows that $1 = \lambda\bar{\lambda}$. If $\lambda \in$ center (A) then we get the usual concept of an involution. Wall in other publications [34]-[36] calls $(\lambda, -)$ an *antistructure* on A. Wall shows in [32] how to extend Tits' definition of a quadratic form to antistructures. We shall mimic Wall to extend our definitions of quadratic and hermitian modules to antistructures. If $(\lambda, -)$ is an antistructure, we define a form parameter exactly as we did in §1B. In order that the definition makes sense, we must check a few things: If $x = a - \lambda\bar{a}$ then $x = -\lambda\bar{x}$, because $-\lambda\bar{x} = -\lambda(\bar{a} - \bar{\bar{a}}\bar{\lambda}) = -\lambda\bar{a} + \lambda\bar{\bar{a}}\bar{\lambda} = -\lambda\bar{a} + a = x$. If $x \in A$ then $x \min \bar{x} \subset$ min, because $x(a - \lambda\bar{a})\bar{x} = xa\bar{x} - x\lambda\bar{a}\bar{x} =$ (because $\lambda\bar{\bar{x}}\bar{\lambda} = x \Longrightarrow \lambda\bar{\bar{x}} = x\lambda$) $xa\bar{x} - \lambda\bar{\bar{x}}\bar{a}\bar{x} = xa\bar{x} - \lambda\overline{xa\bar{x}} \in \min$. If $x \in A$ then $x \max \bar{x} \subset \max$, because if $a = -\lambda\bar{a}$ then $xa\bar{x} = -x\lambda\bar{a}\bar{x} = -\lambda\bar{\bar{x}}\bar{a}\bar{x} = -\lambda\overline{xa\bar{x}}$. We define Λ-quadratic and Λ-hermitian modules exactly as in §1B and C. Many of our results for quadratic and hermitian modules over rings with the usual kind of involution are valid with little or no change in their proofs for quadratic and hermitian modules defined over antistructures.

We record next a handy corollary of 13.2.

COROLLARY 13.10. a) (M, B) *is the trivial* Λ*-quadratic module* $\Longleftrightarrow$ $B = -\lambda\bar{B}$ *and* $B(m, m) \in \Lambda$ *for all* $m \in M$.

b) *Let* A^n *be a free module with basis* $e_1, \cdots, e_n$. *Then* (A^n, B) *is the trivial* Λ*-quadratic module* $\Longleftrightarrow$ $B(e_i, e_i) \in \Lambda$ *for all* i *and* $B(e_i, e_j) = -\lambda\overline{B(e_j, e_i)}$ *for all* i *and* j.

Proof. a) Clear. b) The implication from left to right is clear. Conversely, it is clear that $B = -\lambda\bar{B}$. Now use 13.1 to show that $B(m, m) \in \Lambda$ for all $m \in A^n$.

§14. REMARKS ON NOTATION

The notations used for the K-theory groups of forms are multiplying rapidly. We have chosen very carefully our notations with special regard to simplicity, suggestiveness, and historical development. We have made them flexible enough to contain the major nuances used by most authors.

The notations KQ_i and KH_i are meant to suggest the K-theories of quadratic and hermitian modules. If one takes for granted the meaning of K_i of a category then $KQ_i(A, \Lambda) = K_i\mathbf{Q}(A, \Lambda)$ and $KH_i(A, \Lambda) = K_i\mathbf{H}(A, \Lambda)$ where $\mathbf{Q}(A, \Lambda)$ (resp. $\mathbf{H}(A, \Lambda)$) is the category of nonsingular Λ-quadratic (resp. Λ-hermitian) modules over A. If the results we are quoting are valid for both quadratic and hermitian modules then we replace the letters Q and H by the single letter F. F is meant to suggest form, either quadratic or hermitian. Witt considered first factoring out the hyperbolic elements from $KQ_0(A, \Lambda)$, and so the notation $WQ_i(A, \Lambda)$ is appropriate. If one is interested in modules which lie in a certain subgroup X of $K_0(A)$ then the notations $KQ_0(A, \Lambda)_X$ and $WQ_0(A, \Lambda)_X$ seem appropriate. In the latter definition, we factor out only the hyperbolic elements $\mathbf{H}(M)$ such that M lies in X. We define also groups $KQ_1(A, \Lambda)_X$ and $WQ_1(A, \Lambda)_X$ (see §4). If one is interested in based modules where automorphisms are restricted to lie in a subgroup Y of $K_1(A)$ then we use the notations $KQ_i(A, \Lambda)_{\text{based-}Y}$ and $WQ_i(A, \Lambda)_{\text{based-}Y}$. By varying Λ (between min and max) and the subscripts X and based-Y, one gets most of the K-theory groups used today. In the next section, we show how the surgery obstruction groups of Wall fit into the system above. For handy reference, we shall give now a table which translates the notations found most often in the literature into our notation.

Author or paper	Notation	Translation	Conditions imposed by author
Bak [1], [2]	$KU_i(A, \Lambda)$	$KQ_i(A, \Lambda)$	only in [1], [2]
Bak-Scharlau [9]	$KU_i(A)$	$KQ_i(A,\max)$	
	$KU_i(A,B)$	$KQ_i((A,\max),(B,\max))$	
	$W_0(A)$	$WQ_0(A,\max)$	
	$W_0(A,B)$	$\operatorname{coker} H : K_0(A,B) \to KU_0(A,B)$	
Bass [11], [12]	$KU_i(A,\Lambda)$	$KQ_i(A,\Lambda)$	
	$W_i(A,\Lambda)$	$WQ_i(A,\Lambda)$	
Ranicki [22]	$U_{2n}{}^Y(A)$	$WQ_0{}^{(-1)^n}(A,\min)_Y$	
	$U_{2n+1}{}^Y(A)$	$K_0'\mathrm{Rel}_Y{}^{(-1)^n}(A,\min)/[H(P),P,P^*]$	
	$V_{2n}{}^X(A)$	$WQ_0{}^{(-1)^n}(A,\min)_{\text{based-}X}$	$[\pm 1] \subseteq X$, rank of a free A-module is unique
	$V_{2n+1}{}^X(A)$	$WQ_1{}^{(-1)^n}(A,\min)_X/\begin{bmatrix}0 & (-1)^n\\ 1 & 0\end{bmatrix}$	$[\pm 1] \subseteq X$, rank of a free A-module is unique
	$U_n(A)$	$U_n{}^{K_1(A)}(A)$	
	$V_n(A)$	$V_n{}^{K_1(A)}(A)$	
	$W_n(A)$	$V_n{}^{[\pm 1]}(A)$	
Sharpe [26]	$KU_0{}^\lambda(A)$	$WQ_0{}^{\bar{\lambda}}(A,\max)_{\lvert\lambda\rvert\text{-based-}0}$	the class of λ in $K_1(A)$ has finite order $\lvert\lambda\rvert$
Wall [34], [35]	$L_{2n}{}^X(A)$	$WQ_0{}^{(-1)^n}(A,\min)_{\text{discr-based-}X}$	
	$L_{2n+1}{}^X(A)$	$WQ_1{}^{(-1)^n}(A,\min)_X$	
	$L_n{}^K(A)$	$L_n{}^{K_1(A)}(A)$	
	$L_n{}^S(A)$	$L_n{}^0(A)$	

Let π be a group. Give the integral group ring $\mathbf{Z}\pi$ the involution which sends each element of π to its inverse. Then in the notations of §1B and §4, the surgery obstruction groups $L_n^{s,h}(\pi)$ of Wall [31] are

$$\begin{aligned}
L_{2n}^{h}(\pi) &= WQ_0^{(-1)^n}(\mathbf{Z}\pi, \min)_{\text{based-}K_1(\mathbf{Z}\pi)} \\
L_{2n}^{s}(\pi) &= WQ_0^{(-1)^n}(\mathbf{Z}\pi, \min)_{\text{based-}[\pm\pi]} \\
L_{2n+1}^{h}(\pi) &= WQ_1^{(\ 1)^n}(\mathbf{Z}\pi, \min) \\
L_{2n+1}^{s}(\pi) &= WQ_1^{(-1)^n}(\mathbf{Z}\pi, \min)_{\text{based-}[\pm\pi]}
\end{aligned}$$

where $[\pm\pi] = \text{image}(\pm\pi \to K_1(\mathbf{Z}\pi))$.

BIBLIOGRAPHY

[1] A. Bak, *The stable structure of quadratic modules*, Thesis, Columbia University (1969), available from author.

[2] ————, *On modules with quadratic forms*, Lecture Notes in Math. 108 (1969), 55-66.

[3] ————, *Strong approximation for central extensions of elementary groups*, preprint (1972).

[4] ————, *The computation of surgery groups of odd torsion groups*, Bull. Amer. Math. Soc. (1974), 1113-1116.

[5] ————, *Odd dimension surgery obstruction groups of odd torsion groups vanish*, Topology 14 (1975), 367-374.

[6] ————, *The computation of surgery groups of finite groups with abelian 2-hyperelementary subgroups*, Lecture Notes in Math. 551 (1976), 384-409.

[7] ————, *Surgery and K-theory groups of quadratic forms over finite groups and orders*, preprint (1977).

[8] ————, *The computation of even dimension surgery groups of odd torsion groups*, Communications in Alg. 6(14) (1978), 1393-1458.

[8.1] ————, *Definitions and problems in surgery and related groups*, Gen. Top., Appl. 7 (1977), 215-231.

[8.2] ————, *Arf's theorem for trace neotherian and other rings*, J. Pure, Applied Alg. 14 (1979), 1-20.

[8.3] A. Bak and U. Rehmann, *Le problème des sous-groupes de congruence dans* $SL_{n \geq 2}$ *sur un corps gauche*, C.R. Acad. Sc. Paris, 289, série A, (1979), 151.

[8.4] ————, *The congruence subgroup and metaplectic problems for* $SL_{n \geq 2}$ *of division algebras*, preprint (1979).

[9] A. Bak and W. Scharlau, *Grothendieck and Witt groups of orders and finite groups*, Inventiones math. 23 (1974), 207-240.

[10] H. Bass, *Algebraic K-theory*, Benjamin, New York (1968).

[11] ————, *Unitary algebraic K-theory*, Lecture Notes in Math. 343 (1973), 57-265.

[11.1] H. Bass, *Clifford algebras and spinor norms over a commutative ring*, Amer. J. Math. 96 (1974), 156-206.

[12] ———, *L_3 of finite abelian groups*, Ann. Math. 99 (1974), 118-153.

[13] A. Dress, *On relative Grothendieck rings*, Lecture Notes in Math. 488 (1975), 79-131.

[14] ———, *Induction and structure theorems for orthogonal representations of finite groups*, Ann. Math. 102 (1975), 291-325.

[15] A. Fröhlich and A. M. Mc Evett, *Forms over rings with involution*, J. Algebra 12 (1969), 79-104.

[16] ———, *The representation of groups by automorphisms of forms.* J. Algebra 12 (1969), 114-133.

[16.1] M. Karoubi, *Périodicité de la K-théorie hermitienne*, Springer Lecture Note N° 343 (1973), 301-411.

[16.2] ———, *Localisation de formes quadratiques I et II*, Ann. Sci. Ec. Norm. Sup. 7 (1974), 359-404 et 8 (1975), 99-155.

[17] W. Klingenberg and E. Witt, *Über die Arfsche Invariante quadratischer Formen mod 2*, J. reine angew. Math. 193 (1954), 121-122.

[18] T. Y. Lam, *Induction theorems for Grothendieck groups and Whitehead groups of finite groups*, Ann. Ec. Norm. Sup. (4) 1 (1968), 91-148.

[19] S. Lang, *Algebra*, Addison-Wesley, Reading, Mass. (1965).

[20] J. Milnor, *Introduction to algebraic K-theory*, Ann. Math. Studies 72, Princeton Univ. Press, Princeton (1971).

[21] D. G. Quillen, *The Adams conjecture*, Topology 10, (1971), 67-80.

[22] A. Ranicki, *Algebraic L-theory I. Foundations. II Laurent extensions*, Proc. Lond. Math. Soc. (3) 27 (1973), 101-158.

[23] ———, *Algebraic L-theory III. Twisted Laurent extensions*, Lecture Notes in Math. 343 (1973), 412-463.

[24] ———, *Algebraic L-theory IV. Polynomial extension rings*, Comm. Math. Helv. 49 (1974), 137-167.

[25] J. Shaneson, *Wall's surgery obstruction groups for* $G \times \mathbf{Z}$, Ann. Math. 90 no. 2 (1969), 296-334.

[26] R. Sharpe, *On the structure of the unitary Steinberg group*, Ann. Math. 96 (1972), 444-479.

[27] T. A. Springer, *Note on quadratic forms in characteristic 2*, Nieuw Archief voor Wiskunde (3), 10 (1962), 1-10.

[28] M. Stein, *Relativization functors on rings and algebraic K-theory*, J. Algebra 19 (1971), 140-152.

[29] R. Swan, *K-theory of finite groups and orders*, Lecture Notes in Math. 149 (1970).

[29.1] ———, *Algebraic K-theory*, Lecture Notes in Math. 76 (1968).

[30] J. Tits, *Formes quadratiques, groupes orthogonaux et algèbres de Clifford*, Inventiones math. 5(1968), 19-41.

[31] C. T. C. Wall, *Surgery on compact manifolds*, Academic Press, London-New York (1970).

[32] ———, *On the axiomatic foundations of the theory of Hermitian forms*, Proc. Camb. Phil. Soc. 67(1970), 243-250.

[33] ———, *On the classification of Hermitian forms I. Rings of algebraic integers*, Composito Math., Vol. 22, Fasc. 4(1970), 425-451.

[34] ———, *On the classification of Hermitian forms II, III, IV, V*, Inventiones math. 18(1972), 119-141, 19(1973), 59-71; 23(1974), 241-260, 261-288.

[35] ———, *Classification of Hermitian forms, VI Group Rings*, Ann. Math. 103(1976), 1-80.

[36] ———, *Foundations of algebraic L-theory*, Lecture Notes in Math. 343(1973), 266-300.

SUBJECT INDEX

NOTATION INDEX

Library of Congress Cataloging in Publication Data

Bak, Anthony.
K-theory of forms.

Bibliography: p.
Includes index.
1. K-theory. 2. Modules (Algebra) 3. Forms (Mathematics) I. Title.
QA169.B33 1981 512'.55 81-5176
ISBN 0-691-08274-X AACR2
ISBN 0-691-08275-8 (pbk.)

Milton Keynes UK
Ingram Content Group UK Ltd.
UKHW042323190224
438117UK00001B/95

9 780691 082752